Developments in Plant Genetics and Breeding, 3

Genetics of Scots Pine

Developments in Plant Genetics and Breeding

1A ISOZYMES IN PLANT GENETICS AND BREEDING, PART A
edited by S. D. Tanksley and T. J. Orton
1983 x + 516 pp.

1B ISOZYMES IN PLANT GENETICS AND BREEDING, PART B
edited by S. D. Tanksley and T. J. Orton
1983 viii + 472 pp.

2A CHROMOSOME ENGINEERING IN PLANTS: GENETICS, BREEDING, EVOLUTION, PART A
edited by P. K. Gupta and T. Tsuchiya
1991 656 pp.

2B CHROMOSOME ENGINEERING IN PLANTS: GENETICS, BREEDING, EVOLUTION, PART B
edited by T. Tsuchiya and P. K. Gupta
1991 in preparation

3 GENETICS OF SCOTS PINE
edited by M. Giertych and Cs. Mátyás
1991 280 pp.

Developments in Plant Genetics and Breeding, 3

Genetics of Scots Pine

Edited by

Maciej Giertych

Institute of Dendrology,
Polish Academy of Sciences,
Kórnik, Poland

Co-editor

Csaba Mátyás

Department of Plant Sciences,
Sopron University,
Sopron, Hungary

ELSEVIER

Amsterdam–Oxford–New York–Tokyo 1991

Joint edition published by

Elsevier Science Publishers, Amsterdam, The Netherlands and Akadémiai Kiadó, Budapest, Hungary

Distribution of this book is being handled by the following publishers:

for the U.S.A. and Canada
Elsevier Science Publishing Company, Inc.
655 Avenue of the Americas
New York, N.Y. 10010

for the East European countries, Democratic People's Republic of Korea, People's Republic of Mongolia, Republic of Cuba, Socialist Republic of Vietnam
Kultura Foreign Trading Company, P. O. Box 149, 1389 Budapest, Hungary

for all remaining areas
Elsevier Science Publishers
Sara Burgerhartstraat 25
P.O. Box 211, 1000 AE Amsterdam, The Netherlands

Library of Congress Cataloging-in-Publication Data

Giertych, Maciej.
Genetics of Scots pine / editor, Maciej Giertych : co-editor, Csaba Mátyás.
p. cm. — (Developments in plant genetics and breeding; 3)
Includes bibliographical references and index.
ISBN 0-444-98724-X
1. Scots pine—Genetics. 2. Scots pine—Breeding. I. Mátyás, Csaba. II. Title. III. Series.
SD397.P614G54 1991
634.9'751—dc20 91–6309
CIP

ISBN 0–444–98724–X

Printed in Hungary by Akadémiai Kiadó és Nyomda Vállalat, Budapest.

Preface

In April 1986 Dr. Csaba Mátyás from Hungary, chairman of the IUFRO Working Party S2.03.05 Breeding Scots pine, wrote, asking me to act as editor in an international project aimed at compiling literature on the subject of Scots pine genetics. Since the species is the most important one in Eurasia it was considered necessary to involve authors from as many countries as possible, including Eastern Europe and the USSR where the species dominates all forestry endeavour. Since Scots pine represents the highest proportion of forest cover in Poland (72 per cent), it seemed natural that there should be an editor from Poland. I agreed tentatively and started to discuss the format with prospective authors.

During the 18th IUFRO World Congress in Ljubljana, Yugoslavia, in September 1986, at a business meeting of the Working Party, we discussed the details of the project and agreed on the contents of the book. With minor changes the list of chapters of this book is as originally planned. My nomination as editor of the volume was confirmed. It was agreed that publication be arranged through Akadémiai Kiadó, the Publishing House of the Hungarian Academy of Sciences. In a letter dated 5th March 1987 from the publisher I was formally asked to organise and edit the monograph on 'Genetics of Scots pine'. I accepted.

With help of other members of the Working Party an intensive search for willing and competent authors was started through correspondence. On 4th May 1987 I sent out a circular letter asking prospective authors to write different chapters and also asking them to help with the editing. It was my belief that the best reviewers were to be found among the authors of this monograph, all specialists in Scots pine genetics. Thus I have asked each author to act also as a reviewer for a chapter of the volume. On 27th July 1987 I was able to announce that 19 authors had agreed to my terms and that I expected the agreement of the remaining few shortly, which indeed happened.

With minor modifications the list of authors and chapter titles was fixed in mid-1987. The writing and reviewing proved longer than expected, so that the manuscript was ready for presentation to the publisher by the end of July 1989.

This volume is a truly international effort, prepared in the traditional IUFRO spirit of selfless co-operation. In all, 24 authors from 9 countries are involved. Each chapter was reviewed by two reviewers from two different countries, by myself as the editor of the whole volume, and by Cs. Mátyás as associate editor commissioned by the publisher, who was responsible for the final, co-ordinative editing work and the linguistic revision of the chapters. Needless to say a lot of international correspondence was involved and the process proved time-consuming. I hope it was worth. I leave this to our readers to judge.

I wish to express my sincere gratitude to all authors for their co-operation, for their willingness to submit to the editorial pressure and for their help with the reviewing.

Maciej Giertych

Contents

CHAPTER 11

CHAPTER 12

CHAPTER 13

Contributors

Barnéoud, C., AFOCEL, Direction des Recherches,
2 Place Ferber, 69009 Lyon, France
Boratyński, A., Institute of Dendrology,
62–035 Kórnik, Poland
Borzan, Ž., Forestry Faculty, Department of Forest Genetics and Dendrology,
P.O.Box 178, 41001 Zagreb, Yugoslavia
Chałupka, W., Institute of Dendrology,
62–035 Kórnik, Poland
Ericson, B., SLU Department of Forest Yield Research,
S–770 73 Garpenberg, Sweden
Gerhold, H. D., School of Forest Resources,
The Pennsylvania State University,
University Park, PA 16802, USA
Giertych, M., Institute of Dendrology,
62–035 Kórnik, Poland
Hattemer, H. H., Abteilung Forstliche Biometrie,
Büsgenweg 5, 3400 Göttingen, FRG
van Haverbeke, D. F., USDA Department of Agriculture,
Rocky Mountain Forest and Range Experimental Station,
University of Nebraska,
Lincoln, NE 68583, USA
Kosiński, G., Institute of Dendrology,
62–035 Kórnik, Poland
Koski, V., Department of Forest Genetics,
The Finnish Forest Research Institute,
P.O. Box 18, SF–01301, Vantaa 30, Finland
Lindgren, D., SLU Department of Forest Genetics,
S–901 83 Umeå, Sweden
Mátyás, Cs., Forestry Faculty,
Department of Plant Sciences,
P.O.Box 132, 9401 Sopron, Hungary
Mikola, J., The Foundation for Forest Tree Breeding,
Viljatie 4 A 5, SF–00700 Helsinki, Finland

Molotkov, P. I., Ukrainian Forest and Melioration Research Institute,
Pushkinskaya 86, 310024 Kharkov 24, USSR
Monteuuis, O., AFOCEL, Domaine de l'Etancon,
F–77370 Nangis, France
Müller-Starck, G., Abteilung Forstgenetik und Forstpflanzenzüchtung,
Büsgenweg 2, 3400 Göttingen, FRG
Nilsson, J.-E., SLU Department of Forest Genetics,
S–90183 Umeå, Sweden
Oleksyn, J., Institute of Dendrology,
62–035 Kórnik, Poland
Patlaj, I. N., Ukrainian Forest and Melioration Research Institute,
Pushkinskaya 86, 310024 Kharkov 24, USSR
Pihelgas, E., Eesti Pollumajanduse Akadeemia,
Tartu, Estonia, USSR
Prus-Głowacki, W., UAM Department of Genetics,
Ul. Dąbrowskiego 165, 60–594 Poznań, Poland
Ståhl, E. G., SLU Department of Forest Yield Research,
S–7770 73 Garpenberg, Sweden
Stephan, B. R., Institut für Forstgenetik und Forstpflanzenzüchtung,
Sieker Landstrasse 2, 2070 Grosshansdorf 2, FRG

CHAPTER 1

Range of natural distribution

A. Boratyński ← Poland

– does not read well. Difficult to quickly glean information.

The range of genus *Pinus L.*

The genus *Pinus* occupies an extensive area in the Northern Hemisphere, covering primarily the north temperate zone, particularly the boreal forest belt (taiga). In this zone pines, together with species from other genera of the *Pinaceae* family, such as *Abies, Picea* and *Larix,* are the most frequent trees which, depending on the environmental conditions, locally even dominate the natural vegetation. In spite of their abundance, there are not too many pine species in the boreal zone. The majority of species are found in regions situated much further south, in areas linking the temperate and tropical zones, with a particular increase in diversity in the mountains of the southern part of North America, in Central America, as well as in south-eastern Asia (Critchfield and Little, 1966; Mirov, 1967; Little and Critchfield, 1969; Farjon, 1984). In the last-mentioned region are to be found the southernmost natural stands of the genus *Pinus* (Fig. 1.1). *Pinus merkusii* Jungh et De Vriese crosses the equator on Sumatra (Indonesia) and goes as far south as 2° S (Stein, 1978), whereas *P. caribaea* Morelet only reaches 12° N in Nicaragua (Central America) (Critchfield and Little, 1966).

In the far north some pines spread into the subarctic zone, and at times reach the arctic timberline. On the Eurasian continent *P. pumila* (Pall.) Regel has its northernmost stands at about 72°10′ N along the Lower Lena tributaries in Siberia (Sokolov et al., 1977; Koropachinskij, 1983), and *P. sylvestris* L. reaches about 70°20′ N in northern Norway (Hultén, 1950).

In America the northern limit of pines is at latitudes about 10° lower than in Europe and Asia (see Fig. 1.1). *P. banksiana* Lamb. ranges furthest to the north in America to about 62°30′ N in the middle part of the Mackenzie river basin, while *P. contorta* Dougl. occurs up to latitude 62° N in the Upper Yukon (Little, 1971).

Pines also have a large range with respect to altitude, since they grow from the sea level up to the timberline in the mountains. Of all the representatives of the genus, *P. montezumae* Lamb. occurs at highest elevations, at about 4 000 m in the mountains of Mexico (Mirov, 1967; Farjon, 1984).

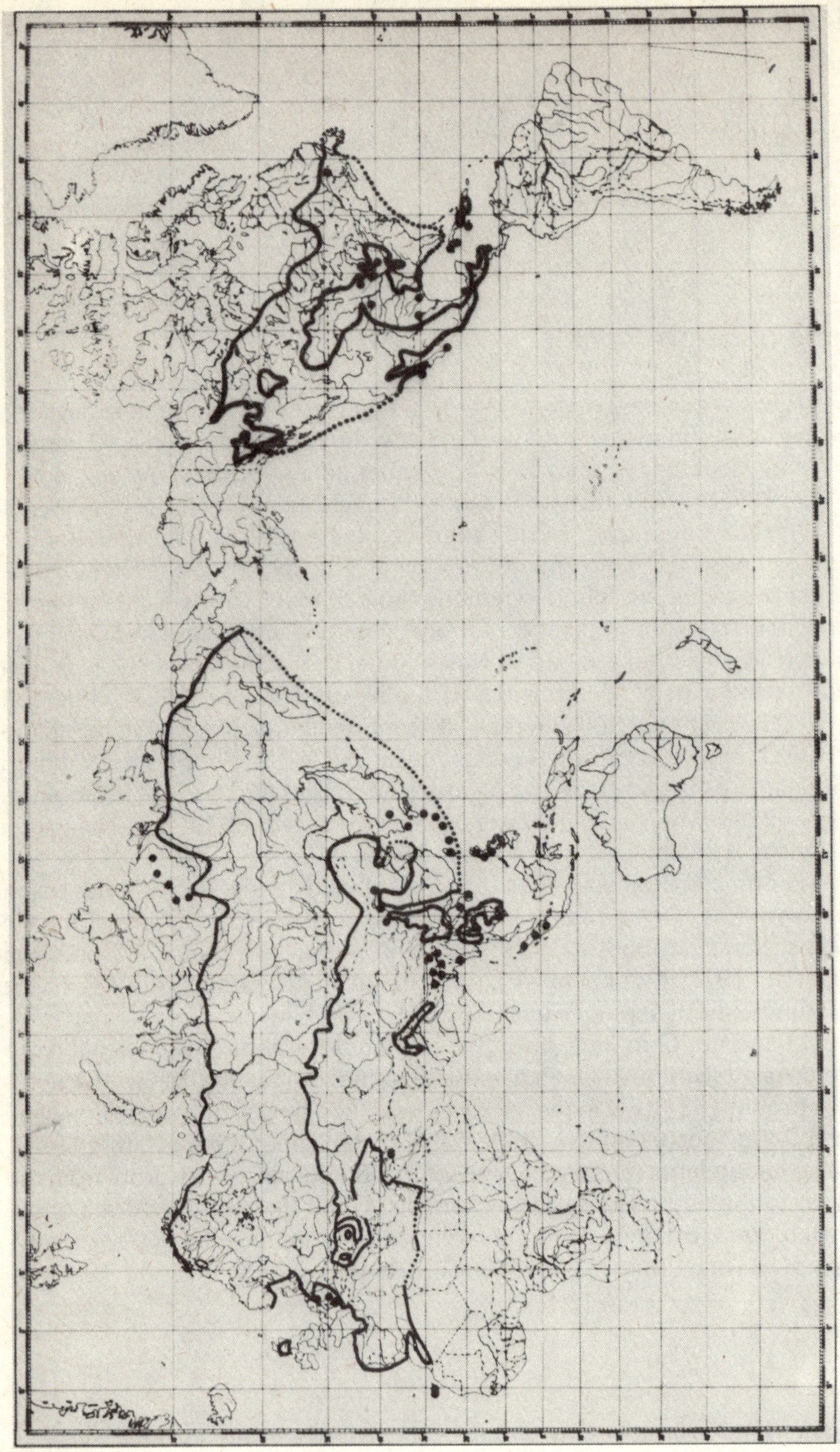

Fig. 1.1. Range of the genus *Pinus*

Natural distribution of *Pinus sylvestris* L.

Scots pine occurs on a very large area and it has formed many geographical races and varieties (Ruby and Wright, 1976). Some of these are relatively easily distinguishable on the basis of their morphology and frequently have been described in terms of lower taxonomical units (varieties or subspecies) of *P. sylvestris* (Gaussen et al., 1964; Pravdin, 1964). Some occurrences, growing most frequently at the limits of the range, have sometimes been discussed as independent species.

In the present account, following the practice of the "Flora Europaea", all of these "small species" are considered only as subspecies or varieties of *P. sylvestris s. l.* (Gaussen et al., 1964), and the range described below (Fig. 1.2) concerns the species understood in this broad sense.

Horizontal distribution

The geographical distribution of *P. sylvestris* is described on the basis of the maps elaborated by Meusel and his co-workers (1965) and by Critchfield and Little (1966), simultaneously taking into account a number of rather insignificant changes and corrections in papers either not considered by the authors mentioned above, or published later (Wu Ch.-L., 1956; Coode and Cullen, 1965; Jasičova, 1966; Eliçin, 1971; Svoboda, 1971; Jalas and Suominen, 1973; Bobrov, 1975, 1978; Fenaroli and Gambi, 1976; Sokolov et al., 1977; Boratyńska et al., 1981; Grubov, 1982; Welten and Sutter, 1982; Barbarich et al., 1986; Christensen, 1986).

Scots pine is characterized by a very extensive range, most extensive of all species of the genus *Pinus,* and even of the whole *Pinaceae* family. The range of Scots pine spreads out longitudinally over a distance of about 14 000 km. The westernmost occurrences of *P. sylvestris* are situated at about 8° W in Spain and the easternmost ones at 141° E in the Soviet Far East. In north-south direction the range of Scots pine covers a distance of about 2 700 km, with the northernmost occurrences at the latitude of 70°20′ N in northern Scandinavia and the southernmost ones at about 37° N in the Sierra Nevada Mountains in Spain (see Fig. 1.2).

In Scandinavia and on the Kola Peninsula *P. sylvestris* reaches the arctic timberline (Hultén, 1950; Sokolov et al., 1977), but further to the east the northern limit of the species stays south of the arctic timberline, which is locally formed there by *Picea obovata* Ledeb., *Larix sibrica* Ledeb. (including *L. sukaczewii* Dyils), *L. gmelini* (Rupr.) Kuzeneva and birch populations.

The course of the northern range limit of *P. sylvestris* is mostly determined by climatic conditions; where continentality is considered to be a restrictive factor, characterized by high temperature amplitudes and low precipitation. It was observed that Scots pine does not grow in extremely continental conditions, where the mean annual temperature amplitude is above 40 °C, the temperature sum of the growing season (above 10 °C) is less than 1 200 °C degree-days and the mean annual precipitation is less than 300 to 350 mm (Polikarpov, 1970, after Koropachinskij, 1983).

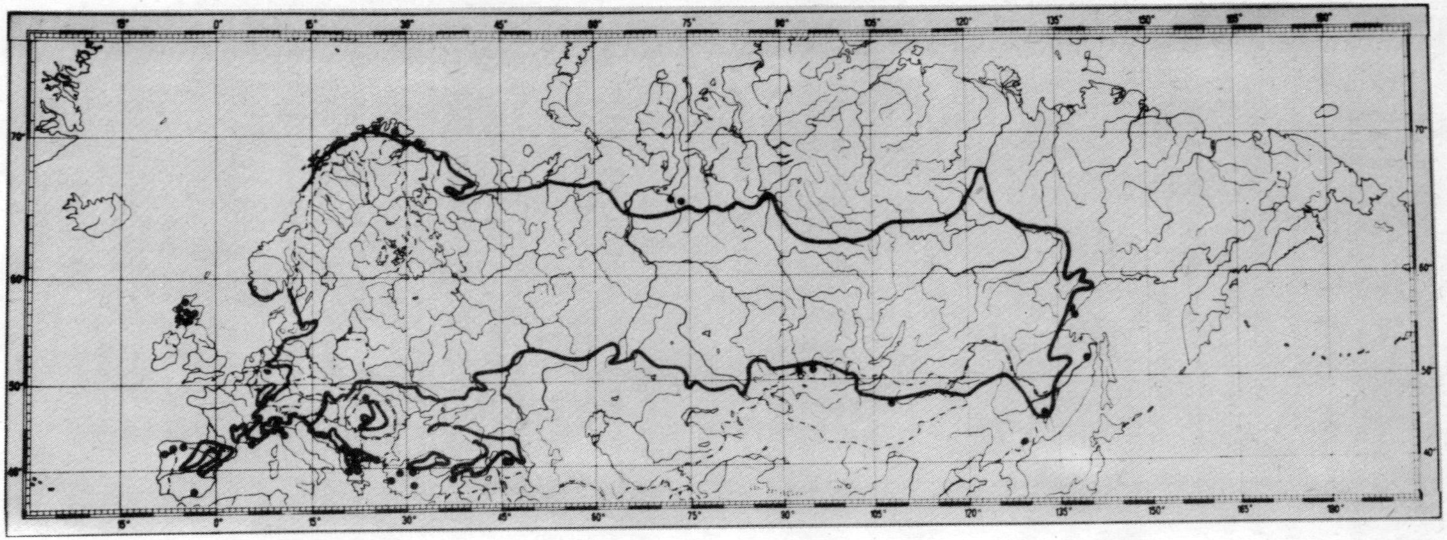
Fig. 1.2. Range of *Pinus sylvestris*

poor map

In north-eastern Europe and on the West-Siberian Lowlands the northern limit of *P. sylvestris* follows occasionally the southern limit of the permafrost zone (Sokolov et al., 1977). It is also considered that the summer isotherm of 13 °C is the northern limit for the maturation of Scots pine seeds in Siberia (Shul'gin, 1975).

On the other hand, the southern limit of distribution range is more closely determined by edaphic and moisture conditions, as well as by the devastating effect of human activities since prehistoric times. This is particularly true in the dry steppe and forest-steppe areas of the Ukraine, Kazakhstan and Mongolia (Polakov and Pavlov, 1956; Svoboda, 1971, Sokolov et al., 1977; Bobrov, 1978). Even so Scots pine is the coniferous species which penetrates most into the steppe zone, and forms the border between forest and steppe over vast areas of the Ukraine, Kazakhstan and Mongolia. In regions, where precipitation is very low and as a rule does not exceed 150–200 mm annually (Dengler, 1944), the species is confined to river valleys and terrain depressions, with a relatively close groundwater level (Polakov and Pavlov, 1956; Svoboda, 1971; Sokolov et al., 1977). Scots pine occurs on hilltops only in areas of loamy soils with better water storage capacity, as on the steppes of Kazakhstan (Polakov and Pavlov, 1956; Koropachinskij, 1975). The very irregular course of the southern border of the range in the steppe region may also be explained therefore by differences in the soil water balance.

In mountainous regions *P. sylvestris* does not reach as far south as some other species of the *Pinaceae* family, but the southernmost natural occurrences are all situated at higher altitudes. These southern Scots pine populations are isolated from the continuous range (see Fig. 1.2), and are considered to be relicts, which have survived from the Ice Age. The classical examples of such relict stands can be found, for instance, in the Grampian Mountains (Scotland), in the Sierra Nevada, the Cordillera Central, the Cordillera Iberica and the Pyrenees on the Iberian Peninsula, in the northern Appenines (Italy), in the Dinarian Alps, the Rodopi, Stara Planina and Pindus Mountains, the eastern and southern Carpathians and Bihor Mountains on the Balkan Peninsula, in the mountains of Crimea and in the Caucasus (USSR), and also in the mountains of northern Anatolia (Asiatic Turkey). Many local varieties and geographical forms of *P. sylvestris* are known from these isolated, mostly mountainous localities.

Altitudinal distribution

At higher altitudes *P. sylvestris* is often part of specific forest associations, scattered over poor sites, where the competition by more demanding tree species is low. Quite frequently the species has its own belt with an upper and a lower limit, determined mostly by climatic conditions. As a rule this forest belt reaches up to the timberline. Scattered, solitary trees may be encountered above the timberline, growing as strongly deformed, dwarf and even prostrate specimens.

The altitudinal range of Scots pine in certain mountain massifs of Europe and Asia is listed in Table 1.1. The highest altitudes are reached in mountains close to

Table 1.1. Altitudinal distribution of *Pinus sylvestris* in the mountains of Europe and Asia

Country	Region	Latitude N [degrees]	Occurrence of Scots pine					References
			lower limit		upper limit			
			scattered [m]	in closed stands [m]	in closed stands [m]	scattered [m]	deform. specim. only [m]	
USSR	Kola Penins.	< 70°				200		Sokolov et al., 1977
	Hibiny Mts	67°30′			360	500	700	
Norway	N part	< 70°				200–250		Rivoli, 1926
	Hardangerwidda and Sogne Mts	60°30′			800	1 040	1 300	Hultén, 1950
Great Britain	Grampian Mts	57°			840			Steven and Carlisle, 1959
France	Vosges Mts	48°				1 200		Rivoli, 1926
	Massif Central	45°		600	1 500	1 600		Chassagne, 1956
Spain	in general		200	500	2 000	2 100		Franco Amaral, 1986
	Pireneos Mts.	42°30′		950	1 600			Willkomm and Lange, 1870
	Sier. Guadarrama	40°40′		1 150	2 000	2 100		Willkomm and Lange, 1870
	Sierra Nevada	37°		1 600	2 100			Willkomm and Lange, 1870; Prieto, 1975
Switzerland	Alps	46–47°		550	1 900	2 250	2 400	Kulesza, 1928; Braun-Blanquet and Rübel, 1932; Hess et al., 1967
Italy	Alps	46°	500			2 000	2 250	Fenaroli and Gambi, 1976
Poland	Tatra Mts	49°10′				1 300	1 510	Pawłowski, 1956; Łysek, 1974
Czechoslovakia	Tatra Mts	49°					1 860	Jasičova, 1966
Yugoslavia	Dinarian Alps	43–44°		1 000	1 600	2 000		Jovanovič, 1970; Šilič, 1973
Bulgaria	Rila and Pirin	42°		850	2 200			Stojanov, 1963

Table 1.1. Cont.

Country	Region	Latitude N [degrees]	Occurrence of Scots pine					References
			lower limit		upper limit			
			scat-tered [m]	in closed stands [m]	in closed stands [m]	scat-tered [m]	deform. specim. only [m]	
Greece	in general	40°30′		1 100	1 800	1 900		Christensen, 1986; Voliotis, 1975
USSR	Caucasus Mts	42°	200	1 000	2 500	2 600	2 700	Sokołowski, 1936; Gulisashvili, 1959
	Altai Mts	51–52°		900	1 150	1 750		Malyshyev, 1965; Sokolov et al., 1977
	Saian Mts	<52°	250	500	1 000	1 850		Koropachinskij, 1975; Sokolov et al., 1977
Mongolia	Bogdo Uul Mts	<48°				1 500		Białobok, 1970
Turkey	Mts of Anatolia	<40°	100	1 250	2 000	2 500		Kayacik, 1954; Coode and Cullen, 1965; Eliçin, 1971

the southern limit of the species, in the Caucasus. Here Scots pine forms more or less close stands up to 2 500 to 2 600 m above sea level, and single, dwarf or prostrate specimens are found even at elevations of 2 700 m (Sokołowski, 1936; Gulisashvili, 1959).

In Europe the altitudinal maximum of Scots pine is reached in the Alps at 2 400 m (Kulesza, 1928; Braun-Blanquet and Rübel, 1932). In other mountains situated at higher latitudes than the Caucasus the upper limit of Scots pine is much lower (see Table 1.1). This is caused, of course, by climatic and environmental conditions, which deteriorate as latitude increases, as well as by the considerably smaller extension of these mountain ranges. It is well known that in less extensive mountainous areas the timberline and the upper limits of tree species reaching the timberline are always lower when compared with the more extensive mountain areas. This is probably the main reason why Scots pine is found at lower altitudes in the mountains south of the Alps.

In the mountains of the southern parts of its range, Scots pine has also a lower altitudinal limit of occurrence (see Table 1.1). This limit lies highest in the southernmost occurrence, the Sierra Nevada in Spain, at 1 600 m altitude (Willkomm and Lange, 1870). To the north the lower limit gradually descends and reaches the sea level in western Europe.

Cultivation

Scots pine has been planted in Central Europe on an ever increasing scale for about 150–200 years (Otto, 1976), leading to its marked presence in the man-made forests of numerous countries (Table 1.2).

In western Europe the planting of Scots pine is being reduced now. A decline of interest is observable first of all in countries where the species has been widely cultivated, as for example in France, Great Britain and the Netherlands.

In Central and eastern Europe, as before, there is still a tendency to increase or at least to maintain the annual area of Scots pine regenerations. About 60% of the total area of Scots pine stands is in the USSR (see Table 1.2).

Besides being planted for timber, Scots pine fulfils also other needs. In Europe Scots pine afforestation is quite frequently used to cover spoil banks in strip-mined areas and gravel, peat or clay pits, dried peat-bogs and maritime or inland dunes, etc. Scots pine is also quite commonly used to establish windbreaks and antierosion belts in arid, mostly steppe regions, as also has been done in Asia, e.g. on the steppes of Kazakhstan (USSR), in Mongolia and in northern China (Anderson and Eiche, 1974; Godnev, 1982; Guo, 1984; Kondratyuk and Polikarpov, 1982; Maksimenko, 1978 and others).

P. sylvestris is also cultivated in North America, mainly in southern Canada and the northern USA. In these countries Scots pine is used for recultivation of stripmine spoils and for windbreaks (Hall, 1977; Davidson, 1979; De Haye et al., 1980), but primarily it is planted for Christmas trees (for further details see Chapters 8 and 21).

Table 1.2. Area of Scots pine forests in selected European countries (after Timber Bulletin 1978, 1979*a*, 1980*a*,*b*, 1981*a*,*b*, 1981*c*, 1982; Rocznik statystyczny leśnictwa 1960, 1971, 1981)

Country (year)	Area (hectares)	Total forest area %
Poland (1956)	4 373 000	73.1
Poland (1967)	4 697 000*	72.5
Poland (1978)	6 035 000*	70.1
The Netherlands (1978)	95 490	45.0
Sweden (1978)	9 982 000	38.8
USSR (1980)	169 131 000**	21.0
Norway (1980)	1 238 000	19.1
Hungary (1983)	129 700	8.8
Switzerland (1979)	45 000***	4.0

* Together with *Larix europaea;*
** Together with *Pinus sibirica* and *P. koraiensis;*
*** Together with *Pinus nigra* and *P. cembra.*

Finally, it should be mentioned that in many countries, there is an interest in growing Scots pine as an introduced exotic species. Examples are Korea, China and even Mexico and New Zealand, where provenance and cultivation trials have been established (Chapa Bezanilla, 1976; Benecke et al., 1978; Benecke and Morris, 1978; Yim, 1980; Chen et al., 1981).

Literature

Anderson, E. and Eiche, V. (1974): Experimental plantations with Scots pine *(Pinus sylvestris)* in the arctic area of Sweden. *Proceedings of the symposium on ecological problems of the circumpolar area, 1971, Luleå.* In: *Forestry Abstracts* (1978) **39:**3667.

Barbarich, A. I. (Ed.) (1986): *Khorologiya flory Ukrainy (Chorology of the Ukrainian flora).* Naukova Dumka, Kiev,

Benecke, U. and Morris, J. Y. (1978): Tree provenance trials. In: Orwin, J. (Ed.) *Revegetation in the rehabilitation of mountain lands.* Symposium, New Zealand Forest Service. In: *Forestry Abstracts,* 1980, **41,** 6363.

Benecke, U., Baker, G. and McCracken, I. J. (1978): Tree growth in Craigeburn Range. In: Orwin, J. (Ed.) *Revegetation in the rehabilitation of mountain lands.* Symposium, New Zeeland Forest Service. In: *Forestry Abstracts,* **41,** 1980, 6362.

Białobok, S. (1970): Rozmieszczenie w Eurazji (Distribution in Eurasia). In: Białobok, S. (Ed.) Sosna zwyczajna *Pinus silvestris* L., 78–85. Państwowe Wydawnictwo Naukowe, Warszawa—Poznań, p. 360.

Boratyńska, K., Boratyński, A., Browicz, K. and Hantz, J. (1981): *Atlas of distribution of trees and shrubs in Poland.* 29. Państwowe Wydawnictwo Naukowe, Warszawa—Poznań.

Braun-Blanquet, J. and Rübel, E. (1932): Flora von Graubünden. 1. *Veröffentlichungen des Geobotanischen Instituts Rübel in Zürich* **7:**1.

Chapa Bezanilla, M. del C. (1976): Principales técnicas de cultivo para 'arboles de Novidad'. Boletin Divulgatio Instituto Nacional de Investigaciones Forestales, Mexico, **41**:34. In: *Forestry Abstracts,* 1977, **38**:5740.

Chassagne, M. (1956): Inventaire Analytíque de la Flore d'Auvergne, 1. Encyclopédie Biogéographique et Écologique. 11, Lechevalier, Paris, p. 460.

Chen, T. Y., Yu, D. F., Utong, Z. A. and Li, X. Y. (1981): A preliminary study on the early growing forecast of *Pinus sylvestris* var. *mongolica. Scientia Silvae Sinicae* **17,1**:74–77. In: *Forestry Abstracts,* 1985, **45:** 5341.

Christensen, K. I. (1986): *Pinus* L. In: Strid, A. (Ed.) *Mountain flora of Greece,* **1**:41–44. Cambridge University Press, Cambridge.

Coode, M. J. E. and Cullen, J. (1965): *Pinus* L. In: Davies, P. H. (Ed.) *Flora of Turkey,* **1**:72–75. Edinburgh University Press, Edinburgh.

Critchfield, W. B. and Little, E. L. jr. (1966): Geographic distribution of the pines of the World. U. S. Department of Agriculture, Forest Service Misc. Publ.

Davidson, W. H. (1979): Results of tree and shrib planting on low pH strip-mine banks. U. S. Department of Agriculture, Forest Service Research Note NE, 285,5.

De Haye, D. H., Dickmann, D. I. and Lemmien, W. A. (1980): Management of exotic timber trees in temperate regions of eastern North America: an assessment. *Forest Ecology and Management* **3,1**:31–44.

Dengler, A. (1944): Waldbau auf ökologischer Grundlage. Springer, Berlin.

Eliçin, G. (1971): Sous-espèce écologique du pin sylvestre (*Pinus sylvestris* L.) en Turquie. Istanbul Üniversitesi, Orman Fakültesi Dergisi, A, **20,9**:189–199 (1970).

Farjon, A. (1984): *Pines.* Brill, Leiden.

Fedorov, A. A. (Ed.) (1974): *Flora evropeiskoi chasti SSSR (Flora partis Europaeae URSS).* Vol. 1. Nauka, Leningrad.

Fenaroli, L. and Gambi, G. (1976): *Alberi.* Museo Tridentio de Scienze Naturali, Trento.

Franco Amaral, J. do (1986): *Pinus* L. In: Lopez Gonzáles, G. (Ed.) *Flora Iberica* **1**:168–174. Real Jardin Botánico, Consejo Superior de Investigaciones Cientificas, Madrid.

Gaussen, H., Heywood, V. H. and Chater, A. O. (1964): *Pinus* L. In: Flora Europaea, **1**:32–35. Oxford University Press.

Godnev, E. D. (1982): Vyrashchivanie kul'tur sosny v sukhikh stepyakh (Growing Scots Pine in the dry steppes). *Lesnoe Khozyaistvo* **1**:31–36.

Grubov, V. I. (1982): *Opredelitel' sosudistykh rastenii Mongolii (Key to the vascular plants of Mongolia).* Nauka, Leningrad.

Gulisashvili, V. Z. (1959): *Dendroflora Kavkaza (Dendroflora of the Caucasus).* 1:132–217. Gruzinskaya Akademia Nauk, Tbilisi.

Guo, L. (1984): Afforestation of sandy deserts in Inner Mongolia (China). *Forstarchiv.* **55,1**:27-29.

Hall, J. P. (1977): Survival and growth of exotic *Pinus* species in central Newfoundland ten years after planting. Informatic Report Newfoundland Forest Research Centre, Canada, N–X–154.

Hess, H. E., Landolt, E. and Hirtzel, R. (1967): *Flora der Schweiz.* Vol. 1. Birkhäuser, Basel und Stuttgart.

Hultén, E. (1950): *Atlas of the distribution of vascular plants in north-west Europe.* Generalstabens Litografiska Anstalt, Stockholm.

Jalas, J. and Suominen, J. (1973): *Atlas florae Europaeae.* Vol. 2. Committee for Mapping the Flora of Europe and Societas Biologica Fennica Vanamo, Helsinki.

Jasičova, M. (1966): *Pinus* L. In: Futák, J. (Ed.) *Flora Slovenska (Flora of Slovakia).* 2:278–294. Vydavatelstvo Slovenskej Akadémie Vied, Bratislava.

Jovanović, B. (1970). *Pinus* L. In: Josifović, M. (Ed.) *Flora SR Srbije (Flora de la République Socialiste de Serbie).* Beograd, **1**:145–155.

Kayacik, H. (1954): Türkiye çadamari ve bunlarin cografi yayilişlari üzerinde araştirmalar (Pines in Turkey and an investigation about their geographical distribution). Istanbul Üniversitesi, Orman Fakültesi Dergisi, **1–2**:44–64.

Komarov, V. L. (Ed.) (1934): *Flora SSSR (Flora of the URSS).* Nauka, Leningrad.

Kondratyuk, E. N. and Polikarpov, A. K. (1982): Puti uskorennoj introduktsii sosny v Donbasse (Ways of accelerating the introduction of pine into the Donbas industrial region (Eastern Ukraine)). *Byul, Glavn. Botan. Sada* **124**:8–13.

Koropachinskij, I. Yu. (1975): *Dendroflora Altaisko–Sayanskoj gornoj oblasti (Dendological flora of the Altai and Saian Mts.).* Nauka, Novosibirsk.

Koropachinskij, I. Yu. (1983): *Drevesnye rasteniya Sibiri (Trees and shrubs of Siberia).* Nauka, Novosibirsk.

Koropachinskij, I. YU. and Skvortsova, A. B. (1966): *Dereva i kustarniki Tuvinskoi ASSR (Trees and shrubs of the Tuvinian ASSR).* Nauka, Novosibirsk.

Kulesza, W. (1928): Doświadczenia z dziedziny proweniencji nasion sosny pospolitej (*Pinus silvestris* L.) w Eglisau i Samaden w Szwajcarii (Provenance trials of the Scots pine (*Pinus silvestris* L.) in Englisau and Samaden in Switzerland). *Sylwan* **46,2**:96–103.

Little, E. L. jr. (1971): *Atlas of United States trees.* Vol. 1. U. S. Department of Agriculure, Forest Service, Misc. Publ. 1146.

Little, E. L. jr. and Critchfield, W. B. (1969): *Subdivision of the Genus Pinus (Pines).* U. S. Department of Agriculture, Forest Service Misc. Publ. 1144.

Łysek, S. (1974): Sosna *Pinus silvestris* L. (Scots pine). In: Myczkowski, S. (Ed.) Rodzime drzewa Tatr (Native trees of the Tatra Mts.). Part 1. *Studia Ośrodka Dokumentacji Fizjograficznej* **3**:87–110.

Maksimenko, A. P. (1978): Sozdanie kultur sosny na peshchano-rakushechnykh pochvakh (Establishment of pine plantations on sand-shell lime soils). *Lesnoe Khozyaistvo* **79**:52–54.

Malyshev, L. I. (1965): *Alpinskaya flora Vostochnogo Sayana (Alpine flora of the East Sayan Mts.).* Nauka, Moskva–Leningrad.

Meusel, H., Jäger, E. and Weinert, E. (1965): *Vergleichende Chronologie der Zentraleuropäischen Flora.* Vol. 1. Parts 1 and 2. Gustav Fischer, Jena.

Mirov, N. T. (1967): *The Genus Pinus.* Ronald Press Company, New York.

Otto, H. J. (1976): Vergangenheit und Zukunft der Kiefernwälder im norddeutschen Tiefland (Past and future of the Pine forests in the North German lowlands). *Unser Wald* **28,2**:44–48.

Pawłowski, B. (1956): *Flora Tatr (Flora of the Tatras).* Vol. 1. Państwowe Wydawnictwo Naukowe, Warszawa.

Polakov, P. P. and Pavlov, I. V. (1956): *Pinus* L. In: Pavlov, I. V. (Ed.) *Flora Kazakhskoj SSSR (Flora of Kazakhstan).* **1**:68–70. Akademiya Nauk Kazakhskoi SSR, Alma-Ata.

Pravdin, L. T. (1964): *Sosna obyknovennaya (Scots pine).* Nauka, Moskva.

Prieto, P. (1975). Los bosques de Sierra Nevada. Algunos consideraciones historicas, ecologicas y fitosociologicas sobre las forestales autoctonas de las regiones mas elevadas de la Peniberica (The forests of the Sierra Nevada. Some historical, ecological and plant sociological considerations on the native forests of the highest altitude regions of the Iberian peninsula). *Ann. Inst. Bot. A. J. Cavanilles* **32,3**:1099–1129.

Rivoli, J. (1926): *Ogólny zarys geografii leśnictwa (General outline of the forest geography).* Fiszer et Krajewski, Poznań.

Rocznik Statystyczny Leśnictwa i Przemyslu Drzewnego (Yearbook of the forestry and forest industry), 1960. Ministerstwo Leśnictwa i Przemyslu Drzewnego, Warszawa.

Rocznik Statystyczny Leśnictwa (Yearbook of forestry), 1971. Główny Urzad Statystyczny, Warszawa.

Rocznik Statystyczny Leśnictwa i Gospodarki Drewnem (Yearbook of forestry and wood economy), 1981. Główny Urzad Statystyczny, Warszawa.

Ruby, J. L. and Wright, J. W. (1976): A revised classification of geographic varieties in Scots pine. *Silvae Genet.* **25,5–6**:169–175.

Shul'gin, V. A. (1975): Geograficheskaya izmenchivost' nekotorykh priznakov i svoistv sosny obyknovennoj v usloviyakh Komi ASSR (Geographical variation of some characters and properties of Scots pine in the Komi ASSR). In: Genetika, selektsiya, semenovodstvo i introduktsiya lesnykh porod, Moskva

Šilić, Č. (1973): *Atlas drveća i grmlja (Atlas of trees and shrubs).* Vol. 1. Sarajevo.

Sokolov, S. Ja., Svyazeva, O. A. and Kubly, V. A. (1977): *Arealy derevev i kustarnikov SSSR (Distribution of trees and shrubs in the USSR).* Vol. 1. Nauka, Leningrad.

Sokołowski, M. (1936): Przyczynek do znajomości reliktowej sosny kaukaskiej (Contribution to the knowledge of the relict pine in the Caucasus). *Sylwan* **54,2:**117–163.

Stein, N. (1978): Coniferen in westlichen Malayischen Archipel. *Biogeographica* **11:**168.

Steven, H. M. and Carlisle, A. (1959): *The native pinewoods of Scotland.* Oliver and Boyd Ltd., Edinburgh.

Stoyanov, V. (1963): *Pinus* L. In: Iordanov, D. (Ed.) *Flora na Narodna Republika B'lgariya (Flora of the Bulgarian People's Republic).* **1**:160–166. Academia Scientiarum Bulgariae, Sofia.

Svoboda, A. M. (1971): Contribution aux recherches sur la variabilitate des cónes du Pin sylvestre (*Pinus silvestris* L.) et sur son extension en Mongolie. *Frag. Flor. Geobot.* **17,3:**367–374.

Timber Bulletin for Europe, Suppl., Forestry and forest products. Country profiles: 1978, Vol. 31, 3, 2. Poland; 1979a, Vol. 31, 3, 3. Sweden; 1979b, Vol. 32, 5, 4. Switzerland; 1980a, Vol. 32, 6, 5. Norway; 1980b, Vol. 33, 5. USSR; 1981a, Vol. 33, 7. Turkey; 1981b, Vol. 33, 10. The Netherlands; 1981c, Vol. 34, 5. USA; 1982, Vol. 35, 7. Hungary.

Voliotis, D. (1975): Die Vegetationsstufung einiger Gebirge in Nordgriechenland (Voras, Vermion, Pieria). Problems of Balkan flora and vegetation. *Proceedings of the first international symposium of Balkan flora, Varna, June 7–14, 1973,* pp. 391–399.

Welten, M., Sutter, R. H. C. (1982): *Verbreitungsatlas der Farn- und Blütenpflanzen der Schweiz.* Vol. 1. Birkhäuser, Basel.

Willkomm, M. and Lange, J. (1870): Prodromus Florae Hispanicae, Vol. 1. *Schweizerbart. Stuttgartiae,* 316.

Wu, Ch.-L. (1956): The taxonomic revision and phyto-geographical study of Chinese pines. *Acta Phytotaxonomica Sinica* **5:**131–164.

Yim, K. B. (1980): Pines of Korea. *Symposium on Scots pine forestry of the future.* Institute of Dendrology, Kórnik, Poland, pp. 48–60.

CHAPTER 2

Systematic position within the genus *Pinus* and intraspecific taxonomy

P. I. Molotkov and I. N. Patlaj

Systematic position within the genus *Pinus*

Scots pine (*Pinus sylvestris* L.) belongs to the class *Coniferae,* order *Pinales* and family *Pinaceae.* The genus *Pinus* includes more than 100 species that are concentrated in the northern hemisphere. Many classifications of this genus exist (Endlicher, 1847; Shaw, 1914; Pilger, 1926; Novak, 1930; Duffield, 1952; Gaussen et al., 1964; Mirov, 1967; Little and Critchfield, 1969).

Pilger's system (1926) is widely used in many countries including the USSR where Maleev (1949) made some revisions. This system is used in this account.

The genus *Pinus* is divided into two subgenera: *Haploxylon,* with a single vascular bundle in the needles, and *Diploxylon* having two bundles.

The subgenus *Diploxylon* includes eight sections. Section *Eupitys* is the most numerous one. Species belonging to this group have 2 (sometimes 3) needles in a fascicle, the resin ducts are either peripheral or internal, the sheaths are persistent, the cones are small and dehiscent, and the seed wing is completely developed. This section includes *P. sylvestris.*

Intraspecific variability and taxonomy

The vast range of Scots pine over extremely changing forest conditions and the long history of the species cause its great variability. On the basis of palaeontological data, the genus *Pinus* appeared on Earth in the Mesozoic era (Florin, 1963), and *Pinus sylvestris* was already growing in Europe in the Tertiary period. It is believed that the species developed in East Asia and then spread into Europe. During the glaciations and interglacial periods of the Pleistocene the range of *P. sylvestris* repeatedly changed in latitudinal, longitudinal and altitudinal directions and numerous new populations were formed. All these factors increased variability and made the creation of a rational and uniform intraspecific taxonomical system very difficult (Staszkiewicz, 1970). Since 1829 repeated attempts have been made to define new species within *P. sylvestris.* To the present at least 11 suggestions of this kind have been made, as for example *P. sosnowskyi* Nakai, *P. hamata* (Stev.) Sosn., *P. armena* Koch and *P. kochiana* Klotzsch in the Caucasus, *P. fomini* Kondr. in Ukraine or *P. lapponum* (Fries ex Hartm.) Mayr in northern Scandinavia and the Kola Peninsula (Komarov, 1934; Czeczott, 1939; Fedorov 1974; Bobrov, 1975, 1978; Eremin, 1976; Barbarich et al., 1986).

The most detailed intraspecific systematics and taxonomy for the species *Pinus sylvestris* were suggested by Pravdin (1964), based on works of numerous Russian and foreign authors. Pravdin's system provides for the separation of: (1) species sensu lato; (2) subspecies; (3) geographic races after Korzhinskij (1892); (4) geographic ecotypes after Turesson (1925); (5) climatic ecotypes; (6) edaphic ecotypes, and (7) populations composed of forms, the diversity and frequency of which determine the degree of polymorphism.

Pravdin analysed his own rich material on the variation of Scots pine that has been gathered from the USSR, Hungary, Sweden and France and he summarized the data published in Switzerland (Engler, 1913), Germany (Schott, 1907; Dengler, 1908; Kienitz, 1911; Münch, 1924; Wiedeman, 1930; Rubner, 1960), Poland (Sokołowski, 1931), Scotland (Steven and Carlisle, 1959) and Sweden (Langlet, 1944; Lindquist, 1956). He concluded that a subdivision into 5 subspecies (*sylvestris, hamata, lapponica, sibirica* and *kulundensis*) is justified. Persistence of the needles, the number and location of resin ducts in the needles, needle length, longevity of the trees, their dimensions, growth habit, size, weight and colour of cones and seeds were used for the diagnosis. A short description of the subspecies is given below.

Pinus sylvestris L. ssp. *sylvestris* grows in the European part of the USSR and in western Europe south of latitude 62° N (with the exception of the Crimea and the Caucasus). Trees live as long as 500 years, attaining up to 40 m height and 1–1.5 m stem diameter. Annual growth is monocyclic. The branches are light brown or yellowish. The flaky bark is reddish brown. The crown may be either narrow or broad. Buds are conical and resinous in winter. The needles are (3–)4–6(–10) cm long and persist on the tree for 2–3, seldom up to 4–5 years, turning yellowish during winter. Resin ducts in the needles vary from eight to twelve (maximum 15) and are directly adjacent to the hypodermis. Cones are (2–)3–5 cm long. Male catkins are mainly yellow, seldom pink. Thousand seed weight is 4–7 g. Seeds are yellow, black, grey, brown or motley.

The distribution of *P. sylvestris* L. ssp. *sylvestris* climatypes (after Schott, 1907; Svoboda, 1953; Klika and Novak, 1953) (Fig. 2.1) may be geographically grouped, as follows:

(a) North European climatypes
— var. *septentrionalis* Schott: Central, South and Southwest Scandinavia;
— var. *scotica* Schott: Scotland;
— var. *rigensis* (Desf.) Aschers. et Graeb.: Estonia, Latvia, Lithuania and northeastern Poland.

(b) Lowland climatypes in western Europe
— var. *batava* Schott: introduced populations of the Rhine delta region;
— var. *borussica* Schott: GDR and Poland;
— var. *haguenensis* Loud. (syn. *superrhenana* Schott): Upper Rhine valley;

Out of the North and western European climatypes var. *borussica, rigensis* and *haguenensis* are the most productive.

(c) Mountain climatypes in the western Mediterranean
— var. *iberica* Svob.: North and Central Spain;

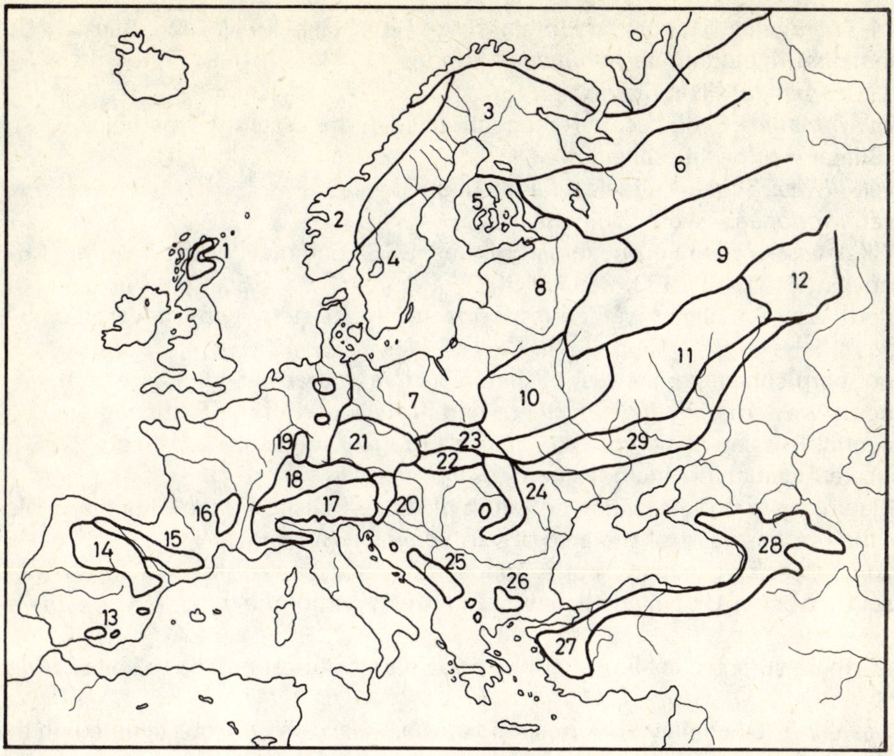

Fig. 2.1. Taxonomic subdivision of the European area of Scots pine according to Svoboda (1953). 1. *P.s. scotica*, 2. *P.s. norvegica*, 3. *P.s. lapponica*, 4. *P.s. suecica*, 5. *P.s. fennica*, 6. *P.s. borealis*, 7. *P.s. borussica*, 8. *P.s. baltica*, 9. *P.s. rossica*, 10. *P.s. ucrainica*, 11. *P.s. sarmatica*, 12. *P.s. bashkirica*, 13. *P.s. nevadensis*, 14. *P.s. iberica*, 15. *P. pyrenaica*, 16. *P.s. aquitana*, 17. *P.s. alpina*, 18. *P.s. vindelica*, 19. *P.s. superrhenana*, 20. *P.s. pannonica*, 21. *P.s. hercynica*, 22. *P.s. carpatica*, 23. *P.s. polonica*, 24. *P.s. romanica*, 25. *P.s. illyrica*, 26. *P.s. rhodopaea*, 27. *P.s. pontica*, 28. *P.s. caucasica* and *P.s. stepposa*, 29. *P.s. scythica*

— var. *nevadensis* Christ: Sierra Nevada Mountains of southern Spain. This occurrence is the southernmost;
— var. *pyrenaica* Svob.: eastern Pyrenees;
— var. *aquitana* Schott: France, Massif Central Mountains.
(d) Climatypes of Central Europe
— var. *alpina* Svob.: including var. *briganthiaca* Gaussen (Western Alps), var. *engadinensis* Heer (Swiss Alps, Tyrol) and var. *vocontiana* Guinier et Gaussen (north-western slopes of the Swiss Alps);
— var. *vindelica* Schott: Jura and Black Forest Mountains;
— var. *pannonica* Schott: West Hungary;

— var. *hercynica* Münch: mountains between Black Forest and Silesia (Ore Mountains), including var. *bohemica* Simon;
— var. *carpatica* Klika: Slovakia;
— var. *romanica* Svob.: scattered occurrences in the eastern Carpathians.
(e) Climatypes of the Balkans:
— var. *illyrica* Svob.: Yugoslavia and Albania;
— var. *rhodopaea* Svob.: Bulgaria.
(f) Climatypes of the European part of the USSR (Samofal, 1925; Pravdin, 1964, Obnovlenski, 1940, 1953, Patlaj, 1965):
— north-eastern climatype: corresponds to the eastern part of var. *borealis* Svob.;
— northern climatype: the rest of the area of var. *borealis* Svob.;
— north-western climatype: northern and eastern part of the Baltic region;
— central Russian climatype: corresponds roughly to var. *rossica* Svob.;
— western climatype: includes partially var. *rigensis;*
— Bielorussian–Polesie climatype: south of White Russia and the Ukrainian Polesie (the northern part of the area of var. *ucrainica* Svob.);
— south Russian climatype: corresponds practically to var. *sarmatica* Zapalowicz;
— south-west or Ukrainian climatype (southern part of the range of var. *ucrainica*);
— southern steppe climatypes: outlier populations corresponding to var. *scythica* Svob.;
— Bashkir–Tatar climatype: corresponds to var. *baschkirica* Svob., situated on the left bank of the river Volga;
— central pre-Ural climatype; and
— southern Ural climatype.

The taxa *Pinus sylvestris* L. var. *cretacea* Kalinicz, growing on Cretaceous hills in the south of the European part of the USSR and var. *nana* Pallas (its synonyms are var. *sphagnicola* Ruprecht, var. *turfosa* Woerlein and several others), exhibit poor growth and occur in various parts of the range. These should be classified as clearly expressed edaphic ecotypes of Scots pine.

Pinus sylvestris L. ssp. *hamata* (Steven) Fomin is concentrated in the Caucasus and in the Crimea. Its life-span is up to 200 years. The height is up to 20 (35) m and the diameter up to 1 m. There is often an irregular thickening at the stem base. The reddish yellow bark is covered with thin flakes. The oval, apiculate buds are resinous during the winter. The bluish green needles are up to 7 cm long, persisting 3–4 years, do not turn yellow in the winter. There are 4–8 resin ducts in the needles, as a rule, adjacent to the hypodermis. There are a few sclerenchyma cells in the central cylinder. Male strobili are mostly yellow, single anthers have a pink tint. Cones are yellowish (brownish) with very convex shields (umbos), lustrous, as if lacquered, like a blunt-pointed hook. Seeds are brown, 5 mm long.

The following climatic (altitudinal) ecotypes have been recognized:

— var. *kochiana* (Klotsch) Fomin: south-western Transcaucasus, Turkey and West Iran;

— var. *subalpina* Fomin (syn. var. *caucasica* Svob., var. *pontica* Svob.): Caucasian subalpine region and the Crimea;
— var. *armena* (Koch) Pravdin, to which the North Caucasian mountain ecotype should also be added.

Pinus sylvestris L. ssp. *lapponica* Fries. grows in Europe and Asia, north of latitude 62° N. The life-span is up to 300 years. The height is up to 20 m, the diameter seldom exceeds 20 cm. Forms with narrow crowns prevail. Needles are broad and thick, greyish green, persisting 6–7 (8–9) years, rather short, 2.8–3.5 cm long. Needles turn yellow in the autumn and winter. There are (4–)8–12(–16) resin ducts per needle. Cones are 3–4 cm long, straw yellow or brown. The apophysis is usually flat (plana type). Seeds are brownish yellow, black or yellowish white, 1000 seed weight is between 2 and 4 g.

Pravdin (1964) identified 2 climatic ecotypes:
— var. *occidentalis* Pravdin: in regions with a more mild climate; and
— var. *orientalis* Pravdin: in regions where conditions seem more continental, roughly in the zone of permafrost (longitude 60° E being the approximate border).

The western climatype of Scandinavia (also Kola Peninsula and Karelia) has been studied to some extent, while the eastern climatype is insufficiently known.

Pinus sylvestris L. ssp. *sibirica* Ledebour grows in the Asian part of the USSR between latitudes 52° and 62° N. The height is up to 25–30 m and the maximum diameter exceeds 1 m. The dark-brown bark covers 2/3 of the stem and is light yellow above. Both narrow and broad crown forms are frequent. The needles are green and turn yellow for the winter. They persist more than 5–(8–9) years. Average needle length is 4 to 5.5 (10) cm. There are (8–)9–14(–16) resin ducts per needle, in both peripheral and parenchymal position.

Cones are (2.5–)4–5(–5.5) cm long. All apophysis forms are common. The colour of the cone varies from light yellow to dark brown. Seeds are yellow, brown, black or motley, 1000 seed weighing 5 to 6 g.

Within the ssp. *sibirica* the following climatypes were proposed by Pravdin (1964) and Iroshnikov (1977):

1. Ural (syn. var. *uralensis* Fischer); *2.* Northern Altai (syn. var. *altaica* Ledebour); *3.* Upper Ob; *4.* Middle Ob (the latter two roughly correspond to var. *objensis* Pravdin); *5.* Chulimo-Yenisei; *6.* Middle Yenisei (climatypes number 5 and 6 seem to correspond to var. *eniseensis* Pravdin); *7.* Sayan; *8.* Angara; *9.* Prebaikal; *10.* Selengin; *11.* South-eastern Transbaikal; *12.* Zeisk; *13.* Upper Lena; *14.* Tungussk; *15.* Vitimo-Olekminsk; *16.* Aldansk; and *17.* Central Yakutia climatype (syn. var. *jakutensis* Sukaczew, var. *annulata* Jurinson).

The enormous range of subspecies *sibirica* and the fact that it has been insufficiently studied yet suggest that in the future more climatic ecotypes will be recognized.

Recent investigations by the Forest Institute of the Academy of Science of the USSR suggest two additional subspecies in Siberia: one in Yakutia, in the permafrost zone (ssp. *jakutensis* Iroshnikov) and another in Transbaikalia in the Dahurian floral region (ssp. *bajkalii* Iroshnikov).

Pinus sylvestris L. ssp. *kulundensis* Sukaczew grows in the steppe zone of the Asian part of the USSR, in isolated forest patches, south of latitude 52° N. Tree height is up to 20–25 m, the diameter may surpass 1 m and the life-span is up to 200 years. The bark is thick, dark grey in the lower part of stem (to the half or the lower two third), and is light brown above. The crowns are either broad or narrow, seldom pyramidal. The needles are of medium length, (6–)8–10(–11) cm, persisting 6 to 7 years, yellowish in winter. Resin ducts average 12 to 14 per needle, maximum 20 (–26). Alongside peripheral resin ducts parenchymal ones are observable. There is a third, additional, vascular bundle that is not always developed. Cones are large, 5 to 6 (7) cm long, either light grey or dark-brown. Seeds are large, 1000 seeds weighing (8–)9–10(–12) g, of black, motley, brown or light colour.

The taxon var. *galofitii* Krupenikov tolerates alkaline soil in the north-east of Kazakhstan, in particular in the Naurzumskij pine forest. It is a typical edaphic ecotype of ssp. *kulundensis* Sukaczew.

Chudny (1982) confirmed the taxonomic range of the above-mentioned subspecies within the USSR borders through the investigation of monoterpenes. (The same investigations supported also the existence of racial differences at subspecies level in Europe.)

The diversity of forms

Unlike races and ecotypes, the forms are separated on the basis of either single or few traits of qualitative or quantitative character that deviate from the typical (crown, stem, needle, etc. forms).

P. sylvestris forms can also be classified on the basis of many other traits, which are unfortunately insufficiently studied, such as anatomical and biochemical characters (Dukharev, 1979) including resin yield (Prokazin, 1959, Chudny, 1977), resistance to fungal diseases and pests.

List of described Scots pine forms

Forms based on crown and stem traits

f. *'Albyns'* Krüssman: prostrate stem, needles greyish green

f. *anguina* hort. Beissner: twigs growing in different directions, stem twisted, needles bluish

f. *ascensa* Carlisle: narrow crown, rapid growth, thick, greyish green needles

f. *colummaris compacta* Croux: columnar dwarf crown with short straight twigs and green, thick needles

f. *columnaris* Croux: columnar crown

f. *compressa* Carrière: slow growing, with branches pressed to the stem, similar to f. *fastigiata*

f. *fastigiata* Carrière: narrow pyramidal crown

f. *fastigiata glauca* Schelle: narrow pyramidal crown, needles bluish

f. *genevensis* Beissner: shrub with twisted spreading stems

f. *globosa* hort. Pardé: spherical crown

f. *globosa nana* hort. Carlisle: dwarf form, globular crown

f. *globosa viridis* hort. Beissner: dwarf form, globular crown, needles bright green

f. *'Hibernia'* Krüssman: dwarf form, buds red, needles short, blue

f. *horizontalis* D. Don: crowns broad, rapid growth, good stem form

f. *pendula* Caspary: weeping, pendulous twigs

f. *plicata* Schenk: dwarf form without main stem

f. *pumila* Beissner: up to 2 m in height, oval crown, needles bluish green

f. *pygmaea* hort. Hornibrook: suppressed, globular crown, needles bluish green

f. *pyramidalis compacta* Hornibrook: pyramidal dense crown, dwarf form

f. *pyramidalis glauca* hort. Beissner: pyramidal crown, needles bluish green

f. *saxatilis* Carrière: dwarf form from rocky sites

f. *tortuosa* Buttner: very twisted stem and twigs

f. *turfosa* Woerlein: dwarfish and crooked form with short needles and small cones, growing in bogs

f. *umbraculifera* Beissner: dwarf form, broad crown, globular and dense, needles glaucescent

f. *virgata* Caspary: crooked stem and twisted twigs

f. *watereri* Beissner: crowns intermediate between pyramidal and columnar

Forms based on needle characters

f. *alba* hort. Krüssman: needles greyish blue, apical shoots reddish brown with intensively blue and curved needles

f. *argentea* Steven: needles silvery, rapid growth

f. *argentea compacta* hort. Beissner: needles silvery grey, bark of young shoots yellowish green, dense crown

f. *argentea* Steven: needles silvery, rapid growth

f. *aurea* Beissner: needles golden yellow, small tree

f. *aureapicta* Krüssman: needle tips golden yellow

f. *baenitzii* Tscherning: needles small, twigs thin

f. *beissneriana* Schwerin: needles first green, later yellow

f. *brachyphylla* Wittrack and Beissner: needles small, twigs thin

f. *brevifolia* Heer: needles small, twigs thin

f. *cristata* Schwerin: needles long, curved, rapid growth

f. *divaricata* Wahleb and Beissner: needles brush-like

f. *glauca compacta* Krüssman: needles dark bluish green, poor growth

f. *glauca* Krüssman: needles intensively blue, crown globular

f. *'Kamon blue'* Krüssman: needles bluish, crown globular

f. *latifolia* Cordon: needles broad, long, glaucous, rapid growth

f. *microphylla* Elwes: needles small, thin branches

f. *monophylla* Hodgins: two needles fused into one
f. *nivea* Schwerin: young needles white, probably a chlorophyll mutant
f. *parvifolia* Heer and Beissner: needles small, thin branches
f. *spiralis* Carrière: needles large, twisted, rapid growth
f. *variegata* Beissner: needles mottled
f. *viridis compacta* hort. Hornibrook: needles grass green, crown compact, narrow and conical

Forms based on bark traits

f. *annulata* Caspary: bark scales arranged in rings around the stem
f. *bonapartei* Seitz: stem covered with small, thin, brown scales
f. *corticans* Domin: bark covered by a cork layer
f. *gibberosa* Kihlmann: 25 to 50 cm broad tuberosities scattered on the stem
f. *kienitzii* Seitz: bark similar to larch with deep furrows
f. *laevigata* Schwerin: bark smooth, without cracks
f. *rubra* Carrière and Beissner: young bark smooth and reddish
f. *seitzii* Schwerin: bark lamellar, like a tortoise shell

Forms based on traits of reproductive organs

f. *baliosperma* Kurdiani: seeds mottled
f. *cinnamomeoconus* Pravdin: mature cones beige
f. *conglomerata* (syn. f. *conis aggregatus* Carrière): cones clustered as needle fascicles
f. *erythroconus* Pravdin: mature cones brown
f. *erythranthera* Sanio: ripe male strobili lilac-red, carmine, or reddish
f. *gibba* Christ: cone apophysis convex
f. *leucosperma* Kurdiani: seeds yellow
f. *macrocarpa* Schröter: cones 2–3 × normal size
f. *melanosperma* Kurdiani: seeds black
f. *microcarpa* Schröter and Betelini: cones much smaller than normal
f. *monticola* Schröter: abundant male strobili, needles persist 8 years
f. *phaeosperma* Kurdiani: seeds brown
f. *plana* Christ: cone apophysis flat and blunt
f. *reflexa* Heer (syn. f. *hamata* Steven): cone apophysis with a hook
f. *sulfuranthera* Kozubov: ripe male strobili yellow
f. *uncinata* D. Don (syn. f. *scariosa* Loddiges): cone apophysis bent backwards and blunt

Forms based on wood quality

f. *alba* Senilis: wood cream-coloured
f. *'Condovaya'* Tkachenko: wood durable, fine-grained
f. *'Mayandovaya'* Tkachenko: wood coarse grained
f. *rubra* Senilis: wood yellowish brown

Literature

Barbarich, A. (Ed.) (1986): *Khorologija flory Ukrainy (Chorology of the Ukrainian flora).* Naukova Dumka, Kiev.

Bobrov, E. G. (1975): *Pinus sylvestris* na Kavkaze, istorija i sistematika (Pinus sylvestris in the Caucasus Mts.: history and systematics). *Botan. Zh.* **60, 10:**1421–1432.

Bobrov, E. G. (1978): *Lesoobrazuyushchie khvojnye USSR (Conifers of the USSR).* Nauka, Leningrad, p. 188.

Chudny, A. V. (1977): 'Izmenchivost' sostava terpentinnykh masel (Variability of terpene oils). *Rastitel'n. resursy* **13 (2)**:291–304.

Chudny, A. V. (1982): Issledovania polimorfizma sosny obyknovennoj (na primere biosinteza monoterpenov) (Studies on Scots pine polymorphism on the example of monoterpene biosynthesis). Avtoreferat, Moskva, VNIILM, p. 48.

Czeczott, H. (1939): Contribution to the knowledge of the flora and vegetation of Turkey. *Feddes Repertorium, Beihefte,* p. 105.

Dengler, A. (1908): Das Wachstum von Kiefern aus einheimischem und nördlichem Saatgut in der Oberforsterei Eberswalde. *Z. Forst. Jagdw.* **40**:137–152.

Duffield, I. W. (1952): Relationships and species hybridization in the genus Pinus. *Z. Forstgenetik* **1**:93–100.

Dukharev, V. A. (1979): Biokhimicheski polimorfizm v populacii sosny obyknovennoj (Biochemical polymorphism of Scots pine populations). Avtoreferat, Moskva, p. 16.

Endlicher, S. (1847): *Sinopsis Coniferarum.* Vol. 1. St. Gallen.

Engler, A. (1913): Einfluss der Provenienz des Samens auf die Eigenschaften der forstlichen Holzgewaechse. 2. Mitteilung. *Mitt. Schweiz, Centralanst. Forstl. Versuchswesen* **10(3)**:81–236.

Eremin, V. M. (1976): K voprosu o sistematicheskom polozhenii *Pinus lapponica* Mayr (On the problem of the systematics of *Pinus lapponica* Mayr. based on bark anatomical data). *Lesnoj Zh.* **19,4**:9–13.

Fedorov, A. A. (Ed.) (1974): *Flora evropeiskoj chasti SSSR (Flora of the European part of the USSR).* Nauka, Leningrad.

Florin, R. (1963): The distribution of Conifer and Taxad genera in time and space. *Acta Horti Berg.* **20 (4)**:121–312.

Gaussen, H., Heywood, V. H. and Chater, A. O. (1964): *Pinus* L. *Flora Europaea* **1**:32–35.

Iroshnikov, A. I. (1977): Geograficheskie kultury khvojnykh v yuzhnoj Sibiri (Conifer provenance studies in southern Siberia). In: *Geogr. kultury i plantacii khvojnykh v Sibiri.* Nauka, Novosibirsk, pp. 4–10.

Kienitz, M. (1911): Formen und Abarten der gemeinen Kiefer (*Pinus silvestris* L.). *Zeitschr. Forst Jagdw.* **1**:4–35.

Klika, I. and Novak, F. A. (1953): *Jehlicnate (Conifers).* Nakladatelsv. CSAV, Praha.

Komarov, V. L. (Ed.) (1934): *Flora SSSR (Flora of the USSR).* Ak. Nauk SSSR, Leningrad.

Korzhinskij, S. I. (1892): Flora vostoka Evropejskoj Rossii v ee sistematicheskikh i geograficheskikh otnosheniakh (Flora of the eastern part of European Russia in its systematic and geographic aspects). Tomsk.

Langlet, O. (1944): Photoperiodismus und Provenienz der gemeinen Kiefer (*Pinus silvestris* L.). Medd. Statens Skogsförsöksanst. **38**:295–327.

Lindquist, B. (1956): *Forstgenetik in der schwedischen Waldbaupraxis.* Vol. 2. Neumann, Radebeul und Berlin.

Little, E. Z. and Critchfield, W. B. (1969): *Subdivisions of the Genus Pinus (Pines).* USDA Misc. Publ. No. 1144.

Maleev, V. N. (1949): Genus 9. *Pinus* L. In: Maleev, V. N. and Sokolov, S. I. (Eds) *Derev'ya i kustarniki SSSR (Trees and shrubs in the USSR).* Ak. Nauk SSSR, Moskva—Leningrad. Vol. 1.

Mirov, N. T. (1967): *The Genus Pinus.* Roland, New York.

Münch, E. (1924–1925): Beitraege zur Kenntnis der Kiefernrassen Deutschlands. *Allg. Forst. Jagdzeitung.* **100**:540–556; **101**:89–123.

Novak, F. A. (1930): *Systematica botanica.* Prague.

Obnovlenski, V. M. (1940): Opyt izuchenia klimatipov i ekotipov sosny v kulturakh Brjanskogo uch.-opytnogo leskhoza (Studies on pine climatypes and ecotypes in the plantations of the Briansk Research Forest District). *Tr. Brjanskogo Leskhoz. Instituta* **2–3**:119–164; **4**:225–254.

Obnovlenski, V. M. (1953): Ob edaficheskikh izmeneniakh sosny obyknovennoj (On the edaphic variation in Scots pine). *Tr. Brjanskogo Leskhoz. Instituta* **4**:19–40.

Patlaj, I. N. (1965): Vlijanie geograficheskogo proiskhozdenia semjan na rost i ustojchivost sosny v kulturach severnoj levobereznoj chasti UkrSSR (Influence of geographic origin of seed on the growth and survival of pine in plantations of the southern left-bank part of the Ukrainian SSR). Ukr. Selskokhozjajstvennaya Akademia, Kiev. Avtoreferat, p. 27.

Pilger, R. (1926): Gymnospermae. In: Engler, A. and Prantl, K. (Eds) *Die natürlichen Pflanzenfamilien.* 2. Aufl. **13**, Leipzig—Berlin.

Pravdin, L. F. (1964): *Sosna obyknovennaya (Scots pine).* Nauka, Moskva.

Prokazin, E. F. (1959): Selekciya smoloproduktivnykh form sosny obyknovennoj (Selection of forms of Scots pine with high resin yield). In: *Opyt i dostizheniya po selekcii lesnykh porod.* Goslezbumizdat, Moskva, p. 38.

Rubner, K. (1960): *Die pflanzengeographischen Grundlagen des Waldbaues.* Neumann, Radebeul und Berlin.

Samofal, S. A. (1925):Klimaticheskie rasy obyknovennoj sosny i rol ikh v organizacii semnnogo khozjajstva USSR (Climatic races of Scots pine and their role in the organisation of seed management in the USSR). *Tr. po lesnomu opytnomu delu* (Moskva) **1**:5–50.

Shaw, G. R. (1914): The genus Pinus. *Arnold Arboretum* **1(5)**.

Schott, P. K. (1907): Rassen der gemeinen Kiefer. *Forstl. Centralbl.* **(4)**: 199–218; **(5)**: 262–279.

Sokołovski, S. (1931): Prace biometryczne nad rasami sosny zwyczajnej (*Pinus silvestris* L.) na ziemiach Polski (Biometric studies of Scots pine in Poland.). *Prace roln.-lesn. PAU (Krakow)* **5.**

Sztaszkiewicz, J. (1970): Systematyka i zmienność (Systematics and variability). In: Białobok, S. (Ed.): *Sosna zwyczajna* (*Pinus sylvestris* L.). Warszawa—Poznan, PWN, pp. 55–76.

Steven, H. M. and Carlisle, A. (1959): *The native pinewoods of Scotland.* Oliver and Boyd, Edinburgh—London.

Svoboda, P. (1953): *Lesni dreviny a jejich porosty (Forest trees and their growth).* Part I. Statni zemedelske nakladatelstvi, Praha.

Turesson, G. (1925): The species in relation to habitat and climate. *Hereditas* **6**:147–236.

Wiedemann, E. (1930): Die Versuche über den Einfluß der Herkunft der Kiefernsamen. (Parts 1 and 2) *Zeitschr. Forst. Jagdw.* **62(7–8)**:496–522; 809–836.

CHAPTER 3

Karyotypes and cytogenetics

Ž. Borzan

Introduction

Pines are suitable objects for cytological research. Papers written at the turn of the 20th century pointed to this. Apart from the predominant interest at that time in discovering and interpreting the process of fertilization in pines, as we can see in the works of Strasburger (1892), Dixon (1894), Lewis (1908) and others, from our present perspective we place special emphasis on the extensive work of Margaret C. Ferguson (1904), who explained in detail the development of the egg cells, fertilization and microsporogenesis of the investigated pine species (*Pinus strobus, P. austriaca, P. rigida, P. resinosa* and *P. montana* var. *uncinata*). Ferguson gave the precise number of chromosomes in the haploid state for these pine species (n=12), while Dixon (1894) gave various numbers of chromosomes for *P. sylvestris* at the time archegonia form, such as 8, 12, and 24, which was probably due to the use of methods, inadequate from the present karyological point of view. Sax and Sax (1933) published their pioneering work on the karyology of 53 gymnosperm species, classified into 16 genera. They emphasized the advantages of using female gametophyte tissue in cytological research of gymnosperms and this opened the way for further karyological research of pines. In spite of this, it was only in the middle of this century that a greater number of significant contributions were made to the cytology of pines.

Materials and methods

For the cytological analysis of the Scots pine, growing at different locations in southern Norway, Aass (1957) used needle meristem tissue and first pollen mitoses showing diagrams of chromosomes.

Most often root tip meristem is used, as in works by Vidaković (1958), Natarajan et al. (1961), Pravdin (1964), Saylor (1964), Fahmy (1966), Simak and Happel (1966). The structure of chromosomes and of interphase nuclei of Scots pine root tips and terminal buds were investigated by Kupila-Ahvenniemi and Hohtola (1977) and Kupila-Ahvenniemi (1978). Female gametophyte tissue in development was used by Sax and Sax (1933) and Borzan (1981, 1986*a, b,* 1988). Karyological anatomy of Scots pine ovules was investigated by Nagl (1962, 1965, 1967).

In these investigations various methods of preparation and making slides were applied. Before fixing, the root tips are usually pretreated with cytostatics, in order to shorten the chromosomes and produce slides with as large a number of cells as possible in the metaphase stage. Then hydrolysis or different enzymes are used to secure the satisfactory separation of cells or of single cell layers when squashing the root tip. The squash methods are described in detail by Darlington and La Cour (1962), or various improvements by Conger and Fairchild (1953) as well as by Simak and Happel (1966).

The developing haploid female gametophyte provides several advantages for the research of the karyotype: it has half the number of chromosomes (Sax and Sax, 1933), a pretreatment is unnecessary, it permits the registration of numerous "tertiary constrictions" (Pederick, 1967, 1970); and it facilitates the use of Giemsa C banding method for the differential staining of chromosomes (Borzan and Papeš, 1978). Differential staining of Scots pine chromosomes is shown in Fig. 3.5.

Measurement of chromosomes

Many authors have investigated the karyotype of Scots pine, often compared with certain other species. The research results are quite difficult to compare, even when the same species is involved, because each author tries to improve the method in use or the presentation of results. It is possible, for example, to compare the number of chromosomes and possibly the total observed number of secondary constrictions but usually it is not possible to compare relative chromosome lengths, arm lengths and positions of secondary constrictions on individual chromosomes. This is due to the fact that in earlier karyological studies statistical methods were not applied, and the karyotype was defined following an analysis of only one or at most several cells. The other reason is the diversity of ways of showing karyotypes. For example, Sax and Sax (1933) presented the haploid karyotype in the descending order of the short arm, Vidaković (1958) in the ascending order of diploid chromosomes, Natarajan et al. (1961) and Saylor (1964) used the descending order of the short arm of diploid chromosomes, Pederick (1967, 1970) showed on idiograms a large number of tertiary constrictions on each chromosome, while Borzan (1977*a*) combined the methods of Saylor and Pederick. Borzan and Papeš (1978) emphasized the need for systematization and standardization of methods in analysing pine karyotypes, also suggesting a more precise method for identification of individual chromosomes by using differential staining methods for chromosomes. It is also necessary to point out that by using female gametophytes, cells suitable for karyotype analysis at any stage of division from the moment of prometaphase even to anaphase are available (see Fig. 3.1).

The morphology of pine chromosomes was well-characterized by Saylor after he had conducted a karyological investigation of a total of 87 pine species (Saylor 1961, 1964, 1972, 1983): "...the karyotypes of all species of pine have a remarkable degree of similarity based on general features of chromosome number, length, and centromere location. Thus, it would appear that this genus, which has been in existence for millions of years and has been subjected to a wide variety of selection forces, has been quite conservative in changing major features of its karyotype. Polyploidy or even the addition or loss of a single chromosome does not exist in the genus at any level except for an occasional individual. Similarly, evolution has proceeded with quite limited involvement of chromosome translocations and inversions that include the centromere. Although it was anticipated initially that more differences would be found, it is nevertheless important to describe the basic species karyotypes and to document the similarities as well as the differences that do exist" (Saylor, 1983).

Table 3.1 presents a confirmation of the basic similarities of pine karyotypes, with a survey of the relative chromosome lengths of certain two-needle pines, which were investigated by various authors. By comparison of the differences between adjacent chromosomes of this species we see that, as a rule, the least differences in lengths of adjacent chromosomes are from chromosome number II to number IX. The largest differences are in the length of adjacent chromosomes I and II, then X and XI, and XI and XII. An exception are relative chromosome lengths of Scots pine in the work by Pravdin (1964), where the largest differences are shown between adjacent chromosomes VI and VII as well as VIII and IX.

A karyotype is defined numerically with corresponding statistical parameters. Examples of numerical karyotypes of Scots pine trees can be found in Tables 3.2 and 3.3. One of the useful statistical parameters for karyotype analysis is the coefficient of variation (C. V.) which is regularly the largest for chromosomes XII and I, i.e. for those which are the easiest to identify. The coefficient of variation actually confirms that the risk in reversal of order when identifying pine chromosomes, as well as the risk of arm reversal for metacentric chromosomes is significantly high. More details on the subject can be found in the papers of Matern and Simak (1968, 1969), Chetty et al. (1970) and Borzan (1988).

Summing up: a uniform approach to karyological research of pines is still lacking. The reasons for this and the subsequent difficulties in comparing the results of various authors are the differences of opinions on the suitability of particular tissues for karyological analysis, the use of different techniques, which may also be related to the character of the research (basic, applied, methodological), as well as the lack of statistical data processing. Standardization of methods and statistical data processing in cytological research of female gametophyte tissue is particularly necessary. This tissue provides more information on the morphology of chromosomes and has greater variability than diploid root tip cells. Conclusions can therefore be drawn only from sufficiently large samples. Borzan (1988) illustrates this with an analysis of anaphase cells (see Fig. 3.1*b*) where chromatids of the same chromosome differ significantly in length after division. The following causes are mentioned for the variability in chromosome length: various types of nuclei dependent on the part of the female gametophyte tissue from which the analysed cells are taken (see Fig. 3.1*c*), differential activity of cells

Table 3.1. Comparison of Scots pine and some other pine chromosome lengths as described by different authors

Species	Chromosome												References
	I	II	III	IV	V	VI	VII	VIII	IX	X	XI	XII	
P. sylvestris	120	114	109	107	107	106	101	101	97	90	79	70	Natarajan et al., 1961*
P. sylvestris	120	114	109	109	106	103	101	98	97	89	82	72	Saylor, 1964*
P. sylvestris	140	132	127	123	122	123	101	96	72	65	53	46	Pravdin, 1964
P. sylvestris	120	115	111	109	107	103	102	99	94	88	80	71	Mihailescu and Dalu, 1972
P. sylvestris, tree No 367	123	116	111	109	106	104	102	98	95	85	79	70	Borzan, 1981
P. sylvestris, tree No 77	124	116	111	108	106	103	101	98	94	87	80	71	Borzan, 1981
P. x nigrosylvis, tree No 410	125	116	112	108	105	102	99	97	94	88	81	72	Borzan, 1988
P. nigra	118	113	110	107	106	105	103	98	98	90	81	71	Saylor, 1964*
P. nigra var. banatica	131	115	115	107	99	99	99	99	90	82	82	82	Tarnavschi and Ciobanu, 1965
P. nigra var. austriaca	131	121	121	111	111	111	97	97	90	90	60	60	Tarnavschi and Ciobanu, 1965
P. nigra ssp. austriaca	116	112	108	107	105	104	102	97	97	91	84	75	Mihailescu and Dalu, 1971
P. nigra var. koekaelare	121	116	112	110	106	104	101	97	95	88	81	69	Mihailescu and Dalu, 1971
P. nigra ssp. calabrica	122	112	110	108	107	104	103	91	95	88	87	72	Mihailescu and Dalu, 1971
P. nigra	119	113	112	110	109	104	106	101	95	85	78	67	Mihailescu and Dalu, 1972
P. nigra	121	115	111	110	105	104	100	97	94	89	84	68	Pederick, 1967
P. nigra, tree No 47	125	118	113	109	106	104	102	98	94	86	79	65	Borzan and Papeš, 1978
P. nigra, tree No 221	125	115	111	108	105	103	100	97	94	89	82	69	Borzan and Papeš, 1978
P. nigra, tree No 336	123	117	113	109	106	103	100	97	93	88	80	69	Borzan, 1988
P. densiflora, tree No 116	127	120	113	109	107	103	100	96	92	85	80	68	Borzan, 1988

* = after Pederick, 1967.

Table 3.2. Numerical Scots pine karyotype according to Saylor (1964)

Chromosome	a	b	a/b
I	13.7*	14.7	0.931
II	13.0	14.2*	0.915
III	12.7	13.1	0.969
IV	12.4	12.8	0.969
V	12.1	13.8	0.877
VI	11.8	12.3	0.959
VII	11.5	13.0	0.885
VIII	11.2	11.8	0.949
IX	10.8	12.4	0.871
X	9.9*	11.2	0.884
XI	7.9	11.6	0.681
XII	6.8	10.2	0.667

Values are mean arm lengths in millimetres of chromosome magnifield 2000×. a= short arm; b=long arm; a/b=arm ratio; *=secondary constriction in the medial portion of the arm.

in that rapidly dividing and developing tissue over a relatively short period, different division stages of the analysed cells; the sticky chromosome phenomenon (Borzan 1977*b*; see Fig. 3.1*d*) and others.

Scots pine karyotype

As the result of statistical analysis of chromosome measurements a numerical karyotype of the species (see Table 3.2) or of the individual tree (see Table 3.3) is obtained, which is the basis for a graphic presentation of the karyotype.

Depending on the centromere position, Scots pine chromosomes can be either metacentric or submetacentric. According to Saylor (1961, 1964) the centromere position on pine chromosomes is defined by the short/long (S/L) arm ratio. The chromosome is metacentric if this ratio is between 0.75 and 1.00, and submetacentric between 0.50 and 0.75. Often the classification of Levan et al. (1964) is used in cytology. Saylor's classification is more suitable for pines because of the easily noticeable difference in arm lengths on chromosome XII and frequently on some other chromosomes, particularly chromosome XI. Most of the authors working on pines use Saylor's classification.

Table 3.3. Numerical karyotype of the Scots pine tree sy 77 according to Borzan (1988)

Chromosome number	$\bar{x}$	d	s	$s_{\bar{x}}$	C.V. %
(S+L) Chromosome length					
1	123.9739	7.89	8.3164	1.9602	6.7082
2	116.0796	5.43	4.4278	1.0436	3.8145
3	110.6449	2.57	2.9153	0.6871	2.6348
4	108.0727	2.15	2.6048	0.6140	2.4102
5	105.9224	2.63	2.7627	0.6512	2.6082
6	103.2879	2.22	2.4248	0.5715	2.3476
7	101.0662	2.66	3.4137	0.8046	3.3777
8	98.4110	3.92	3.3890	0.7988	3.4437
9	94.4889	7.58	4.1208	0.9713	4.3611
10	86.9059	6.84	3.9162	0.9231	4.5062
11	80.0680	8.99	4.4165	1.0410	5.5159
12	71.0785		6.0041	1.4152	8.4472
(S) Short arm					
1	58.3464		4.8334	1.1392	8.2840
2	55.1547		3.1698	0.7471	5.7471
3	52.1101		2.9462	0.6944	5.6539
4	51.4108		2.5617	0.6038	4.9829
5	49.5821		2.4539	0.5784	4.9492
6	48.9807		2.0823	0.4908	4.2514
7	47.1899		2.3421	0.5520	4.9631
8	46.3743		3.0312	0.7145	6.5363
9	44.2933		3.9882	0.9400	9.0042
10	40.7389		3.4555	0.8145	8.4820
11	34.1637		2.7094	0.6386	7.9306
12	28.0366		3.2589	0.7681	11.6237
(L) Long arm					
1	65.6275		6.0812	1.4334	9.2663
2	60.9249		3.6242	0.8542	5.9487
3	58.5348		2.1413	0.5047	3.6581
4	56.6619		2.5687	0.6054	4.5333
5	56.3402		2.9886	0.7044	5.3046
6	54.3072		2.6131	0.6159	4.8117
7	53.8763		3.5977	0.8480	6.6777
8	52.0367		2.4991	0.5890	4.8026
9	50.1956		3.3485	0.7892	6.6709
10	46.1670		2.8716	0.6768	6.2199
11	45.9044		2.9770	0.7017	6.4852
12	43.0419		4.1301	0.9735	9.5954
(S/L) Arm ratio					
1	0.8952		0.0973	0.0229	10.8632
2	0.9087		0.0785	0.0185	8.6413
3	0.8920		0.0683	0.0161	7.6518
4	0.9100		0.0734	0.0173	8.0634
5	0.8835		0.0784	0.0185	8.8716
6	0.9048		0.0704	0.0166	7.7761
7	0.8803		0.0788	0.0186	8.9565
8	0.8939		0.0809	0.0191	9.0480
9	0.8878		0.1104	0.0260	12.4319
10	0.8866		0.0980	0.0231	11.0530
11	0.7464		0.0677	0.0160	9.0748
12	0.6551		0.0832	0.0196	12.7025

Table 3.3. Cont.

Chromosome number	$\bar{x}$	d	s	$s_{\bar{x}}$	C.V. %
		(S/S+L) · 100 Centromere index			
1	47.0963		2.8976	0.6830	6.1524
2	47.5233		2.2220	0.5237	4.6756
3	47.0791		1.9541	0.4606	4.1507
4	47.5690		2.0633	0.4863	4.3374
5	46.8201		2.2369	0.5272	4.7776
6	47.4317		1.9528	0.4603	4.1171
7	46.7226		2.3647	0.5574	5.0612
8	47.1047		2.3174	0.5462	4.9196
9	46.8403		3.4057	0.8027	7.2710
10	46.8492		2.9586	0.6974	6.3152
11	42.6569		2.2102	0.5210	5.1814
12	39.4371		3.0155	0.7108	7.6463

For each chromosome characteristic the following statistical parameters are given: standard deviation (s), standard error ($s_{\bar{x}}$) and coefficient of variation (C. V. %). For chromosome length the difference (d) between the mean values of two adjacent chromosomes are given.

According to Saylor (1983), the chromosomes XI and XII of Scots pine and of all other pine species of the *Sylvestres* subsection are heterobrachial (submetacentric), which distinguishes them from other pines, where only chromosome XII is heterobrachial. On this point opinions of various authors differ. Fahmy (1966) identifies four pairs of short chromosomes and two pairs of the longest chromosomes of Scots pine as submetacentric chromosomes. In contrast Pravdin (1964) shows a Scots pine idiogram without submetacentric chromosomes. I considered (Borzan, 1981) Scots pine chromosomes XI and XII as submetacentric in terms of their average lengths, but after a detailed analysis of the length of chromosome XI in each of 18 analysed female gametophyte cells it seems that chromosome XI may be heteromorphic (below or above 0.75, the limiting value of arm ratio), i.e. submetacentric in some ovules or metacentric in others, on the same tree (Borzan, 1986*b,* 1988).

Secondary constrictions

For a karyotype presentation by idiograms, it is necessary to show the positions of secondary constrictions along the arms. Mehra and Khoshoo (1956) have already demonstrated the existence of a varying number of secondary constrictions in pine species. On chromosomes of Scots pine the number and position of secondary constrictions is still subject to discussion. It is characteristic that most authors stress the inconsistency in the appearance of secondary constrictions in their own material. Saylor (1961) included secondary constrictions in karyotype descriptions, but they were not used as main criteria for identification because of their unreliability. Comparing karyotypes of various pine species, only Pederick (1967, 1970) was able to identify chromosomes based on the location of secondary and tertiary constrictions. Such reliability and precision in studying the haploid female gametophyte chromosomes could not be repeated in my work (Borzan, 1977*a*, 1981, 1988). Vidaković (1958), analysing the root tip meristem of seeds of *Pinus nigra, P. sylvestris* and their hybrids, could not establish a valid difference in secondary constrictions either. Natarajan et al. (1961) found all constrictions in Scots pine on the short arms of chromosomes I, II, VI, VII and X. They found from several to 8–9 nucleoli in the interphase nuclei.

Saylor (1964) determined the Scots pine karyotype based on 13 cells from the root tips of 8 seedlings and obtained different results from those of Aass (1957) and Natarajan et al. (1961) in the appearance of secondary constrictions. He found that secondary constrictions were on the short arms of chromosomes I and X, on the long arms of chromosome II, and that all constrictions were on the medial portion of the arms.

Pravdin (1964) identified secondary constrictions on chromosomes I, II and VI. Fahmy (1966) found "four pairs of short chromosomes, 7–10 μm in length, all with submedian centric-constrictions, and with no secondary constrictions (chromosomes IX, X, XI and XII). Six pairs of medium-sized chromosomes, 11–13 μm in length; two pairs with median centric-constrictions, and the rest with submedian ones. Of these chromosomes, three pairs have secondary constrictions. Two pairs of fairly long chromosomes, 13–15 μm in length, with submedian centric-constrictions. One pair of these chromosomes has secondary constrictions (chromosomes I and II). Many of the interphasic nuclei contain six to eight nucleoli."

Mihailescu and Dalu (1972) analysed two Scots pine provenances, a mixed sample of Scots pine and a sample of European black pine *(P. nigra)*. They concluded that the chromosomes of European black pine are, as a rule, shorter than chromosomes of Scots pine. They also found differences in length of certain arms for chromosomes I, II, VI and XII. In two provenances of Scots pine they identified secondary constrictions on short arms of the chromosomes I, II, III, V, VI, IX and X, and on the long arm of chromosome IV. In a mixed sample of Scots pine they found the same constrictions, except for chromosome V, which had none.

Kormutak (1975) wrote that the European black pine has no secondary constrictions, while Scots pine has two. Martinović (1978) claimed that from among the species *Pinus sylvestris, P. nigra, P. mugo, P. banksiana* and *P. flexilis,* only the last has secondary constrictions on one pair of chromosomes. Muratova (1978*a*) reported that the number of secondary constrictions in *Pinus sibirica, P. pumila, P. koraiensis,* and *P. sylvestris* varies from 4 (in *P. koraiensis*) to 7 (in *P. sibirica*). Borzan (1981) found 1 to 7 clearly visible secondary constrictions on chromosomes in female gametophyte cells originating from two Scots pine trees.

Figure 3.1*a* shows that Scots pine may have two constrictions on the same chromosome. Chromosome II shows a clear, prominent secondary constriction on the long arm, and a less expressed constriction on the short arm. The figure illustrates that many questions remained unsettled in pine karyology. For example:

how can secondary constrictions be recognized and consistently distinguished from "tertiary" ones; which tissue should be used for karyological research; how reliable are the research results, even when using the same methods?

A lack of consistency among various authors in registration of constrictions on a certain chromosome results from similar lengths of adjacent chromosomes and variations in the length of each chromosome, which both contribute to reversal of chromosome order or arm reversals, i.e. to inaccurate identification. An example of this can be seen in Fig. 3.1*a*, where chromosome II displays two clearly visible constrictions, while on the idiogram of Scots pine tree sy 77 (see Fig 3.3*d*) two constrictions are marked on chromosome I. The reason for this is that in Fig. 3.1*a*, chromosome II is shorter than chromosome I, while in the rest of 17 analysed cells, two constrictions can be found mostly on chromosome I (see Fig. 3.2).

Difficulties in registering constrictions are also due to the various techniques of making slides, overlapping of chromosomes, varying degrees of chromosome contractions in various stages of differential tissue activities, i.e. variation in length of the part of chromosome which contributes to the registration of a constriction. The identification of chromosomes on the basis of secondary constrictions only is impossible because they appear regularly on the medial part of an arm. It seems that some other methods are necessary to provide a reliable and exact chromosome identification.

Graphic presentation of karyotypes

It is theoretically possible to identify chromosomes with constrictions on the basis of the position and the frequency distribution of secondary and tertiary constrictions, if the risk of reversal of order is less than 50%, and the analysed sample is large enough.

Borzan (1988) considers that the registration of constrictions based on statistics (Fig. 3.2) would attribute them to the chromosomes they most frequently appear on, and would restrict their display on the idiogram only to those constrictions corresponding to the number of observed nucleoli in interphase nuclei. Such a presentation would minimize the risk of reversal in order and arm reversal. In this way the constriction which appears most frequently in some position can be considered as a characteristic of the chromosome in question.

On the basis of the numerical karyotype and the frequency distribution of the constrictions, it is easy to represent the karyotype with an idiogram (Fig. 3.3*d*). Apart from the idiogram presentation, the karyotype can also be drawn as a polykaryogram according to Iljcenko (1975; after Muratova, 1978*a, b*) in which values of the centromere indices have been entered for each chromosome on the abscissa, and values of the length of the corresponding chromosome on the ordinate. In the idiogram, the morphology of each of the chromosomes is presented graphically, while in the polykaryogram all chromosomes from all investigated cells are represented by points. When grouped, these points clearly provide the characteristic of each karyotype. Borzan (1988) proposes the marking of the values of the mean chromosomes (values from the numerical karyotype) with small circles or larger points in the polykaryogram (Fig. 3.4).

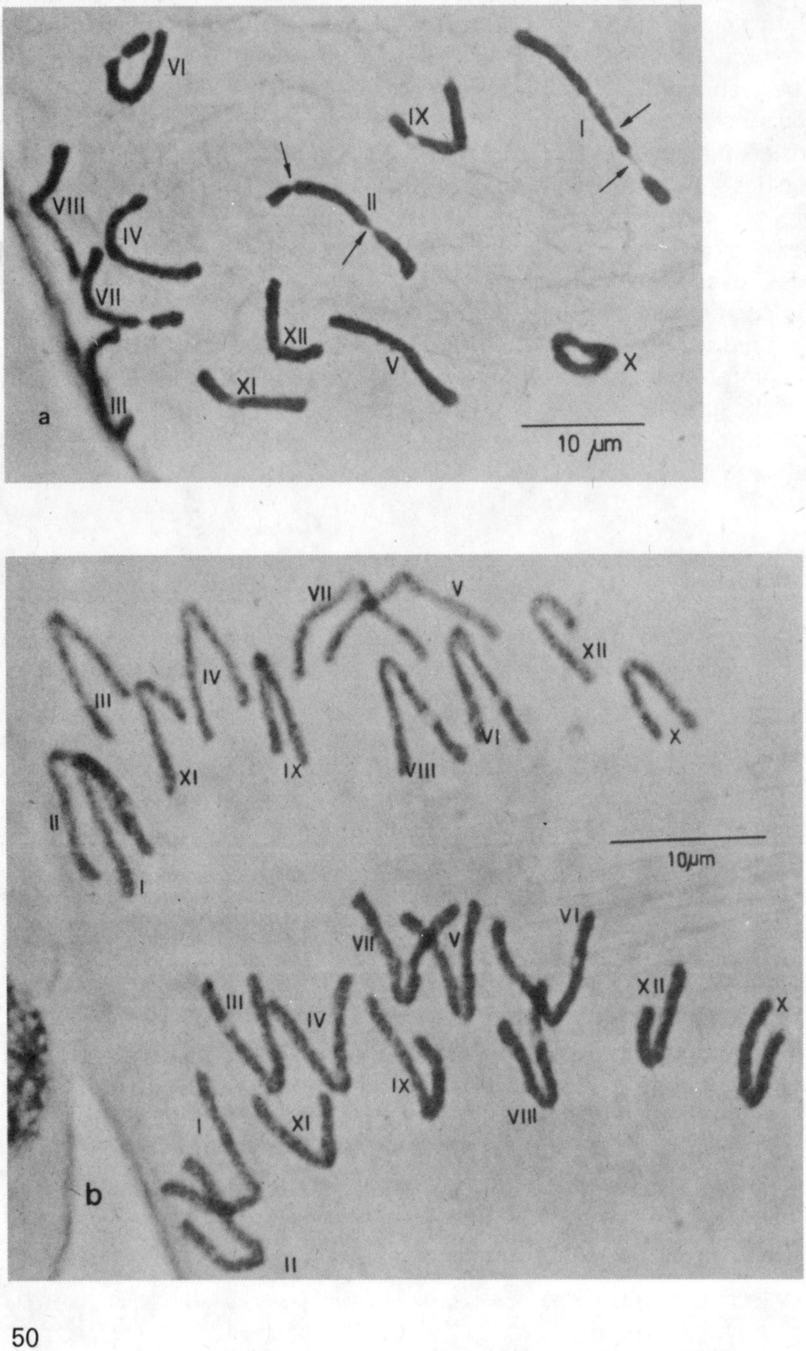

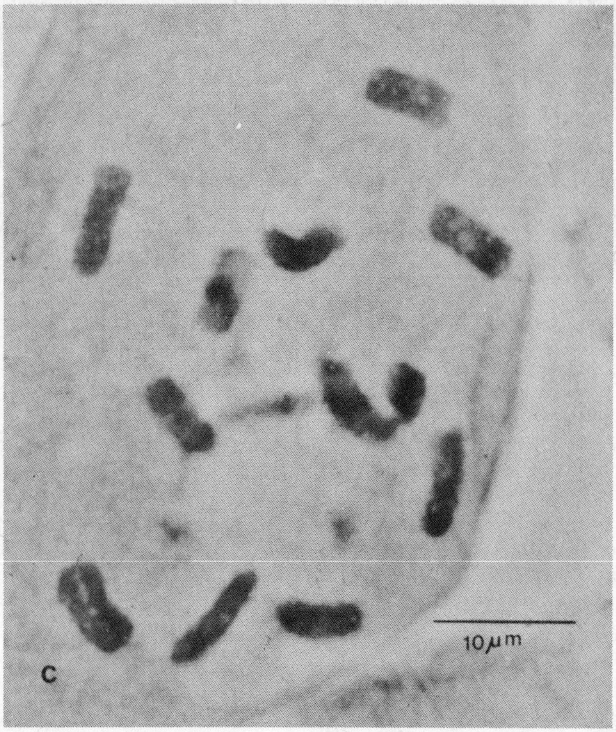

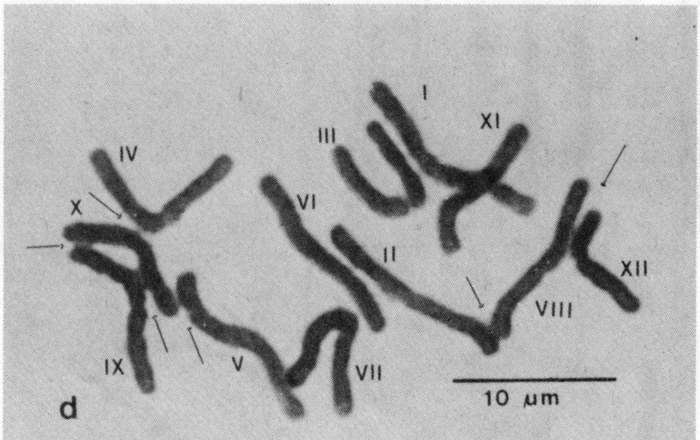

Fig. 3.1. Scots pine chromosomes in the female gametophyte tissue: *(a)* metaphase; *(b)* anaphase; *(c)* extremely short chromosomes; *(d)* metaphase chromosomes with sticky connections shown by arrow

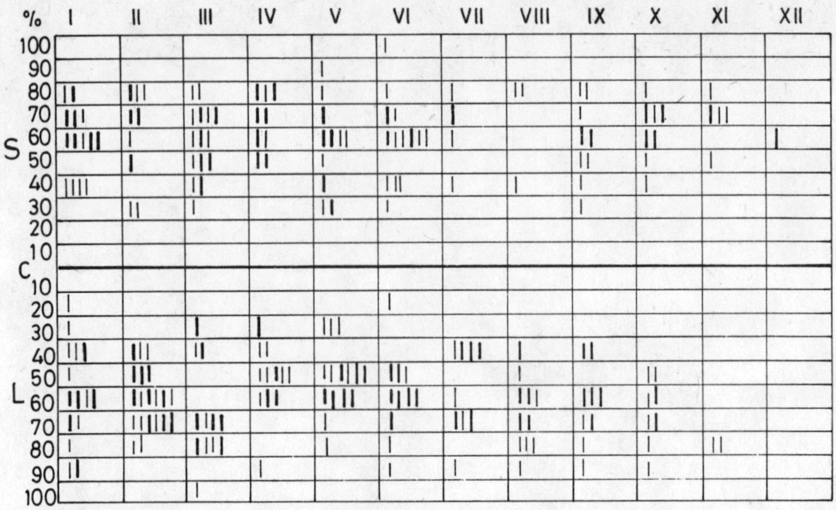

Fig. 3.2. Frequency distribution of secondary and tertiary constrictions of chromosomes in the female gametophyte of Scots pine tree sy 77. The bars in bold-face represent the prominent constrictions

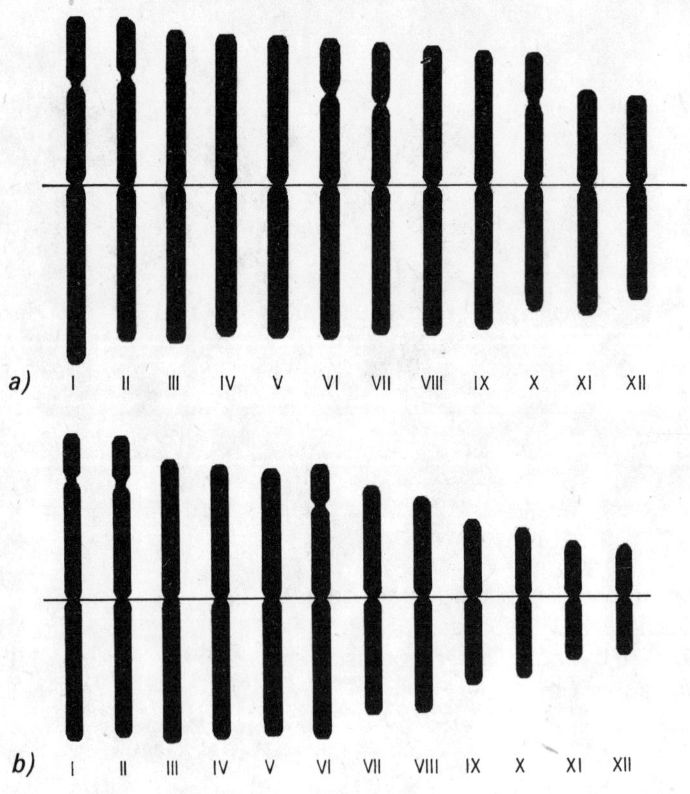

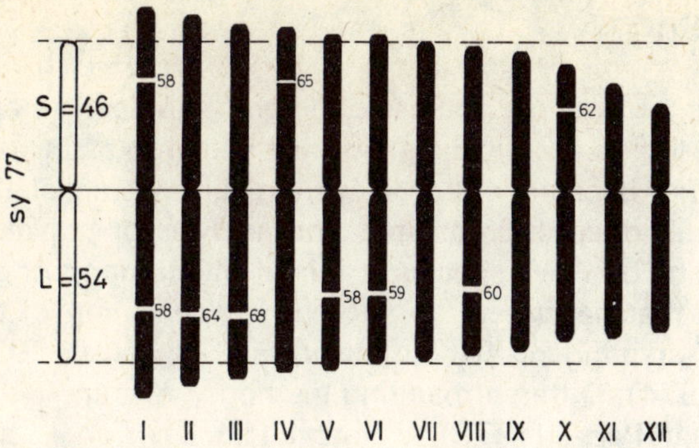

c)

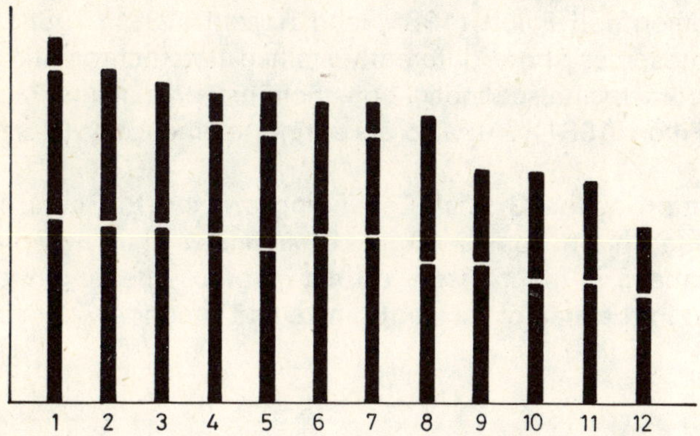

d)

Fig. 3.3 Idiograms of the Scots pine according to *(a)* Natarajan et al. (1961); *(b)* Pravdin (1964); *(c)* Fahmy (1966); and *(d)* Borzan (1981)

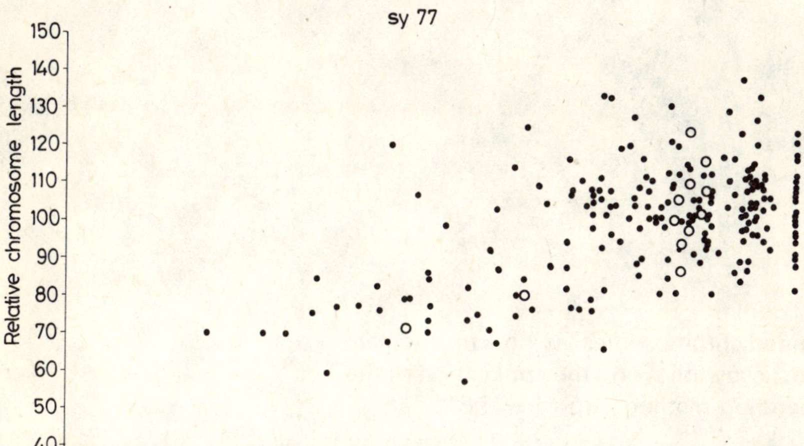

Fig. 3.4. A polykaryogram of Scots pine tree sy 77, according to Borzan (1988)

Chromosome banding methods

Numerous staining methods, as for example, the use of different quinacrin stains that fluoresce under ultraviolet light, as well as staining with Giemsa stain were developed in the early 1970s. They became routine in cytological research, enabling precise identification of individual chromosomes and analysis of polymorphism of chromosomes within and between species in animal and human cytogenetics. Studies soon appeared that described the successful application of the Giemsa C banding method of staining plant chromosomes (Schweizer, 1973, 1974; Marks and Schweizer, 1974). Different banding methods were applied to pine chromosomes by Borzan and Papeš (1978), Muratova (1978*a*), Tanaka and Hizume (1980), MacPherson and Filion (1981) and Borzan (1981, 1988). In these studies, pine chromosomes show differently stained heterochromatic regions not only at the centromere and secondary constrictions (Borzan and Papeš, 1978; MacPherson and Filion, 1981), but also appearing telomerically (Borzan, 1981, 1988).

The successful application of the Giemsa C banding method to Scots pine female gametophyte tissue chromosomes (Fig. 3.5) indicates that they have relatively little heterochromatin in comparison to total chromosome length, and that the bands are located in the area of the centromere and secondary constrictions. The Giemsa C banding method is still inadequately precise in application, because there are not enough successful slides to allow the use of statistical analysis to results. It is clear that a greater number of slides is needed with

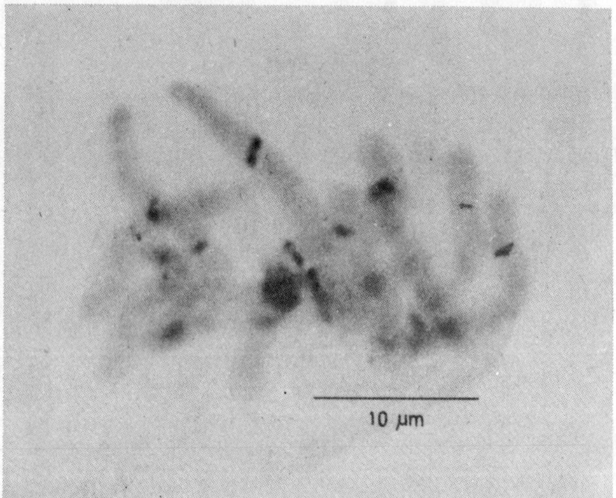

Fig. 3.5. Banded chromosomes in Scots pine female gametophyte tissue following the application of the Giemsa C banding method (Borzan, 1981).

well-spread chromosomes and clear bands, in order to facilitate the reliable location of bands on each chromosome, thus making it possible to identify each chromosome precisely (particularly from the group of II to X). This is a precondition for cytological research towards mapping of individual chromosomes. In that respect other studies which have shown successful application of certain methods for differential staining of the gymnosperm chromosomes deserve attention. The Japanese investigators, Kondo and Hizume (1982), Hizume et al. (1983) and Kondo et al. (1985), have used fluorescent staining with chromomycin A_3 and 4'-6-diamidino-2-phenylindol on chromosomes of the species *Cryptomeria japonica, Pinus nigra* var. *maritima* and *P. densiflora.*

In Finland, Kupila-Ahvenniemi and Hohtola (1977) stained the chromosomes of Scots pine with a simplified Giemsa procedure and revealed differentially stained patterns, different in shape and quantity from those found by other researchers, sometimes with one chromosome arm fully and intensively stained.

The question is still open regarding the advantage of using ovules or root tips for differential staining of pine chromosomes. It is worth quoting Natarajan and Natarajan (1972) who consider that one must have well-spread chromosomes in nuclei free of cytoplasm in order to successfully apply the Giemsa method or the quinacrin fluorescence technique. This, however, is difficult to manage without hydrolysis or the enzyme treatment of the root tip cell walls.

In pines, dividing free nuclei in a developing female gametophyte tissue are easier to expose to chemicals used for the banding of chromosomes, because there are no cell walls. This is also a drawback, however, because of the aggressive nature of chemicals in use, the time required and the complexity of the method, which results in few successful slides.

Chromosome mapping

The next step towards mapping of pine chromosomes will be possible after applying various differential staining methods which will make identification of individual chromosomes much more reliable. Along with allozyme studies on Scots pine (Szmidt, 1984), molecular research of conifers (Kriebel, 1987) and application of various methods that enable isolation of individual chromosomes from the cells (Schlarbaum, 1986) it will become possible to determine the position of particular genes on the identified chromosomes.

Application of Scots pine karyological research

Simak (1986) studied chromosome aberrations in seeds of 15 Scots pine provenances, stored between 0 and 24 years. There was a tendency for an increased frequency of mitotically defective cells with the extension of the storage period.

The European black pine and Scots pine are incompatible species, yet rarely able to produce F1 hybrids (*Pinus x nigrosylvis* Vid.) due to the special combining ability of individual trees (using European black pine as female and Scots pine as male crossing partners). Borzan (1984, 1986*a, b,* 1988) explored the possibility of using cytology towards understanding and resolving the problem of incompatibility of these two species. He investigated the karyotypes of individual European black and Scots pine trees that show special combining ability and the karyotypes of trees which do not produce hybrids when crossed. It has been shown that significant differences in morphological characteristics of chromosomes of the investigated trees do exist, as well as between the entire karyotypes. However, it was not possible to establish a correlation between the ability of certain trees to cross and the morphology of their chromosomes. Borzan (1988) concludes that incompatibility factors between European black pine and Scots pine operate at the molecular level in various phases of female gametophyte development from the moment of pollination to embryo development, and that they are not visible in chromosome morphology.

Literature

Aass, I. (1957): A cytological analysis of Scots pine (*Pinus silvestris* L.) from Skjak, Norway. *Det norske Skogførsoksvesen* **48**:93–109.

Borzan, Ž (1977*a*): Contribution to the karyotype analysis of the European black pine (*Pinus nigra* Arn.). *Ann. Forest.* **8/3**:29–50.

Borzan, Ž. (1977*b*): Sticky chromosomes in the endosperm of European black pine and Scots pine. *Genetika* **9(2)**:139–147.

Borzan, Ž. (1981): Karyotype analysis from the endosperm of European black pine and Scots pine. *Ann. Forest.* **10/1**:1–42.

Borzan, Ž. (1984): Hybridization between *Pinus nigra* and *Pinus sylvestris* and morphology of chromosomes. In: *Proceedings of the 8th international symposium on sexual reproduction in seed plants, ferns and mosses, 20–24 August 1984, Wageningen.* PUDOC, The Netherlands, pp. 181.

Borzan Ž. (1986*a*): Citotaksonomske značajke kariotipova nekih borova (Cytotaxonomical properties of karyotypes of some pines). *Glasnik za šumske pokuse, posebno izdanje* **2**:91–103.

Borzan, Ž. (1986*b*): The importance of cytology for improvement of pines. In: *Proceedings of the 18th IUFRO World Congress, Ljubljana,* Division 2, Vol. II, pp. 536–545.

Borzan, Ž. (1988): Kariotipovi nekih borova podsekcije *Sylvestres* (Karyotypes of some Pines of the subsection *Sylvestres*). *Glasnik za šumske pokuse* **24**:1–100.

Borzan, Ž. and Papeš, D. (1978): Karyotype analysis in *Pinus:* A contribution to the standardization of the karyotype analysis and review of some applied techniques. *Silvae Genet.* **27(3–4)**:144–150.

Chetty, C. K. R., Upadhaya, L. P. and Kedharnath, S. (1970): Estimation of risk of reversal of arms and order of chromosomes in karyotype analysis of *Pinus roxburghii* Sarg. *Indian Forester* **96**:811–816.

Conger, A. D. and Fairchild, L. M. (1953): A quick-freeze method for making smear slides permanent. *Stain Technology* **28(6)**:281–283.

Darlington, C. D. and La Cour, L. F. (1962): *The handling of chromosomes.* George Allen and Unwin Ltd., London, pp. 263.

Dixon, H. H. (1894): Fertilization of *Pinus silvestris. Ann. Bot.* **8**:21–34.

Fahmy, T. Y. (1966): Chromosome number and morphology in certain conifers. *Alexandria J. Agric. Res.* **14**:205–221.

Ferguson, M. (1904): Contribution to the knowledge of the life history of *Pinus* with special reference to sporogenesis, the development of the gametophytes and fertilization. *Proc. Wash. Acad. Sci.* **6**:1–202.

Hizume, M., Ohgiku, A. and Tanaka, A. (1983): Chromosome banding in the Genus *Pinus.* I. Identification of chromosomes in *P. nigra* by fluorescent banding method. *Bot. Mag. Tokyo* **96**:273–276.

Kondo, T. and Hizume, M. (1982): Banding for the chromosomes of *Cryptomeria japonica* D. Don. *J. Jap. For. Soc.* **64 (9)**:356–358.

Kondo, T., Hizume, M. and Kubota, R. (1985): Variation of fluorescent chromosome bands of *Cryptomeria japonica. J. Jap. For. Soc.* **67(5)**:184–189.

Kormutak, A. (1975): Kariological structure of some *Pinus* species. *Biologia (Bratislava)* **30 (7)**:545–550.

Kriebel, H. B. (1987): Applying molecular biology to forest tree improvement. *Ann. Forest.* **13/1–2**:3–12.

Kupila-Ahvenniemi, S. (1978): Structure of interphase nuclei and scarcity of nucleoli in Scots pine from northern locations. *Ann. Bot. Fennici* **15**:92–100.

Kupila-Ahvenniemi, S. and Hohtola, A. (1977): Structure of the chromosomes of Scots pine. *Hereditas* **87**:185–188.

Levan, A., Fredga, K. and Sandberg, A. A. (1964): Nomenclature for centromeric position on chromosomes. *Hereditas* **52**:201–220.

Lewis, I. M. (1908): The behaviour of the chromosomes in *Pinus* and *Thuja*. *Ann. Bot.* **22**:529–556.

MacPherson, P. and Filion, W. G. (1981): Karyotype analysis and the distribution of constitutive heterochromatin in five species of *Pinus. J. Hered.* **72**:193–198.

Marks, G. E. and Schweizer, D. (1974): Giemsa banding: Karyotype differences in some species of *Anemone* and in *Hepatica nobilis. Chromosoma (Berl.)* **44**:405–416.

Martinović, V. (1978): Predbežné výsledky štúdia karyologických pomerov pri vybraných druhoch rodu *Pinus* (Preliminary results of studying of karyological relationships in chosen species of the genus *Pinus*). *Fol. Dendrol.* **4/78**:149–157.

Matern, B. and Simak, M. (1968): Statistical problems in karyotype analysis. *Hereditas* **59**:280–288.

Matern, B. and Simak, M. (1969): On some statistical problems connected with the identification of chromosomes. Res. Note 11, from Dept. of For. Biometry, Royal Col. of For. Stockholm, pp. 81.

Mehra, P. N. and Khoshoo, T. N. (1956): Cytology of Conifers. I. *J. Genet.* **54**:165—180.

Mihailescu, A. and Dalu, M. (1971): Comparative study on the karyotype of different provenances of *Pinus nigra* of the Şeţu Sinaia nursery. *Rev. Roum. Biol. Ser. Bot.* **16**:319–327.

Mihailescu, A. and Dalu, M. (1972): Comparative karyotype study in different provenances of *Pinus silvestris* and *Pinus nigra. Rev. Roum. Biol. Ser. Bot.* **17**:343–353.

Muratova, E. N. (1978*a*): Polykariogram analysis and differential staining of chromosomes of some representatives of *Pinus* species. *Abstract. Proc. XIV Int. Congress of Genetics,* Moscow, pp. 300.

Muratova, E. N. (1978*b*): Kariotipy kedrovykh sosen. I. Kariotip *Pinus sibirica* Du Tour (Karyotypes of Pines of the group *Cembra.* I. Karyotype of *Pinus sibirica* Du Tour). *Citologija* **20 (8)**:972–976.

Nagl, W. (1962): Über Endopoliploidie, Restitutionskernbildung und Kernstrukturen im Suspensor von Angiospermen und einer Gymnosperme. *Österr. Bot. Z.* **109**:431–494.

Nagl, W. (1965): Karyologische Anatomie der Samenanlage von *Pinus silvestris. Österr. Bot. Z.* **112**:359–370.

Nagl, W. (1967): Mikrophotometrische DNS-Messungen an Interphase- und Ruhekernen sowie Mitosen in der Samenanlage von *Pinus silvestris. Z. Pflanzenphysiol.* **56**:40–56.

Natarajan, A. T., Ohba, K. and Simak, M. (1961): Karyotype analysis of *Pinus silvestris. Hereditas* **47**:379–382.

Natarajan, A. T. and Natarajan, S. (1972): The heterochromatin of *Rhoeo discolor. Hereditas* **72**:323–330.

Pederick, L. A. (1967): The structure and identification of the chromosomes of *Pinus radiata* D. Don. *Silvae Genet.* **16**:69–77.

Pederick, L. A. (1970): Chromosome relationships between *Pinus* species. *Silvae Genet.* **19**:171–180.

Pravdin, L. F. (1964): Sosna obyknovennaya. Izmenchivost, vnutrividovaya sistematika i selekciya (Scots pine. Variability, systematics and improvement). Nauka, Moskva, pp. 191.

Sax, K. and Sax, H. J. (1933): Chromosome number and morphology in the Conifers. *J. Arnold Arb.* **14**:356–375.

Saylor, L. C. (1961): A karyotypic analysis of selected species of *Pinus. Silvae Genet.* **10**:77–84.

Saylor, L. C. (1964): Karyotype analysis of *Pinus* group *Lariciones. Silvae Genet.* **13**:165–170.

Saylor, L. C. (1972): Karyotype analysis of the genus *Pinus*- subgenus *Pinus*. *Silvae Genet.* **21**: 155–163.

Saylor, L. C. (1983): Karyotype analysis of the genus *Pinus*- subgenus *Strobus*. *Silvae Genet.* **32**: 119–124.

Schlarbaum, S. L. (1986): Cytogenetic manipulations in forest trees through tissue culture. *18th IUFRO World Congress, Ljubljana,* pp. 23.

Schweizer, D. (1973): Differential staining of plant chromosomes with Giemsa. *Chromosoma (Berl.)* **40**:307–320.

Schweizer, D. (1974): An improved Giemsa C-banding procedure for plant chromosomes. *Experientia* **30**:570–571.

Simak, M. (1986): Chromosome aberrations in stored seeds of *Pinus silvestris* and *Picea abies* and the consequences on plant properties. Sveriges lantbruksuniversitet, Inst. för skogsskötsel, *Rapporter* **20**:1–42.

Simak, M. and Happel, C. (1966): Vorbehandlung der Koniferensamen für Chromosomenuntersuchungen. *Silvae Genet.* **15 (2)**:38–41.

Strasburger, E. (1892): Über das Verhalten des Pollens und die Befruchtungsvorgänge bei den Gymnospermen. *Hist. Beitr.* **4**:1–156.

Szmidt, A. E. (1984): Genetic studies of Scots pine (*Pinus sylvestris* L.) domestication by means of isozyme analysis. The Swedish University of Agricultural Sciences, Umeå, pp. 194.

Tanaka, R. and Hizume, M. (1980): C banding treatment for the chromosomes of some Gymnosperms. *Bot. Mag. (Tokyo)* **93**:167–170.

Tarnavschi, I. T. and Ciobanu, I. (1965): Kariologische Untersuchungen an *Pinus nigra* Arn. sspp. *nigricans* Host. var. *banatica* Georg. et Ionescu in vergleiche mit *Pinus nigra* Arn. var. *austriaca* Moess. *Rev. Roum. Biol. Ser. Bot.* **10**:371–375.

Vidaković, M. (1958): Investigations on the intermediate type between Austrian and the Scots pine. *Silvae Genet.* **7**:12–18.

CHAPTER 4

Generative reproduction and genetic processes in nature

V. Koski

— Plays down the problems of selfing.

Finland

— describes gene flow in native stands.

Introduction

Most seed plants, even many conifers, possess some ability to produce asexual offspring in addition to those produced by generative reproduction.

Natural regeneration of *Pinus sylvestris,* however, is based exclusively on sexual reproduction. In Scots pine the formation of gametes and their fusion to a new zygote follow the basic pattern of sexual reproduction. In general, conifers have preserved the original mechanism of generative reproduction, whereas modifications are quite common among angiosperms *(Magnoliophytina).*

Many features of the regeneration of *Pinus sylvestris* are seemingly primitive or inefficient. However, the genetic system of Scots pine has been advantageous and maintained the species throughout tens of millions of years and a diversity of environmental changes (Mirov, 1967; Strasburger, 1983).

The structure and functioning of reproductive organs of pine were described by many early students of plant anatomy and embryology (Strasburger, 1869, 1871, 1892; Ferguson, 1904; Buchholz, 1918, 1920, 1926; Goebel, 1933; Schnarff, 1933; Doyle and O'Leary, 1935).

In his comprehensive study on flowering and seed crop of *Pinus sylvestris* Sarvas (1962) thoroughly cited most of the earlier descriptive works. The great merit of Sarvas' work was the introduction of a quantitative and casual approach. Furthermore, in his later studies Sarvas (1967; 1972) incorporated the timing of reproductive processes with temperature sum accumulation. To my mind, sound understanding of genetic processes must be based on the knowledge of the anatomic structure of reproductive organs as well as of the flowering ecology. A very brief overview of the reproductive cycle is given here, but for details reference should be made to the original studies mentioned above.

Flower primordia of *Pinus sylvestris* are initiated almost a year before flowering proper (Kupila-Ahvenniemi, 1985). Male strobili are formed at basal parts of terminal buds of lateral branches. During the latter part of the growing season male flower buds proceed so far that in a longitudinal section anthers are clearly visible. The actual generative tissue overwinters at the archesporial stage. The female primordia are formed in apical buds of topmost branches. The primordia remain rather tiny and undifferentiated. In longitudinal sections they are hardly visible.

The anatomical development of flowers continues in the beginning of the following growing season. The actual rate of progress depends on temperature and thus the dates of specific stages vary from year to year. In southern Finland male meiosis begins in mid-May and is completed usually by the end of May (Luomajoki, 1982). On female side the meiosis does not take place until July, a month after pollination. Female and male strobili reach their maturity simultaneously, of course.

In each tree the flowering starts with the opening of female strobili, which in a closed stand are mainly located on the upper part of the crown. When the receptiveness of female strobili has begun, the pollen sacs of male strobili start releasing pollen. Pollen grains are transported by turbulent air currents from anthers to female strobili. The pollen grains impacting receptive female strobili enter them through fissures between ovuliferous scales. During the pollination stage female strobili stand in an upright position. Pollen grains glide downwards on the scales and adhere on the sticky micropylar filaments.

Pollen grains of *Pinus sylvestris* carry two air sacs. In textbooks they are said to have a positive effect on the flight characteristics of pollen grains. This effect may be negligible. Pollen grains of *Pseudotsuga menziesii,* for instance, drift over considerable distances, even though they are larger than those of Scots pine and they do not have air sacs (Silen, 1962). Doyle and O'Leary (1935) suggested another function to air sacs. During the pollination period the micropyle of an ovule opens downwards. In daytime pollen grains enter the female strobili and stick on the adhesive strips of ovules. When air humidity rises, usually in the night, the micropylar canal is filled with fluid, the so-called pollination drop. Then the air sacs lift the pollen grains into the pollen chamber and give them a favourable position. Sarvas (1962) supported the same explanation. Among conifers pollen grains with air sacs are always associated with the pollination drop and a downwards oriented micropyle.

After some days the micropylar canal is closed through extension of the surrounding cells. Simultaneously the strobilus bends down and pollen grains germinate. As a rule, there are more than one pollen grain in each pollen chamber. All other pollen grains outside the pollen chamber are excluded from later development. The actual fertilization does not take place until the following summer. The development of the female gametophyte starts with the meiosis of the megagametophyte mother cell after the germination of pollen grains. Unpollinated ovules collapse soon after the flowering period. In pollinated ovules the remaining haploid cell begins to develop through repeated mitotic divisions of the nucleus. After reaching the free nuclear stage of the megagametophyte the ovule becomes dormant and grows further after overwintering. In spring, the haploid tissue continues to grow and also cell walls are formed. Some weeks after the flowering of the following cycle archegonia, usually more than one per ovule, are formed at the micropylar end of the endosperm. Then, about 13 months after pollination, the fusion of female and male gametes takes place. More than one zygote per ovule is formed. In all embryos of one and the same ovule the maternal genotypes are identical, whereas the paternal ones are different.

Polyembryony is characteristic of the sexual reproduction of *Pinus sylvestris.* In addition to polyzygotic polyembryony, monozygotic (i.e. cleavage), polyembryony also occurs, where each embryo is subdivided into four genetically identical proembryos. During embryogeny and seed ripening all but one embryo disappears, having only one fully developed embryo in a normal seed. The anatomical development of the seed proceeds roughly till the end of the growing season. Physiological maturation still continues after a chilling period in the autumn. Natural seed fall takes place in late winter and early spring.

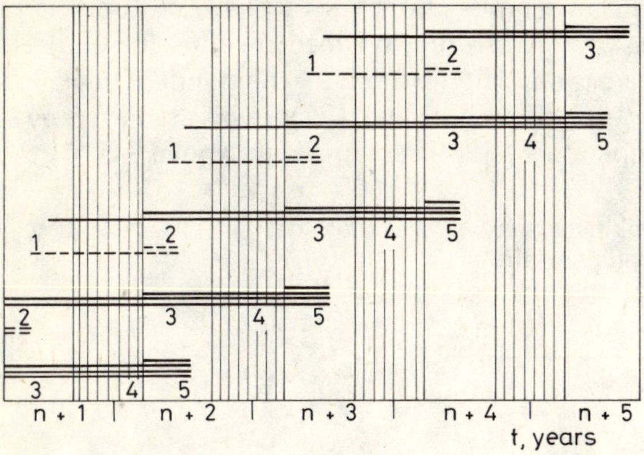

Fig. 4.1. A schematic diagram of overlapping reproductive cycles of *Pinus sylvestris.* The horizontal axis describes time in years. The dormant periods are shown by the shaded columns. The development of female organs is described with continuous lines and that of male organs with dashed lines. The number of lines in each pathway shows the "age" of reproductive organs. Marked phenological stages: 1: formation of flower primordia, 2: pollination; 3: fertilization, 4: seed dispersal, 5: abscission of empty cones

Each generative cycle consists of numerous events and stages, and lasts for almost three years. As pine flowers and sets seed every year, succeeding reproductive cycles overlap. Consequently, female organs of three cycles are observable almost any time in each tree (Fig. 4.1). In the following some of the components of the system are discussed in more detail with special emphasis on the genetic consequences.

Flowering ecology

Flowering time

The flowering period of *Pinus sylvestris* occurs rather regularly in early summer. In South Finland, for instance the first week of June sees dense pollen clouds and pollen deposition everywhere. When dealing with flowering studies or controlled crossings one certainly observes that the calendar date is not a reliable indicator of the onset of flowering every year. A continuous series of measurements has been collected with registering pollen samplers from 1961 to 1987 at Punkaharju in southeastern Finland (Table 4.1). A scrutiny on median dates of anthesis shows that the average for 27 years is day 160 (=June 9), the standard deviation is 7 days, and the range 29 days (from day 142 to day 171). In his studies, Sarvas (1967) found that in units of temperature sum (i.e. degree days above +5 °C) the

Table 4.1. The flowering time of *Pinus sylvestris,* Punkaharju, South Finland. (The median of the total pollen catch, measured with registering pollen samplers.)

Year	Date	Day	d.d.
1961	June 6	157	229
1962	June 20	171	245
1963	May 29	149	234
1964	June 15	167	242
1965	June 13	164	206
1966	June 12	163	256
1967	June 8	159	199
1968	June 13	165	212
1969	June 15	166	217
1970	June 10	161	211
1971	June 14	165	226
1972	June 13	165	214
1973	June 4	155	215
1974	June 19	170	205
1975	May 22	142	183
1976	June 12	164	187
1977	June 11	162	211
1978	June 6	157	224
1979	June 2	153	199
1980	June 9	161	200
1981	June 7	158	211
1982	June 12	163	184
1983	May 31	151	214
1984	May 28	149	212
1985	June 17	168	213
1986	June 8	159	242
1987	June 15	166	243
Average	June 9	160	216
St. dev.		7	19

scatter is essentially smaller than in calendar days. From the data in Table 4.1 we obtain an average of 216 d.d., and SD of 19 d.d. and the range of 184 to 256 d.d. Converted to days with average temperature of +15 °C we obtain SD=2 and a range of 7 days, respectively. Efforts have been made to refine the calculation of thermal time (Sarvas, 1972; Chung, 1981), but relative to the work needed the improvement is not substantial.

The linkage between the accumulation of a constant temperature sum and a certain phenological stage, e.g. flowering, was first suggested by Reaumur (1735; cited by Wang, 1960) and again by Linsser (1867). They also concluded that the characteristic temperature sum is proportional to the total temperature sum of the growing season. In other words pines of northern origin would reach flowering stage with a lower temperature sum than those of southern origin. Sarvas (1967) came to the same conclusion and he gave 17% of the total average temperature sum as the parameter for the maximal pollen shedding in *Pinus sylvestris.* This pattern has been confirmed later in Turkey (Boydak, 1975) and Hungary (Koski, 1987).

Northern and probably alpine marginal populations are an exception to the 17% rule. In regions where the average summer is too short and cool for seed ripening, the proportion of flowering temperature sum is higher than 17%.

For the sake of clarity: even though northern populations reach flowering stage at a lower temperature sum than the southern ones, they normally flower later than southern populations. The growing season begins earlier and temperature sum accumulation is faster in the South. For instance, in Hungary the flowering time is one month earlier than in South Finland, and again in North Finland one month later than in South Finland. In provenance trials, seed orchards and other transferred stands the opposite phenomenon can be observed.

Besides the geographical trend, flowering time exhibits variation at several levels. Within each tree every bud responds autonomously to temperature. Depending on exposure, height and shading temperature differences of several degrees may occur within a tree. Consequently, the rate of progress varies within the population of buds. We can only roughly say that this "topophysis" variation is a few days, but the true range depends on actual circumstances. Anyway, it causes an extension of the effective flowering period at an individual level.

The second level of variation is that among the trees in a stand. There are genetic differences and microclimatic variation within a stand. These factors are extremely difficult to distinguish. As this kind of observations is very tedious at canopy level, there are very few data on the receptiveness of female strobili or pollen release from male flowers of individual trees. Chung (1981) carried out careful surveys in a clonal archive and his figures provide perhaps the most representative information. According to him the receptive period of a female strobilus is around one week, on the average. As the duration is proportional to temperature and also depends on the origin of the population, it can vary from 3 to 10 days in South Finland. Besides temperature pollen shedding depends on air humidity and wind speed. During ideal flowering conditions the variation of pollen shedding is similar to that of female flowering, but rain and windstill can postpone pollen shedding considerably.

The general pattern of within-population variation of flowering time follows a normal distribution. Sarvas (1970) calculated 6% as an average coefficient of variation expressed in thermal time. When this figure is combined with the earlier mentioned 17% of the average temperature sum of the total growing season, the flowering time of any autochthonous population can be predicted. For instance, if the total temperature sum is 1 500 d.d., the mean of flowering will be at 255 d.d. and the whole range between 209 and 301 d.d. The numerical values are not to be considered as exact as constants of physics, because biological systems always display flexibility, but they are accurate enough for operational purposes.

Pollination ecology

It is important to realize that effective pollination is determined by the structure and function of both female and male organs. Thus, the knowledge of pollen dispersal alone is not enough for genetic conclusions.

Sarvas (1962, p. 105) studied the number of fertile ovuliferous scales in female strobili of *Pinus sylvestris.* The average of 10 sample trees and 4 years was 12.8. As each scale carries 2 ovules, there are normally 24 or 26 ovules per strobilus. A strobilus has no mechanism to catch passing pollen grains from the surrounding air. It is only a target of the random trajectories of pollen grains. When pollen grains hit a strobilus, they are not evenly distributed among all ovules. The very first pollen grain deposits on an unpollinated ovule, but as the pollen catch accumulates, the probability of reaching an already pollinated ovule increases.

The number of required pollen depositions per strobilus is much higher than the number of ovules. The number of pollen grains per ovule must be at least two in order to benefit from advantages of polyembryony. In the random process of wind pollination, hundreds or thousands of pollen grains must hit each strobilus during the few days of receptiveness. This requirement makes the observed pollen catch values, such as 200 per mm^2, understandable.

On the other hand, there is an upper limit of pollen grains that can enter each pollen chamber. Sarvas (1962, pp. 76–77) reported that "in most cases there simply has not been room for more than 2–3 pollen grains". In the case of abundant total pollination only the early fraction is effective.

Pollen transportation by air currents from male to female strobili is not channelled. Thus, an adequate pollination necessitates a high density of pollen grains throughout the canopy layer. A high density is only achieved with a large number of pollen grains. Sarvas (1962, p. 23) reported average pollen crops of 9.0 to 34.5 kg per hectare in mature stands of South Finland. The processed data of Sarvas, published by Koski and Tallqvist (1978) reveal that in years of heavy flowering the pollen crop is still higher, from ca. 30 to 130 kg per hectare.

In relation to spores and pollen grains of most flowering plants the pollen grains of *Pinus sylvestris* are large. Their diameter is ca. 40 μm and their fresh weight ca. 10^{-8} g. The sedimentation rate is 2.5 to 5 cm/s (Eisenhut, 1961).

After maturation the male strobilus releases pollen when evaporation from the anther walls opens the pollen sacs with a cohesion mechanism. This takes place only when the relative air humidity is low. However, the anthers have no mechanism to release pollen actively. The launching of pollen needs an external impulse, which is wind. Pollen is released when wind velocity exceeds 0.5 m/s through friction and above all through shaking of branches.

Air currents exceeding 0.5 m/s in a forest stand are always turbulent. Instead of a laminar flow, the wind is gusts and eddies. There are also vertical air movements where velocities are up to 1 m/s. During warm days there are convection currents in pine stands with an ascending component of several metres per sec. Under these circumstances the released pollen grains float in air rather than sink continuously towards the ground.

Plentiful pollen production and turbulent winds make up a mechanism permiting sufficient pollination of female strobili. The effectivity of pollination depends on the relation between distance and pollen density. Pollen is mixed with increasing amounts of air by eddies as soon as it is released from the anthers. The decline of density can be described with a leptocursic curve, the change beeing steepest within the very first metres. The density is highest within the tree itself, which favours self-pollination. The proportion of self-pollination depends on the abundance of male flowering of the tree itself and even more on the magnitude of foreign pollination. The range of self-pollination may be from 0 to 100%, but a normal proportion in a mature stand is 10 to 20%.

The pollen density of an individual tree falls so low within 50 metres that it is difficult to measure with pollen samplers. But beware! This does not indicate that all pollen grains are sedimented within 50 metres, but on the contrary, they mainly rose, out of the reach of samplers. During the flowering time the pollen cloud is dense, regardless of the negligible contribution on any single tree. The pollen cloud must be a mixture from vast numbers of trees.

Dispersion models clearly show that pollen of Scots pine obviously travels over considerable distances, that is tens of kilometres. Empirical evidence has been published since 1919 (Hesselman, 1919). (For further references see, e.g. Koski, 1970.)

The abundant pollen crop of a stand is mixed by the wind to form a general pollen cloud whenever pine forests cover large areas. Nothing but rainfall can bring all travelling pollen grains to the ground. The general pollen cloud or background pollination causes effective gene flow between adjacent stands. The author (Koski, 1970) calculated from empirical data that the proportion of background pollination is ca. 60% in ordinary stands ($<$5 ha) in South Finland. Differences in flowering time restrict gene exchange over very long distances. Overlapping populations and repeated gene flow in each generation form a chain along which gradual changes in gene pool are possible over hundreds of kilometres.

There is every reason to assume that the gene flow is effective mainly from the South towards the North. Reaching of flowering stage is accelerated by high temperatures. At least in Finland warm weather is brought by south winds, and in most years pine flowering takes place during southerly winds.

Polyembryony

Most ovules of *Pinus sylvestris* have more than one archegonium (egg cell). The normal number is 2 but 3 is quite common too (Sarvas, 1962, p. 88). As the number of pollen grains per ovule usually is 2–3, normally more than one zygote is formed after fertilization. The match between pollen grains and archegonia is not perfect, and therefore the average number of zygotes per ovule is below 2, (1.75 according to Sarvas, 1962 and Koski, 1971).

The existence of polyembryony is obviously linked with self-pollination and the elimination of inbreeding. Partial self-pollination occurs normally, and there is no self-incompatibility mechanism to prevent self-fertilization. However, most embryos originating from selfing abort due to homozygosity of embryonic lethals. There are various models explaining the nature of embryonic lethals, e.g. model of independent recessive factors (Bramlett and Popham, 1971; Koski, 1971), of lethal equivalents (Sorensen, 1969) and that of epistatic interactions (Griffin and Lindgren, 1985). As regards elimination of inbreds it does not matter which of the above-mentioned models is most appropriate.

The number of independent recessive lethals per tree has been estimated to be as high as 8 to 10 (Koski, 1971; Koski and Muona, 1985), which means 90 to 95% probability of abortion after self-fertilization. As the total number of loci is limited, even outcrossing results in lethal homozygotes at a low frequency. Sarvas (1962) found that ca. 24% of all embryos abort after open pollination. In case of one embryo per ovule this would result in at least 24 empty seeds out of a hundred, whereas in case of two embryos less than 6% of seed would be empty. This kind of preference of outcrossed seed may be a sufficient evolutionary justification to preserve polyembryony.

The existence of a real competition among surviving embryos and a consequent positive selection for better fitness is hard to confirm. The reliability of early selection at seedling stage is similarly doubtful; it is difficult to consider the embryonal stage more predictive.

Quantity of reproduction

Sarvas conducted extensive measurements on the flowering and seed crop in pure, mature stands of *Pinus sylvestris* (Sarvas, 1962; Koski and Tallqvist, 1978). The average seed crop in a ten years'period was around 100 seeds per m^2 in South Finland, around 50 in North Finland, and more than 150 in southern Scandinavia. In most years the seed crop was 300–400 seeds per m^2. In other words the magnitude is 10^6 seeds per hectare and year. The average pollen crop was 3–4 g per m^2, whereas in years of heavy flowering the figure was 10–15 g per m^2, or 80–120 kg per ha. On a hectare basis the number of pollen grains reaches a magnitude of 10^{12}.

The capacity of pollen and seed production is proportional to the dominant height of the stand. In this connection magnitudes rather than exact figures are meaningful. Especially on the paternal side the immense number offers a chance for rather unlimited recombinations in formation and fusion of gametes. The total number of seeds is so high that even after random perishing of most, the emerging seedlings are numerous enough to provide a starting material for natural selection.

Pinus sylvestris has a relatively long juvenile period after which the female flowering gets started first. Male flowers appear gradually, several years later. Individual differences in the age of maturation and intensity of flowering are large. The variation is not levelled out at later ages. Even among dominant trees there are consistent differences both in the abundance of flowering and in the sex ratio. The empirical material on the variation of trees in natural stands is scarce. Clonal seed orchards and tree banks have been studied to some extent (Jonsson et al., 1976; Bhumibhamon, 1978; Ross, 1984; Müller-Starck and Ziehe, 1984; Koski and Muona, 1986).

The results show large variation among genotypes both in female and male flowerings. Three main issues have to be mentioned: *(1)* The distribution of trees is not a symmetric normal distribution as in many other characteristics. The frequency of scarcely flowering trees is quite high, whereas really profusely flowering trees are infrequent. *(2)* The tendency to flowering is rather consistent, in other words, differences in reproductive capacity do not level out over succeeding years. *(3)* Female and male flowerings vary independently of each other. There are predominantly female trees or predominantly male trees. Without doubt these features cause deviations from the presumed panmixia and Hardy–Weinberg equilibria.

The differences are large enough to cause undesirable bias in the genetic composition of breeding populations and seed orchards with small number of genotypes (Jonsson et al., 1976; Bhumibhamon, 1978; Shen et al., 1981; Müller-Starck, 1982). In large natural populations all the earlier described mechanisms counteract this kind of random changes.

A great deal of viable seeds and seedlings are destroyed by various pests and calamities. Nevertheless, the density of seedlings is tens per m^2 and hundreds of thousands per hectare. As the density of mature stands is less than a thousand per hectare, 99.9% of the individuals forming the natural regeneration die before maturity. The process of self-thinning is obviously coupled with the large within-population variation (Wallace, 1981, p. 486). It is advantageous to a population that in the case of high-density a few individuals take over while the others loose. Another advantage of the high reproductive capacity and the genetic diversity in the offspring is the possibility of adaptation to an environmental change.

Genetic consequences

Rate of outcrossing

Monoecism and synchronization of female and male flowering in Scots pine lead without doubt to partial self-pollination. In large natural stands abundance of foreign pollen dilutes the proportion of own pollen. The opening of female strobili begins a little earlier than that of male flowers. In case of heavy pollination pollen chambers of ovules can be occupied by foreign pollen grains and own pollen is excluded from effective pollination.

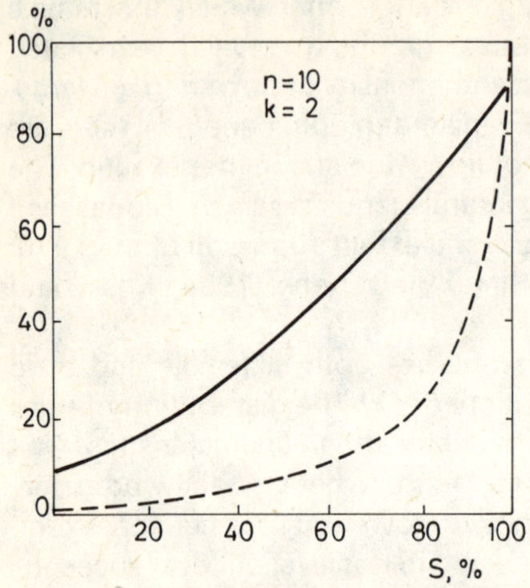

Fig. 4.2. An example of predicted proportions of empty seed (continuous line) and selfed filled seed (dashed line) after partial self-pollination (S). The number of embryonic lethals (n) is 10, and the number of fertilizations per ovule (k) is 2

The most significant elimination of selfing arises from homozygosity of embryonic lethals. The latest, refined models of embryonic lethals (Bishir and Namkoong, 1987) show that with close to 10 lethals per tree, ca. 90% of selfed embryos would be eliminated. Koski and Muona (1986) suggested a formula for estimating the proportion of selfed embryos among filled seeds at any proportion of self-pollination versus any number of ebryonic lethals.

The model is expressed by the two following equations. The probability of embryo abortion after cross-pollination is 0.3. An application to a typical case is given in Fig. 4.2.

$$E = P_n(2)S^2 + 2 \times 0.3P_n(1)\,SC + 0.09\,C^2 \tag{1}$$

$$s = \frac{(1-P_n(2))S^2 + 2[(1-P_n(1)) \times 0.3\,SC + 0.5\,(1-P_n(1)) \times 0.7\,SC]}{(1-E)} \tag{2}$$

where

E	=	Probability of empty seed formation
$P_n(k)$	=	Probability of abortion of all embryos after self pollination
n	=	Number of embryonic lethals
k	=	Number of fertilizations per ovule
S	=	Proportion of self-pollination
C	=	Proportion of cross-pollination ($=1-S$)
s	=	Probability of a selfed embryo in a filled seed.

This model predicts a fairly low frequency of inbreds at normal proportions (<20%) of self-pollination. Electrophoretic studies on isozymes have, however, consistently resulted in higher estimates of inbreeding in natural populations (Müller-Starck, 1977; Rudin and Lindgren, 1977; Rudin et al., 1986; Koski and Muona, 1986; Muona et al., 1987). This discrepancy is, perhaps, caused by overestimates of the number of lethals per tree. This, in turn, is caused by empty seeds formed due to other reasons (Bishir and Namkoong, 1987). Thus, in natural populations some 5% of the filled seeds are selfed according to isozyme studies. On the other hand, values exceeding 10% would require 60% of own pollen (see Fig. 4.2) according to the model of Koski and Muona (1986).

Due to their reduced vigour selfed seedlings disappear during seedling and sapling stages, and the fixation index does not differ from zero in natural stands (Gullberg et al., 1985). From the viewpoint of population genetics it does not make any difference at which stage selfing is eliminated. Outbreeding flowering plants can prevent self-fertilization by means of self-incompatibility mechanisms. *Pinus sylvestris* does not have the same system, but several subsystems together lead to the same result.

All the evidence of flowering ecology and pollen dispersal suggests that the number of pollen parents of the progeny of one tree is very large. The closest neighbours contribute relatively most, but one half of the total pollination consists of small fractions originating from beyond 50 metres' distance (Koski, 1970).

Genetic structure of stands

A high outcrossing rate together with effective gene flow over considerable distances suggests a certain population structure (Koski, 1970; pp. 72–73). Genetic variation within population will be large, and within uniform regions the genetic differentiation of sub-populations is restricted. This kind of pattern has more recently been encountered by means of electrophoretic techniques (Gullberg et al., 1985), so that at many loci >95% of variation is found within populations.

In terms of population genetics this indicates polymorphism in a large proportion of loci and the existence of Hardy-Weinberg equilibria at genotype frequencies. Curiously enough, selection pressure reinforces the preservation of genetic polymorphism. The astonishingly large genetic load of embryonic lethals can only be caused by the advantage of heterozygotes. In the general formulae the equilibrium frequencies are denoted with $\hat{p}$ and $\hat{q}$, and the respective selection pressures with s and t:

$$\hat{p}=\frac{s}{s+t}, \quad \hat{q}=\frac{t}{s+t}$$

The following example shows that even in the case of a recessive lethal genotype both alleles are saved if the heterozygote has the highest fitness value.

If we denote alleles by L and l, ll being lethal we can write:

Genotype	Fitness	Selection coefficient
LL	0.9	0.1
Ll	1.0	0.0
ll	0.0	1.0

$$\hat{p} = \frac{1.0}{1.1} = 0.91$$

$$\hat{q} = \frac{0.1}{1.1} = 0.09$$

Probably the heterozygosity of lethal loci *per se* is not decisive but that of closely linked loci would be.

Adaptive characteristics have a polygenic background. During the formation of gametes gene combinations segregate. Effective outcrossing, gene flow and large offspring number result in a still larger variation in the progeny population than in the parent generation. In an unchanging environment stabilizing selection culls off the extremes from both tails of the distribution, but it neither moves the mean nor removes polymorphism. If the environment permanently changes due to a climatic change, for instance, the gene frequencies can follow the change. In other words the population is able to adapt.

This kind of process has occurred during the millions of years that Scots pine has existed. Quite recently, after the latest glaciation Fennoscandia has gone through essential alterations of climate. The effective mechanism of recombination together with the high reproductive potential seem to provide means for a rather rapid adaptation. In fact, one generation has to be enough for a successful species!

The potential of genetic flexibility is more easily seen in the geographical dimension. Provenance trials and studies on the annual rhythm have revealed the main south-north trends in climatic adaptation. On the other hand, morphological traits or isozyme markers do not necessarily have the same pattern. Thus, only a difference in selection pressure may and does cause genetic differentiation. Due to effective gene flow a fine grain differentation does not occur. As the climate changes gradually, gene frequencies follow the same gradient, and a clinal population structure is maintained.

The scrutiny of generative reproduction and genetics of *Pinus sylvestris* given in the preceding chapters refers to autochthonous populations within the continuous area of distribution. It is evident that in populations of isolated occurrences and in areas where the species is cultivated as an exotic, the genetic processes do not function any more in the described way.

Literature

Bhumibhamon, S. (1978): Studies on Scots pine seed orchards in Finland with special emphasis on the genetic composition of the seed. *Comm. Inst. For. Fenn.* **94,4**:1–118.

Bishir, J. and Namkoong, G. (1987): Unsound seeds in Conifers: Estimation of numbers of lethal alleles and of magnitudes of effects associated with the maternal parent. *Silvae Genet.* **36**:180–185.

Boydak, M. (1975): Eskisehir-Catacik mintikasi ormanlarinda saricam (*Pinus silvestris* L.) in tohum verimi uzerine arastirmalar [Researches on the seed crop of Scots pine (*Pinus silvestris* L.) in Eskisehir-Catacik forest region]. Istanbul Universitesi Orman Facultesi, Diss., p. 82.

Bramlett, D. and Popham, T. (1971): Model relating unsound seed and embryonic lethals in self-pollinated pines. *Silvae Genet.* **20**:192–193.

Buchholz, J. T. (1918): Suspensor and early embryo of *Pinus. Bot. Gaz.* **66**:185–228.

Buchholz, J. T. (1920): Embryo development and polyembryony in relation to the phylogeny of Conifers. *Amer. Journ. Bot.* **7**:125–145.

Buchholz, J. T. (1926): Origin of cleavage polyembryony in Conifers. *Bot. Gaz.* **81**:55–70.

Chung, M-S. (1981): Flowering characteristics of *Pinus sylvestris* L. with special emphasis on the reproductive adaptation to local temperature factor. *Acta For. Fenn.* **169**:1–69.

Doyle, J and O'Leary, M. (1935): Pollination in *Pinus. Scient. Proc. Roy. Dublin Soc.* **20**:181–190.

Eisenhut, G. (1961): Untersuchungen über die Morphologie und Ökologie der Pollenkörner heimischer und fremdländischer Waldbäume. *Fortsw. Forsch.* **15**:1–68.

Ferguson, M. (1904): Contributions to the knowledge of the life history of *Pinus* with special reference to sporogenesis, the development of gametophytes and fertilization. *Proc. Washington Acad. Sci.* **6.**

Goebel, K. (1933): Organographie der Pflanzen. III Teil Samenpflanzen. Jena.

Griffin, A. and Lindgren, D. (1985): Effect of inbreeding on production of filled seed in *Pinus radiata* — Experimental results and model of gene action. *TAG* **71**:334–348.

Gullberg, U., Yazdani, R. and Nyman, N. (1985): Allozyme variation in Scots pine (*Pinus sylvestris* L.) in Sweden. *Silvae Genet.* **34**:193–201.

Hesselman, H. (1919): Iakttagelser över skogsträdens spridningsförmaga (Observations on the ability of dispersal of forest trees). *Medd. Skogsförsöksanst.* **16**:27–60.

Jonsson, A., Ekberg, I. and Eriksson, G. (1976): Flowering in a seed orchard of *Pinus sylvestris* L. *Studia For. Suec.* **135**:1–38.

Koski, V. (1970): A study on pollen dispersal as a mechanism of gene flow in Conifers. *Comm. Inst. For. Fenn.* **70,4**:1–78.

Koski, V. (1971): Embryonic lethals of *Picea abies* and *Pinus sylvestris. Comm. Inst. For. Fenn.* **75,3**:1–30.

Koski, V. (1987): Long geographic transfers, a possible way of eliminating pollen contamination in advanced generation seed orchards of *Pinus sylvestris. Forest Ecol. and Manag.* **19**:267–271.

Koski, V. and Muona, O. (1986): Probability of inbreeding in relation to clonal differences in male flowering and embryonic lethals. *Proc. Joint IUFRO Conference, Williamsburg, Virginia 1986,* pp. 391–400.

Koski, V. and Tallqvist, R. (1978): Tuloksia monivuotisista kukinnan ja siemensadon mittauksista metsäpuilla (Results of long-term measurements of flowering and seed crop of forest trees). *Folia For.* **364**:1–60.

Kupila-Ahvenniemi, S. (1985): Wintertime changes in the fine structure and the ribosome content of the buds of Scots pine. In: Kaurin, A., Junttila, O. and Nielsen, J. (Eds), *Plant Production in the North.* Norw. Univ. Press Tromsö, pp. 171–180.

Linsser, G. (1867): Die periodischen Erscheinungen des Pflanzenlebens in ihrem Verhältnis zu den Wärmeeincheinungen. Mem. de l'Acad. Imp. d. sc. de St Petersbourg XI, **7**:1–44.

Luomajoki, A. (1982): Temperature and dates of male meiosis in trees. *Hereditas,* **97**:167–178.

Mirov, N. T. (1967): *The Genus Pinus.* The Ronald Press, New York.

Muona, O., Yazdani, R. and Rudin, D. (1987): Genetic change between life stages in *Pinus sylvestris:* Allozyme variation in seed and planted seedlings. *Silvae Genet,* **36**:39–42.

Muona, O., Harju, A. and Kärkkäinen, K. (1988): Genetic comparison of natural and nursery grown seedlings of *Pinus sylvestris* using allozymes. *Scand. J. For. Res.* **3**:37–46.

Müller-Starck, G. (1977): Untersuchungen über die natürliche Selbstung in Beständen der Fichte (*Picea abies* (L) Karst) und Kiefer (*Pinus sylvestris* L). *Silvae Genet.* **26**:207–217.

Müller-Starck, G. (1982): Sexually asymmetric selection and partial self-fertilization. 2. Clonal gametic contributions to the offspring of a Scot pine seed orchard. *Silva Fenn.* **16**:99–106.

Müller-Starck, G. and Ziehe, M. (1984): Reproductive systems in Conifer seed orchards 3. Female and male fitnesses of individual clones of *Pinus sylvestris* L. *TAG* **69**:173–177.

Pukacki, P. (1980): Temperature of Norway spruce and Scots pine buds. *Arboretum Kórnickie* **24**:277–287.

Ross, M. D. (1984): Frequency-dependent selection in hermaphrodites: the rule rather than the exception. *Biol. Journ. Linn. Soc.* **23**:145–155.

Rudin, D. and Lindgren, D. (1977): Isozyme studies in seed orchards. *Studia For. Suec.* **139**:1–23.

Rudin, D., Muona, O. and Yazdani, R. (1986): Comparison of the mating system of *Pinus sylvestris* in natural stands and seed orchards. *Hereditas* **104**:15–19.

Sarvas, R. (1962): Investigations on the flowering and seed crop of *Pinus silvestris. Comm. Inst. For. Fenn.* **53,4**:1–198.

Sarvas, R. (1967): The annual period of development of forest trees. *Proc. Finn. Acad. Sci. Lett.* **1965**: 211–231.

Sarvas, R. (1970): Establishment and registration of seed orchards. *Folia For.* **89**:1–24.

Sarvas, R. (1972): Investigations on the annual cycle of development of forest trees. Active period. *Comm. Inst. For. Fenn.* **76,3**:1–110.

Schnarff, K. (1933): *Embryologie der Gymnospermen. Handbuch der Pflanzenanatiomie X/2.* Berlin.

Shen, H-H., Rudin, D. and Lindgren, D. (1981): Study on the pollination in a Scots pine seed orchard by means of isozyme analysis. *Silvae Gen.* **30**:7–15.

Silen, R. (1962): Pollen dispersal considerations for Douglas fir. *J. For.* **60**:790–795.

Sorensen, F. (1969): Embryonic genetic load in coastal Douglas-fir, *Pseudotsuga menziesii* var. *menziesii. Am. Naturalist.* **103**:389–398.

Strasburger, E. (1869): Die Befruchtung bei den Coniferen. Jena.

Strasburger, E. (1871): Die Bestäubung der Gymnospermen. *Jenaisch. Zeitschr. Medic. Naturwiss.* **6**:247–262.

Strasburger, E. (1892): Ueber das Verhalten des Pollens und die Befruchtungsvorgänge bei den Gymnospermen. Jena.

Strasburger, E. (1983): Lehrbuch der Botanik, 32. Auflage, neubearbeitet von v. Denffer, Ziegler, Ehrendorfer, Bresinsky. Gustav Fischer Verlag, Stuttgart.

Wallace, B. (1981): *Basic population genetics.* Columbia University Press, New York.

Wang, J. (1960): A critique of heat unit approach to plant response studies. *Ecology* **41**:785–790.

CHAPTER 5

Biochemical polymorphism

W. Prus-Głowacki

Isoenzymes

Inheritance of isoenzymes and genetic linkage

The present state of studies on isoenzyme inheritance in *Pinus sylvestris* is presented in Table 5.1. So far an inheritance system has been defined for 26 enzymes. In the course of their studies various authors have distinguished varying numbers of loci for the same enzymatic systems. This reflects ways in which the isoenzymatic patterns are interpreted, as well as various techniques used in electrophoretic separation. The total number of enzymatic loci described for Scots pine is between 37 and 54, and numbers of alleles coded by the loci, between 90 and 110.

The first results on linkages between four enzymatic loci (ADH A, ADH B, LAP B and GOT B) in *Pinus sylvestris* were published by Rudin and Ekberg (1978), who also provided a distance map for these loci. Studies by Szmidt et al. (1984) using 27 loci, demonstrated four linkage groups in Scots pine: group A includes 8 loci (GOT A, ADH A, ADH B, GOT B, PGI B, DIA D, LAP B and DIA A); group B two loci (LAP A and F EST); group C three loci (G6PD A, ACO, G6PD B); and group D two loci (GOT C and SDH B). Recently Niebling et al. (1987), studying the linkage for 29 enzymatic loci, have shown six linkage groups in *Pinus sylvestris.* A first (I) large group included seven loci (DIAC, PGIB, PEPLGG C, ADHA, PGK B, LAP B and GOT B), a second (II) group four loci (GOT C, 6PGDA, DIAB and MDHA). The third (III), fourth (IV), fifth (V) and sixth (VI) group each included two loci, LAPA–MDHC, EstA–APHB, APHA–GAPA, DIAA–PEPGLC, respectively. The linkage groups described for Scots pine correspond to linkage groups of enzymatic loci described by Conkle (1981) for other pine species belonging to different subgenera. This suggests that differentiation within this plant group has not been associated with more extensive chromosomal rearrangement.

Rasmuson (1979) was first to draw attention to the linkage disequilibrium existing between loci Est B and GOT B in *Pinus sylvestris,* based on the data of Rudin and Ekberg (1978). Muona (1982) reviewed the polygenic theory of multilocus structure in predominantly outcrossing populations with regard to gametic and zygotic disequilibria. She also discussed the possibility of taking advantage of the phenomenon in tree breeding. Studies of Muona and Szmidt (1985) on the genetic structure of three natural populations of Scots pine from northern Sweden have shown small but significant gametic disequilibria, the importance of which is, however, limited.

Table 5.1. Enzymatic gene markers described for *Pinus sylvestris*

Enzyme system	Number of loci	Number of alleles per locus	Tissue	References
Esterase E.C. 3.1.1.1. (Est)	3	5, 4, 2	n	Rudin and Rasmuson, 1973
Glutamate-oxalate-transaminase E.C. 2.5.1.1. (GOT)	2	3, 5	m, n	Rudin, 1975
	3	3, 5, 2	m, n	Rudin and Ekberg, 1978
	3	1, 2, 2	m	Krzakowa, 1979
	4	3, 5, 3, 1	m, e, b	Chung, 1981
	3	3, 7, 4	m, e	Müller-Starck, 1982*a*
Leucine amino-peptidase E.C. 3.4.1.1. (LAP)	2	3, 6	m	Rudin, 1977
	2	2, 2	m	Krzakowa, 1979
	2	5, 6	m	Mejnartowicz, 1979*a*
Acid phosphatases E.C. 3.1.3.2. (Aph)	2	3, 6	m	Rudin, 1977
	1	15	m	Mejnartowicz, 1978
	3	2, 2, 2	m	Krzakowa et al., 1977
Ribonuclease E.C. 3.1.4.23. (RNS)	2	np	m	Mejnartowicz and Bergmann, 1977
Phosphodiesterase E.C. 3.1.4.1. (PDE)	1	np	m	Mejnartowicz and Bergmann, 1977
Catalase E.C. 1.11.1.6. (Cat)	1	7	m	Szmidt, 1979*a*
Alcohol dehydrogenase E.C. 1.1.1.1. (ADH)	2	2, 3	m	Rudin and Ekberg, 1978
	4	2, 2, 2, 2	m	Krzakowa et al., 1977
Malate dehydrogenase E.C. 1.1.1.37. (MDH)	2	3, 4	m	Rudin and Ekberg, 1978
6-phospho-gluconate dehydrogenase E.C. 1.1.1.44. (6 PGD)	3	3, 2, 4	m	Krzakowa and Szweykowski 1979
	2	2, 3	m, e, b	Szmidt and Yazdani, 1984*a*
Glutamate dehydrogenase E.C. 1.4.1.3. (GDH)	3	2, 2, 2	m	Krzakowa, 1979
	1	3	m	Gullberg et al., 1982
Fluorescent esterase E.C. 3.1.1.2. (Fest)	1	4	m, n, b	Yazdani and Rudin, 1982
Fluorescent β-galactosidase E.C. 3.2.1.23. (βGal)	1	2	m, n, b	Yazdani and Rudin, 1982
Shikimate dehydrogenase E.C. 1.1.1.25. (ShDH)	2	3, 2	m, e, b	Szmidt and Yazdani, 1984*a*

Aconitase E.C. 4.2.1.3. (Aco)	1	3	m, e, b	Szmidt and Yazdani, 1984*b*
Phosphoglucose isomerase E.C. 5.3.1.9. (PGI)	2	1, 4	m, e, b	Szmidt and Yazdani, 1984*b*
Phosphoglucomutase E.C. 2.7.5.1. (PGM)	2	3, 2	m, e, b	Szmidt and Nagasaka, 1984
Formate dehydrogenase E.C. 1.2.1.2. (FDH)	1	3	m	Mejnartowicz and Bergman, 1985
Aldolase E.C. 4.1.2.13. (ALD)	1	np	m	Niebling et al., 1987
G-lyceraldehate-3-phosphate dehydrogenase E.C. 1.2.1.12. (GAP)	1	np	m	Niebling et al., 1987
Hexokinase E.C. 2.7.1.1. (HK)	1	np	m	Niebling et al., 1987
Phosphoglycerate kinase E.C. 2.7.2.3. (PGk)	2	np	m	Niebling et al., 1987
Glutamic-pyruvic transaminase E.C. 2.6.1.2. (GPT)	1	np	m	Niebling et al., 1987
Superoxide dismutase E.C. 1.15.1.1. (SOD)	1	np	m	Niebling et al., 1987
Peptidase (glycyl-leucine) E.C. 3.4.13.1. (PEPGL)	2	np	m	Niebling et al., 1987
Peptidase (leucyl-glycyl-glycine) E.C. 3.4.13.1. (PEPLGG)	1	np	m	Niebling et al., 1987

m = macromegatophytes;
n = needles;
b = buds;
e = embryos;
np = not published

Genetic structure and geographic differentiation

The study of Rasmuson and Rudin (1971) was one of the first reports of the use of isoenzymes of alpha-naphthyl acetate esterase in the identification of Scots pine clones and for the examination of population genetic structure in relation to geographic variability. Since then esterase variability has also been examined by Dukharev (1978*a, b*), Polozova and Dukharev (1978) and Pravdina et al. (1982) in pine populations from different regions of USSR. Rudin et al. (1974) have demonstrated marked genetic similarity in three enzymatic loci (EST, LAP B, and GOT B) for two adjacent populations of central Sweden and the genetically distant character of pine populations from the Kiruna area of northern Sweden. Krzakowa et al. (1977) examined the genetic structure of 22 plus trees. Variability of 4 enzymes (ADH, GOT, APh, LAP) revealed high heterozygosity in the trees.

Genetic variability of 8 Scots pine populations from Poland has been examined by Krzakowa (1979, 1982) and Krzakowa and Szweykowski (1979). The authors have concluded that populations from northern Poland constitute a separate group. Szmidt (1980) has also examined genetic variability of 11 Polish populations using catalase as the genetic marker. He demonstrated a high level of heterozygosity in those populations which show advantageous economic traits. The author found no correlation between isoenzymatic variability of catalase and the geographic distribution of the populations.

The studies of Mejnartowicz (1979*a*) on the variability of LAP and APh allozymes, performed on 19 populations of *P. sylvestris* from Poland, FRG, USSR, Hungary and Turkey, demonstrated that southern populations differ significantly from the northern ones. Genetic variability of 17 populations in relation to climatic and geographic conditions has been studied by Kieliszewska-Rokicka (1981) on the basis of esterase, peroxidase and glutamate dehydrogenase polymorphism. The author demonstrated that southern outliers of the species range (Spain and Turkey) exhibit a wider spectrum of peroxidase isoenzymes. Studies by Gullberg et al. (1982) and by Mejnartowicz and Bergmann (1985) have documented an evident relationship between the climatic and environmental conditions and the genetic variability of populations and subpopulations. Gullberg et al. (1985), studying variability of allozymes coded by 11 loci in 9 native stands of *P. sylvestris* in Sweden, have detected significant differences in allozyme frequencies between the studied populations, but at the same time have failed to demonstrate evident geographic or racial differences.

Reproductive systems

The investigation of the reproductive processes (mating system, selfing level, gene flow and influx of foreign pollen in seed orchards) is greatly facilitated by the isozyme technique. Details and results of these investigations are summarized in Chapter 11.

Hybridization and introgression

Only limited studies have been carried out on hybridization and introgression in *P. sylvestris* using isoenzyme techniques. Djurbabić et al. (1977) have dealt with incompatibility problems in *P. sylvestris, P. nigra* and *P. densiflora.* They have demonstrated variation in the number of protein fractions in the pollen of these species, as well as differences in the amount of protein within individual fractions. On the basis of the patterns obtained a correlation has been found between electrophoretic patterns and ability to form hybrids.

Kormutak (1984) has studied composition and activity of peroxidase, esterase MDH and ribonuclease isoenzymes in artificial *P. sylvestris x P. nigra* and *P. sylvestris x P. contorta* hybrids during ontogenesis. The author has noted significant variation in isoenzyme spectra during ontogenesis. The results indicate that protein similarity exists between *P. sylvestris* and *P. nigra,* but *P. contorta* is distinct. He suggested that peroxidases might serve as incompatibility indicators in *P. nigra.*

Hybrid swarms of *P. sylvestris* and *P. mugo* from peat bogs in the Sudety Mountains have been studied by Prus-Głowacki and Szweykowski (1977, 1983) with regard to the genetic structure of putative parents of the hybrid swarm populations (*P. sylvestris, P. mugo* and *P. uliginosa*), as well as with regard to the genetic structure of the hybrid swarm population itself. The results indicate that only limited gene flow takes place between the species in nature.

Demographic genetics and studies on adaptive processes

Reconstruction of the phenomena of adaptive processes in populations of long-lived organisms with overlapping generations, such as forest trees can be achieved by the examination of allozyme and isoenzymatic genotype frequencies in different age groups of naturally regenerating populations. As far as studies on the genetic structure of naturally regenerating Scots pine populations are concerned, heterozygosity has been observed to increase with age. Younger populations exhibit significant deviations from Hardy–Weinberg proportions (Dukharev, 1978*b*; Polozova and Dukharev, 1978; Polozova et al., 1978). Studies by Rudin and Tigerstedt (1979) and Tigerstedt et al. (1982) have demonstrated differences in the genetic structure between parental populations and their progeny and the latter has been found to include a significant number of homozygotes. The authors have suggested that selection against homozygotes represents an important factor restricting the effects of self-fertilization in naturally regenerating populations. Prus-Głowacki (1982) studied three age groups of a naturally regenerating population and has noted a tendency in the genetic structure of the younger age groups to converge towards the genetic structure of the parental group, which may indicate that stabilizing selection operates in the population studied. Studies by Yazdani et al. (1985) on the genetic structure of a naturally regenerating population of *P. sylvestris,* have shown an increase in the average heterozygosity level for

7 loci, observed from the embryo subpopulation through young trees to mature trees. Elimination of homozygous individuals takes place during the early stages of population development. Results of Prus-Głowacki and Nowak (1989) obtained for 4 age groups of naturally regenerating populations are consistent with the results of Yazdani et al. (1985).

Mejnartowicz (1984) reviewed literature on the effects of industrial emissions on enzyme activities in plant tissues and the isoenzyme differences between individuals sensitive and tolerant to industrial pollution. No differences have been found between tolerant and sensitive individuals by Kieliszewska-Rokicka (1979), and Szmidt (1979*b*) in peroxidase spectra.

Allelic frequencies at a few enzymatic loci in groups of trees either tolerant or sensitive to pollution by SO_2 and HF, in the vicinity of a phosphorus fertilizer plant have been observed by Mejnartowicz (1978, 1979*b*, 1983, 1984, 1986; Mejnartowicz and Okoniewska, 1982; Mejnartowicz et al., 1983). These studies have shown allozyme frequencies to be suitable as indicators of tolerance to environmental pollution. Resistant trees show a distinct genetic structure and, in general, a greater heterozygosity than sensitive trees. Studies of Geburek et al. (1987) on the genetic structure of two *P. sylvestris* groups from northern Westphalia, one relatively sensitive and the other relatively resistant, have demonstrated significant differences between the groups in GDH and GOT alleles and genotype frequencies. In tolerant trees the average heterozygosity is 10 to 30% higher. Also, variability at individual gene loci and multilocus diversities are approximately 2.5 times greater in tolerant trees. Studies carried out by Prus-Głowacki and Nowak (1989) on the genetic structure of a naturally regenerating population of a region polluted with heavy metals and SO_2, have demonstrated 20% higher heterozygosity in the group of mature individuals than in populations of embryos and seedlings.

The review presented above, relating to the present position of research on various genetic and population problems of *P. sylvestris* using isoenzymatic markers show that our knowledge in the subject is still fragmentary.

Scots pine has been found to be a highly polymorphic species as far as enzymatic loci are concerned, exhibiting low interpopulation variability; so far evidence from European populations indicates that 95–98% of gene diversity in the species will be observed within an average population. The average multilocus heterozygosity of such a population amounts to approximately $H_0 = 0.340$. Populations of the species are close to the Hardy–Weinberg equilibrium (Table 5.2).

Unfortunately, data concerning isoenzymatic differentiaton of Scots pine as related to geographic and racial aspects still remains largely incomplete, due to few published studies and the enormous territorial range of the species.

As far as future studies are concerned, extensive reviews of research perspectives associated with the use of isoenzymes in forest tree populations have been given by Adams (1983), Mitton (1983) and Rudin (1986).

Table 5.2. Some genetic parameters of *Pinus sylvestris* populations

Number of populations	Origin of populations	Number of loci	A	$\bar{H}$	H_{max}	H_{min}	F	References
19	Poland, Turkey, USSR, FRG	3	8.60	0.312	0.413	0.180	—	Mejnartowicz, 1979*a*
9	Sweden	9	4.36	0.307	–	–	−0.005	Gullberg et al., 1985
18	Western Europe	8	4.25	0.361	0.427	0.250	0.033	Prus-Głowacki et al. (in preparation)
14	Soviet Union	8	4.00	0.366	0.471	0.271	0.017	Prus-Głowacki et al. (in preparation)
7	FRG	8	3.36	0.339	0.419	0.295	−0.009	Prus-Głowacki et al. (in preparation)
13	Poland	8	3.50	0.385	0.446	0.335	0.033	Prus-Głowacki et al. (in preparation)
Average			4.67	0.345			0.014	

$\bar{H}$ = heterozygosity of the investigated population;
H_{max}, H_{min} = maximal and minimal heterozygosity values observed;
A = average numbers of alleles per locus;
F = fixation index.

Antigens

Immunochemical techniques permit the demonstration of a single amino acid substitution localized within the antigenic determinant. Any protein molecule inducing an immune response, carries on its surface a few to a few hundred antigenic determinants, which are unique and capable of binding the antibodies in a highly specific way. Because of this phenomenon immunochemical techniques permit precise detection of homologous proteins in individuals either of the same population or of the same species, or at a higher systematic level (Jensen and Fairbrothers, 1983).

The pioneering work by Hagman (1967, 1975, 1977) on antigen polymorphism in *Pinus* with respect to intra- and interspecific differentiation and incompatibility, have shown that the antigen spectrum in various pine species (including *P. sylvestris*), may prove useful not only for taxonomic studies but also for the identification of populations, provenances and forest tree clones. Similarly, comparative studies by Saito (1968) have provided an additional rationale for the use of protein antigenic properties for chemotaxonomic and phylogenetic tests. Studies by Prager et al. (1976) on the antigenic homology of seed storage proteins in conifers, have permitted closer determination of the divergence time for these genera, and offers a new phylogenetic approach to the taxa.

Serological studies of incompatibility in a few pine species, including *P. sylvestris,* have demonstrated that individual species are characterized by specific pollen antigen spectra (Hagman, 1975; Petricević et al., 1977; Kormutak, 1984). A very clear differentiation has been observed for needle proteins in *P. sylvestris, P. mugo* and *P. uliginosa* individuals and in individuals of hybrid swarms, using various serological techniques (Prus-Głowacki et al., 1978, 1981; Prus-Głowacki and Szweykowski, 1979, 1980; Prus-Głowacki, 1983).

Studies on antigenic differentiation in needle proteins of 6 Swedish populations have revealed differences between individual trees within a given population, as well as interpopulation differences, both in immunodiffusion spectra and in degree of antigen polymorphism (Prus-Głowacki and Rudin, 1981; Prus-Głowacki, 1983).

Terpenes

Terpenes are widely spread throughout the plant world and are particularly frequent in conifers. They are compounds synthesized along a complicated biosynthetic pathway, and frequently two different terpenes may originate from the same precursor. Common precursors, common intermediate compounds and the long biosynthetic pathway make these biochemical markers modifiable to a certain extent by environmental factors.

In *Pinus sylvestris,* 11 monoterpene compounds have been identified: alpha-pinene, beta-pinene, myrcene, limonene, beta-phellandrene, sabinene, delta-3-carene, tricyclene, camphene, terpinolene and trans-beta-ocimene. Of these, the first four are present in greater amounts, but relative levels of the compounds differ significantly (Yazdani et al., 1982). The use of terpenes as markers for genetic investigations was reviewed by Squillace (1976).

Inheritance of terpenes

Hiltunen et al. (1975*a, b*) have shown that the distribution of delta-3-carene in a population of *Pinus sylvestris* is bimodal and that synthesis of the compound is controlled by a single locus with two alleles. Similar results for delta-3-carene inheritance have been obtained by Chudnyi (1979) who has demonstrated that beta-pinene synthesis is also under control of a single locus. The inheritance pattern of five monoterpenes; beta-phelladrene, beta-pinene, myrcene, limonene and delta-3-carene, has been examined in *P. sylvestris* by Yazdani et al. (1982). By analysing results of controlled crosses, the authors have demonstrated that segregation patterns are in general consistent with a simple Mendelian inheritance, although some exceptions to the rule are noted.

The inheritance of the diterpenes pinifolic acid and isoabienol, occurring in needles of *P. sylvestris,* has also been examined. In controlled crosses the amount of pinifolic acid has shown no signs of Mendelian inheritance and environmental effects on the level of compound synthesis cannot be excluded (Gref and Lindgren, 1984). On the other hand, the isoabienol content segregated in the majority of cases according to a simple Mendelian inheritance (Gref et al., 1985).

Geographical variation in monoterpene content

Studies on differences in monoterpene content in 108 populations from the entire range of Scots pine, performed by Tobolski and Hanover (1971) have demonstrated that the compounds may serve as means of identification for populations. Alpha-pinene and 3-carene have proven to be most variable. In general, alpha-pinene is present in a higher proportion of trees from the southern limits of the tree range, where 3-carene was not detected, while 3-carene reaches high levels in the northern populations. Populations from Central Europe exhibit intermediate levels of the compounds. Isolated populations from southern Europe (Spain, France, Greece) and Asia Minor (Turkey, Georgia) show characteristic monoterpene spectra of their own, allowing the separation of varieties and races. The authors have also shown that the monoterpene spectrum patterns may serve for the reconstruction of species evolution and the history of pine migrations. On the other hand, studies of Chudnyi (1979) and Yazdani (1986) have shown that changing ecological conditions had no effect on monoterpene spectra. Hiltunen et al. (1975*a*) and Tigerstedt et al. (1979) have examined monoterpene variability in 18 Scots pine populations from Finland and have observed that 3-carene and alpha-pinene distribution patterns exhibit a discontinuity at 63°–64° N. On each side of the discontinuity clinal variability has been observed.

Studies by Forrest (1979, 1980) on monoterpene differences in relic Scots pine populations from Scotland have shown that such data may be used for the identification of populations. Groups of young trees regenerating naturally show marked similarities to parental populations in monoterpenes and exhibit extensive variability. Yazdani et al. (1985) have described geographic variability in monoterpene content in 26 *P. sylvestris* populations originating from seeds collected from natural stands in Sweden. In populations originating from southern Sweden a higher proportion of individuals show lower contents of limonene and beta-pinene and higher ones of delta-carene content than in northern populations. Limonene shows a clearly clinal variability from south to north: a response to the climate according to the authors. Further studies on terpene variability in 10 natural populations of *P. sylvestris* from central and northern Sweden have also demonstrated a significant clinal interpopulation variability in monoterpene composition (Yazdani and Nilsson, 1986).

Concluding remarks

Each of the biochemical markers and each of the techniques demonstrates some advantages and limitations. Isoenzymatic techniques are most suitable for characterization of genetic parameters of natural and artificial populations (e.g. seed orchards) such as: gene diversity, heterozygosity, fixation index, level of selfing, gene flow and changes of all these factors in space and time. Isoenzymes are less suitable for studies on the geographic differentiation of Scots pine. Immunochemical techniques give valuable data for taxonomy, hybridization and introgression and may be useful also for identification of populations, provenances, seed samples and clones.

Terpenes are useful compounds in the search for intra- and interpopulation markers, as well as for the identification of clones, provenances and geographic races.

Literature

Adams, W. T. (1983): Application of isozymes in tree breeding. In: Tanksley, S. D. and Orton, T. J. (Eds) *Isoenzymes in plant genetics and breeding.* Part A. Elsevier Science Publishers, Amsterdam, pp. 381–400.

Chudnyi, A. V. (1979): Population structure of Scotch pine under different ecological conditions. (On the example of biosynthesis of monoterpenes.) *Ekologiya* **1**:37–42; In: *Soviet Journal of Ecology* **10**:1, 25–30.

Chung, M. S. (1981): Biochemical methods for determining population structure and mating system in *Pinus sylvestris* L. *Acta For. Fenn.* **173**:1–28.

Conkle, T. M. (1981): Isozyme variation and linkage in six conifer species. In: Conkle, T. M. (Ed.) *Isoenzymes of North American forest trees and forest insects. Proceedings of the symposium. Berkeley, California, July 27, 1979,* pp. 11–17.

Djurbabić, B., Vidaković, D., Kolbah, D. and Borzan, Ž. (1977): Quantitative analysis of soluble proteins extracted from the pollen of some incompatible pine species. *Genetica* **9**:149–157.

Dukharev, V. A. (1978*a*): Poliformizm esteraz u sosny obyknovennoj v raznovozrastnykh populyatsiakh *Pinus sylvestris* L. (Investigation of esterase polymorphism in the population of *Pinus sylvestris* L.) *Genetika* **14**:1789–1793.

Dukharev, V. A. (1978*b*): Sostav esteraz u sosny obyknovennoj v raznovozrastnykh drevostoyakh (Composition of esterases in *Pinus sylvestris* in uneven-aged stands). *Lesovedenie* **6**:45–50.

Forrest, G. I. (1979): Monoterpene variation in lodgepole pine *(Pinus contorta)* and Scots pine *(Pinus sylvestris).* In: Rudin, D. (Ed.) *Biochemical genetics of forest trees. Proceedings of the Conference, Umeå. Sweden 1978.*

Forrest, G. I. (1980): Genotypic variation among native Scots pine populations in Scotland based on monoterpene analysis. *Forestry* **53**:101–128.

Geburek, Th., Scholz, F., Knabe, W. and Varnweg, A. (1987): Genetic studies by isozyme gene loci on tolerance and sensitivity in an air polluted *Pinus sylvestris* field trial. *Silvae Genet.* **36**:49–53.

Gref, R. and Lindgren, D. (1984): The inheritance of pinifolic acid in Scots pine (*Pinus sylvestris* L.) needles. *Silvae Genet.* **33**:235–237.

Gref, R., Rudin, D. and Lindgren, D. (1985): Segregation patterns of isoabienol in *Pinus sylvestris* L. *Silvae Genet.* **34**:12–14.

Gullberg, U., Yazdani, R. and Rudin, D. (1982): Genetic differentiation between adjacent populations of *Pinus sylvestris. Silva Fenn.* **16, 2**:205–214.

Gullberg, U., Yazdani, R., Rudin, D. and Ryman, N. (1985): Allozyme variation in Scots pine (*Pinus sylvestris* L.) in Sweden. *Silvae Genet.* **34**:193–200.

Hagman, M. (1967): Serological studies of pollen and the incompatibility in forest trees. *14th IUFRO Congress, München* **3**:346–365.

Hagman, M. (1975): Incompatibility in forest trees. *Proc. Roy. Soc. (London)* **188**:313–326.

Hagman, M. (1977): The use of serological methods for the identification of species, provenances and clones of forest trees. *E. E. C. Symposium on forest tree biochemistry. Brussels, January 1977,* pp. 227–262.

Hiltunen, R., Juvonen, S. and Tigerstedt, P. M. A. (1975*a*): Geographical variation in some monoterpenes in Scots pine (*Pinus sylvestris* L.) in Finland. *Farm. Notisbl.* **84**:73–82.

Hiltunen, R., Tigerstedt, P. M. A., Juvonen, S. and Pohjola, J. (1975*b*): Inheritance of 3-carene quantity in *Pinus sylvestris* L. *Farm. Notisbl.* **84**:69–72.

Jensen, U. and Fairbrothers, D. E. (Eds) (1983): *Proteins and nucleic acids in plant systematics.* Springer, Berlin, p. 408.

Kieliszewska-Rokicka, B. (1979): Peroxidase activity in varieties of *Weigela* and *Pinus sylvestris* resistant and susceptible to SO_2. *Arboretum Kórnickie* **24**:313–320.

Kieliszewska-Rokicka, B. (1981): Geographic and climatic differentiation of electrophoretic forms of esterase, glutamate dehydrogenase and peroxidase in Scots pine tissues. *Acta Soc. Bot. Pol.* **50, (3)**:481–492.

Kormutak, A. (1984): Some cytological and biochemical aspects of interspecific incompatibility in pines (*Pinus* sp.). *Acta Dendrobiol.* **7**:7–90.

Krzakowa, M. (1979): Enzymatyczna zmienność miedzypopulacyjna sosny zwyczajnej (Interpopulational differentiation of Scots pine populations revealed by isozyme electrophoresis). *Wydawnictwo UAM (Poznan) Ser. Biol.* **17**:1–43.

Krzakowa, M. (1982): Genetic differentiation of Scots pine populations. I. Genotypes. *Silva Fenn.* **16, 2**: 200–205.

Krzakowa, M., Szweykowski, J. and Korczyk, A. (1977): Population genetics of Scots pine (*Pinus sylvestris* L.) forests. Genetic structure of plus-trees in Bolewice near Poznan (West Poland). *Bull. Ac. Pol. Sci.* **2, 25**:583–590.

Krzakowa, M. and Szweykowski, J. (1979): Variation of 6-PGO in the populations of Polish Scots pine *(Pinus sylvestris).* In: Rudin, D. (Ed.) *Proceedings of the conference on biochemical genetics of forest trees. Umeå, Sweden, 1978.*

Mejnartowicz, L. E. (1978): Genetic characteristic of some Scots pine trees susceptible or somewhat resistant to the action of SO_2. In: Białobok, S. (Ed.) *4th annual report PG-Po-326-studies on the effect of sulfur dioxide and ozone on the respiration and assimilation of trees and shrubs in order to select individuals resistant to action of these gases,* pp. 26–33.

Mejnartowicz, L. E. (1979*a*): Genetic variation in some isoenzyme loci in Scots pine (*Pinus sylvestris* L.) populations. *Arboretum Kórnickie* **24**:91–104.

Mejnartowicz, L. E. (1979*b*): Changes in genetic structure of Scots pine populations affected by industrial emissions of sulphur and fluorine oxides. First Annual Report, Institute of Dendrology, Kórnik.

Mejnartowicz, L. E. (1983): Changes in genetic structure of Scots pine (*Pinus sylvestris* L.) population affected by industrial emission of fluoride and sulphur dioxide. *Genet. Polon.* **24**:41–50.

Mejnartowicz, L. E. (1984): Enzymatic investigations on tolerance of forest trees. In: Koziol, M. J. and Whitley, F. R. (Eds) *Gaseous air pollutants and plant metabolism.* Butterworths, London–Boston, pp. 381–398.

Mejnartowicz, L. E. (1986): Genotypes of Scots pine trees differing in resistance to the action of fluoride and sulphur dioxide described on the basis of glutamic-oxalacetic-transaminase isoenzymes. *Acta Physiol. Cytol. Genet.* **1**:139–146.

Mejnartowicz, L. E. and Bergmann, F. (1977): Variation and genetics of ribonucleases and phosphodiesterases in conifer seeds. *Can. J. Bot.* **55**:711–717.

Mejnartowicz, L. E. and Bergmann, F. (1985): Genetic differentiation among Scots pine populations from the lowlands and the mountains in Poland. In: Gregorius, H.-R. (Ed.) *Population genetics in forestry. Lecture notes in biomathematics,* 60. Springer, Berlin–Heidelberg–New York, pp. 253–266.

Mejnartowicz, L. E. and Okoniewska, B. (1982): Genetische Struktur von Kiefernpopulationen nach Einwirkung von Industrie-Immissionen. *Proceedings of 2nd Workshop on Forstgenetik und Forstpflanzenzüchtung, Forstliche Biometrie und Informatik der Universität Göttingen, 29 September—1 October 1982.*

Mejnartowicz, L. E., Bergmann, F. and Okoniewska, B. (1983): Effect of fluorides and SO_2 on activity of malate dehydrogenases (E.C. 1.1.1.37) in Scots pine *(Pinus sylvestris)* needles. *Proceedings of 19th meeting of Polish Biochemical Society Szczecin 1983, 26–28 Sept.*, p. 46.

Mitton, J. B. (1983): Conifers. In: Tanksley, S. D. and Orton, T. J. (Eds) *Isoenzymes in plant genetics and breeding.* Part B. Elsevier Science Publishers B. V., Amsterdam, pp. 443–472.

Muona, O. (1982): Potential causes for multilocus structure in predominantly outcrossing populations. *Silva Fenn.* **16**:107–114.

Muona, O. and Szmidt, A. E. (1985): A multilocus study of natural populations of *Pinus sylvestris.* In: Gregorius, H. R. (Ed.) *Population genetics in forestry. Lecture notes in biomathematics,* 60, Springer, Berlin–Heidelberg–New York, pp. 226–240.

Müller-Starck, G. (1982): Sexually asymmetric fertility selection and partial self-fertilization. 2. Clonal gametic contributions to the offspring of a Scots pine seed orchard. *Silvae Fenn.* **16 (2)**:99–106.

Niebling, C. R., Johnson, K. and Gerhold, H. D. (1987): Electrophoretic analysis of genetic linkage in Scots pine (*Pinus sylvestris* L.). *Biochem. Genet.* **25**:803–814.

Petriević, S., Vidaković, M., Bilić, I. and Borzan, Ž. (1977): Immunological identity of pollen-wall proteins in some incompatible pine species. *Genetika* **3**:271–280.

Polozova, L. Y. and Dukharev, V. A. (1978): Izmenchivost' fermentov esteraz v dvukh-yarusnej populatsii sosny obyknovennoj (Variation in the esterase enzymes in a two-storied population of Scots pine). *Lesovedenie* **5**:30–35.

Polozova, L. Y., Pravdin, L. F. and Dukharev, V. A. (1978): Biokhimicheskij polimorfizm v dvukhyarusnykh populyatsiyakh sosny obyknovennoj (Biochemical polymorphism in two-storied populations of *Pinus sylvestris*). *Doklady Academii Nauk SSSR* **243**:778–781.

Prager, E. M., Fowler, D. P. and Wilson, A. C. (1976): Rate of evolution in conifers (Pinaceae). *Evolution* **30**: 637–649.

Pravdin, L. F., Altukov, J. P., Dukharev, V. A. and Zivotovskij, L. A. (1982): Polifmorfizm esteraz v populyatsiakh sosny obykhnovennoj (Polymorphism of Scots pine populations in esterase loci). *Ann. Ac. Sci. USSR.* **262**:998–1000.

Prus-Głowacki, W. (1982): Badania nad zmiennoscia genetyczna w klasach wiekowych naturalnie odnawiajacej sie populacji *Pinus sylvestris* (Studies on genetic variation in age classes of naturally regenerating population of *Pinus sylvestris*). *Wyd. Nauk. UAM Poznan,* pp. 5–88.

Prus-Głowacki, W. (1983): Serological investigation of a hybrid swarm population of *Pinus sylvestris* L. x *Pinus mugo* Turra and the antigenic differentiation of *Pinus sylvestris* L. in Sweden. In: Jensen, U. and Fairbrothers, D. E. (Eds) *Proteins and nucleic acids in plant systematics,* Springer, Berlin, pp. 353–361.

Prus-Głowacki, W. and Nowak, R. (1989): Demographic processes in *Pinus sylvestris* populations from regions under strong and weak anthropogenous pressure. *Silvae Genet.* **38**:55–62.

Prus-Głowacki, W. and Rudin, D. (1981): Studies on antigenic proteins of *Pinus sylvestris* from six Swedish provenances. A pilot study. *Silvae Genet.* **30**:200–203.

Prus-Głowacki, W., Sadowski, J., Szweykowski, J. and Wiatroszak, I. (1981): Quantitative and qualitative analysis of needles antigens of *Pinus sylvestris, Pinus mugo, Pinus uliginosa* and *Pinus nigra* and of some individuals from a hybrid swarm population. *Genetica Polon.* **22**:447–454.

Prus-Głowacki, W. and Szweykowski, J. (1977): Studies on isoenzyme variability in Scotch pine *(Pinus sylvestris)* and mountain dwarf pine *(Pinus mugo)* populations. *Bull. Sci. Amis Poznan. D.* **17**:15–27.

Prus-Głowacki, W. and Szweykowski, J. (1979): Studies on antigenic differences in needle proteins of *Pinus sylvestris* L., *P. mugo* Turra, *P. uliginosa* Neumann and *P. nigra* Arnold. *Acta Soc. Bot. Pol.* **48**:217–238.

Prus-Głowacki, W. and Szweykowski, J. (1980): Serological characteristics of some putative hybrid individuals from a *Pinus sylvestris x Pinus mugo* hybrid swarm population. *Acta Soc. Bot. Pol.* **49**:127–142.

Prus-Głowacki, W. and Szweykowski, J. (1983): Studies on isoenzyme variability in populations of *Pinus sylvestris* L., *Pinus mugo* Turra, *Pinus uliginosa* Neumann and individuals from a hybrid swarm population. *Bull. Soc. Amis Poznan, Serie D.* **22**:107–122.

Prus-Głowacki, W., Szweykowski, J. and Nowak, R. (1985): Serotaxonomical investigation of the European pine species. *Silvae Genet.* **34**:162–170.

Prus-Głowacki, W., Szweykowski, J. and Sadowski, J. (1978): Studies on serological similarity of *Pinus sylvestris* L., *Pinus mugo* Turra and individuals from a hybrid swarm population. *Genetica Polon.* **19**:321–338.

Rasmuson, M. (1979): Interaction between loci-models and observations. In: Rudin, D. (Ed.) *Proceedings of the conference on biochemical genetics of forest trees. Umeå, 1978.*

Rasmuson, B. and Rudin, D. (1971): Variation in esterase zymogram patterns in needles of *Pinus sylvestris* from different provenances in northern Sweden. *Silvae Genet.* **20**:39–41.

Rudin, D. (1975): Inheritance of glutamate-oxalate-transaminases (GOT) from needles and endosperms of *Pinus sylvestris. Hereditas* **80**:298–300.

Rudin, D. (1977): Leucine-amino-peptidases (LAP) from needles and macrogametophytes of *Pinus sylvestris* L. Inheritance of allozymes. *Hereditas* **85**:219–226.

Rudin, D. (1986): Developmental trends in the field of biochemical genetics of forest trees. In: *Proceedings of the 18th IUFRO World Congress Jubljana 1986.* Vol. II. pp. 583–588.

Rudin, D. and Ekberg, I. (1978): Linkage studies in *Pinus sylvestris* using macrogametophyte allozymes. *Silvae Genet.* **27**:1–11.

Rudin, D. and Rasmuson, B. (1973): Genetic variation in esterases from needles of *Pinus silvestris* L. *Hereditas* **72**:89–98.

Rudin, D. and Tigerstedt, P. (1979): Genetic structure of a naturally regenerated stand of *Pinus sylvestris* L. at Oulanka in Finland. Preliminary report. *Aquila, Ser. Zool.* **20**:92–93.

Rudin, D., Eriksson, G., Ekberg, I. and Rasmuson, M. (1974): Studies of allele frequencies and inbreeding in Scots pine populations by the aid of the isozyme technique. *Silvae Genet.* **23**:10–13.

Saito, A. (1968): The phyloserological study on the phylogenic relationship in *Pinus. Bull. Kyushu University Forest* **42**:235–240.

Shaw, D. V., Yazdani, R. and Muona, O. (1982): Methods for analyzing data on the relative proportions of monoterpenes in conifers. *Silva Fenn.* **2**:235–240.

Squillace, A. E. (1976): Biochemical genetics and selection. Composition of volatile terpenes. In: *Proceedings of IUFRO joint meeting on advanced generation breeding.* Cestas. INRA, pp. 167–178.

Szmidt, A. E. (1979*a*): Inheritance of catalase multiple forms in Scots pine (*Pinus silvestris* L.) endosperm. *Arboretum Kórnickie* **24**:105–110.

Szmidt, A. E. (1979*b*): Peroxidase variation and activity in Scots pine (*Pinus sylvestris* L.) needles from polluted forest areas. *Annual Report,* Institute of Dendrology, Kórnik.

Szmidt, A. E. (1980): Genetic variation among eleven populations of Scots pine (*Pinus sylvestris* L.) as determinated by catalase polymorphism. *Arboretum Kórnickie* **25**:181–187.

Szmidt, A. E., Muona, O. and Yazdani, R. (1984): Linkage relationship in Scots pine (*Pinus sylvestris* L.). In: Szmidt, A. E. Ph. D. Thesis, S. K. A. S., 1984, Umeå.

Szmidt, A. E. and Nagasaka, K. (1984): Phosphoglucomutase polymorphism in Scots pine (*Pinus sylvestirs* L.). In: Szmidt, A. E. Ph. D. Thesis, S. K. A. S., Umeå, Part III, pp. 1–7.

Szmidt, A. E. and Yazdani, R. (1984*a*): Electrophoretic studies of genetic polymorphism of shikimate and 6-phosphogluconate dehydrogenases in Scots pine *(Pinus sylvestris* L.). *Arboretum Kórnickie* **29**:63–72.

Szmidt, A. E. and Yazdani, R. (1984*b*): Inheritance of aconitase and phosphoglucoisomerase electrophoretic variants in Scots pine (*Pinus sylvestris* L.) tissues. In: Szmidt, A. E. Ph. D. Thesis, S. K. A. S., Umeå, Part II, pp. 1–10.

Tigerstedt, P. M. A., Hiltunen, R., Chung, M. S. and Moren, E. (1979): Inheritance and genetic variation of monoterpens in Scots pine (*Pinus sylvestris* L.). In: Rudin, D. (Ed.) *Biochemical genetics of forest trees.* Umeå, pp. 29–39.

Tigerstedt, P. M. A., Rudin, D., Niemelä, T. and Tammisola, J. (1982): Competition and neighbouring effect in a naturally regenerating population of Scots pine. *Silva Fenn.* **16**:122–129.

Tobolski, J. J. and Hanover, J. W. (1971): Genetic variation in the monoterpens of Scotch pine. *Forest Sci.* **17**:293–299.

Yazdani, R. (1986): Monoterpene methods for identification of populations of *Pinus sylvestris.* In: Lindgren, D. (Ed.) *Proceedings of the Frans Kempe Symposium, Umeå, June 10–11. 1986.*

Yazdani, R. and Nilsson, J.-E. (1986): Cortical monoterpene variation in natural populations of *Pinus sylvestris* in Sweden. *Scan. J. For. Res.* **1**:85–93.

Yazdani, R. and Rudin, D. (1982): Inheritance of fluorescence esterase and β-galactosidase in haploid and diploid tissues of *Pinus sylvestris* L. *Hereditas* **96**:191–194.

Yazdani, R., Rudin, D., Aldén, T., Lindgren, D., Harbom, B. and Ljung, K. (1982): Inheritance pattern of five monoterpenes in Scots pine (*Pinus sylvestris* L.). *Hereditas* **97**:261–272.

Yazdani, R., Nilsson, J. E. and Ericsson, T. (1985): Geographical variation in the relative proportion of monoterpenes in cortical oleoresin of *Pinus sylvestris* in Sweden. *Silvae Genet.* **34**:201–207.

CHAPTER 6

Provenance variation in growth and phenology

M. Giertych

First trials

The first forest provenance experiment was established in 1820 by the French nurseryman Vilmorin who used several seed sources of Scots pine for the purpose. He published his results in 1857, that is, after 36 years' observation. From these results it appears that the best pine race is the "Riga pine" (after Wright and Bull, 1963) which name was applied to all pine lots originating from the Baltic countries. This experiment is, of course, only of historical interest today, since it had no replicates and there was no strict control of seed collection procedures. However, it needs to be stressed that the high opinion about the value of Baltic pine has not been invalidated by later studies.

At the end of the 19th and at the beginning of the 20th centuries several similar studies were conducted, but unfortunately mostly without replicates. All of these were reviewed by Kalela (1937/38). His general conclusions were as follows. The warmer and more maritime the climate of origin, the more sensitive are the trees to frost and snow damage and to fungal diseases in the nursery, the more stout the branches and knots, the more crooked the stem, the wider the crown, and the lower the specific gravity of wood, however, the growth increment is greater. On the other hand, the more cold and continental is the climate of origin, the more resistant are the trees to early, late and winter frosts, to snow damage, to drought, to fungi and insects, the lower is seedling mortality, the fewer the trunk and wood deformations and the less disturbed is height growth by the changed conditions.

Few of these early studies are still maintained. Perhaps the most notable exceptions are the experiments of M.K. Turskij established in Moscow in the 1880s and 1890s which are still being regularly measured (Timofeev, 1974), currently by G. F. Mihal'chenko. It is consistently reported here that local provenances are most productive and that the further away is the origin of a provenance, the poorer is its performance.

First IUFRO international trial (1907)

Another of these older experiments worth mentioning is the first IUFRO international provenance experiment on Scots pine made in 1907, which though devoid of replicates is still of interest because the same material was outplanted in several places throughout Europe. These different locations constitute replicates of a sort. There are several reviews available of this study (Wiedemann, 1930; Kalela, 1937/38; Giertych, 1979). (In this chapter reviews rather than original papers will be quoted wherever possible.)

The latest review (Giertych, 1979) indicates that the best provenances originate from lowlands of Central Europe. The outlier or fringe populations were generally very poor on all sites. Among these the montane Spiś (Slovakia) provenance appears best. The fairly good Hasselt and Hoenderlo provenances are believed to be from domiciled stands since Scots pine is not indigenous in Belgium and the Netherlands. Judging by their adaptability to various sites they probably originate from Germany (possibly Rheinpfalz) or the Baltic region. The best provenance is Guzianka from the Mazury region in N. Poland. For some reason it is well below the average on one site (Gießen, FRG), a result which is generally considered to be an artefact (there are no replicates, and the plot there is on a different, very wet site). Apart from the Guzianka provenance in Gießen most genotype x environment interactions are provided by the Likavka (Slovakia) planting site on which all provenances appear to behave differently than elsewhere. This is the southernmost location and at the highest altitude. Since most of the planting sites from which data are available are within or close to the region of origin of the best provenances the general conclusion might be drawn that local provenances are best or at least above the average. This is particularly underlined by the Likavka planting site where the nearest Spiś and Bratislava provenances grew well. The disqualification of outlier or fringe populations may be the result of lack of planting sites in these regions.

Ogievskij's series

Almost simultaneously with the IUFRO 1907 trial a great experimental series was performed by V. D. Ogievskij in the Russian Empire from 1910 to 1916. This study did not use specific provenances but seed collections pooled from several stands within administrative provinces (gubernias). Seeds were provided by 42 provinces though not all were included at the various planting sites. There were originally at least 20 planting sites, but due to wars, revolution, border changes, etc. data have been available from only 8 of these (Wiedemann, 1930; Kalela, 1937/38; Giertych and Oleksyn, 1981; Oleksyn and Giertych, 1984). This series also lacks replicates. It was a one-country effort, but, since sampling covers the major portion of the range of Scots pine in eastern Europe and Asia, including parts which at present belong to Poland or Turkey, it is of international interest and complements well the IUFRO 1907 series.

On the basis of the latest figures concerning volume production per unit area it is clear (Giertych and Oleksyn, 1981) that generally local provenances have had relatively satisfactory growth. Provenances from the pre-revolutionary provinces of Volynia, Minsk, Mogilev and Chernigov proved to be of exceptionally high value. Also good were provenances from the Baltic region and from a narrow belt between latitudes 54° and 57° N (provinces Moscow, Vladimir, Nizhni Novgorod, Kazan, Samara). The outlier provenances from the southern fringe of the range exhibited poor survival and growth. Northern provenances, particularly from beyond 60° lat. N, are useful only locally because of poor growth in spite of good survival. The worst provenances were those from the provinces of Vologda, Viatka and Perm. The northwestern and southeastern provenances have growth generally below the average and those from central Poland were around the average.

The most adaptable provenances, performing reliably well on a variety of locations, appear to originate from regions referred to as the mixed forest type, which corresponds to the west-central part of the former Russian Empire and the narrow belt through Moscow, Vladimir to Kazan and Samara, as distinct from the taiga further north or the forest steppe in the south (Giertych and Oleksyn, 1981; Oleksyn and Giertych, 1984). An adaptation to greater competition appears to be involved here. Adaptability is a very important characteristic of a race, since it determines its transfer possibilities, but, it is of little taxonomic value. Therefore, assigning subspecies or varietal rank to races recognized on the basis of provenance experiments (Pravdin, 1964; Ruby and Wright, 1976) does not appear to serve any useful purpose.

Second IUFRO international trial (1938)

In 1938 IUFRO initiated another series of international provenance studies on Scots pine. Seed was collected from 53 places throughout the range of Scots pine in western and Central Europe and distributed to all participating countries and also to North America. Another smaller series with roughly the same seed lots was established in 1939, but only in Canada and Turkey. Jointly these two series provide the most complete information so far on provenance variation of the species in western and Central Europe.

The latest review of height growth results from these series (Giertych, 1979) indicates that the best growing races of Scots pine originate from the lowland of Central Europe, particularly from the Baltic countries and from north and western Poland. They display superior growth on practically all sites from Scandinavia to Turkey and from eastern Europe to Michigan in the USA. If seed transfer is planned, these fast-growing, highly adaptable populations appear to be the most suitable.

The northern races from Scotland and Scandinavia are generally very poor on most sites, especially on the continent. Most of the genotype x environment interactions are to be observed in Scandinavia. Some local provenances grow equally well as the best continental ones. This is particularly true in southern Finland. In SE Norway the local Vitsand provenance is very good and also the one from Tromsø, from lat. 69° N. This is the only example of a successful long distance (9° lat.) southward transfer. In Sweden southward transfers of only about 2° lat. appear justified.

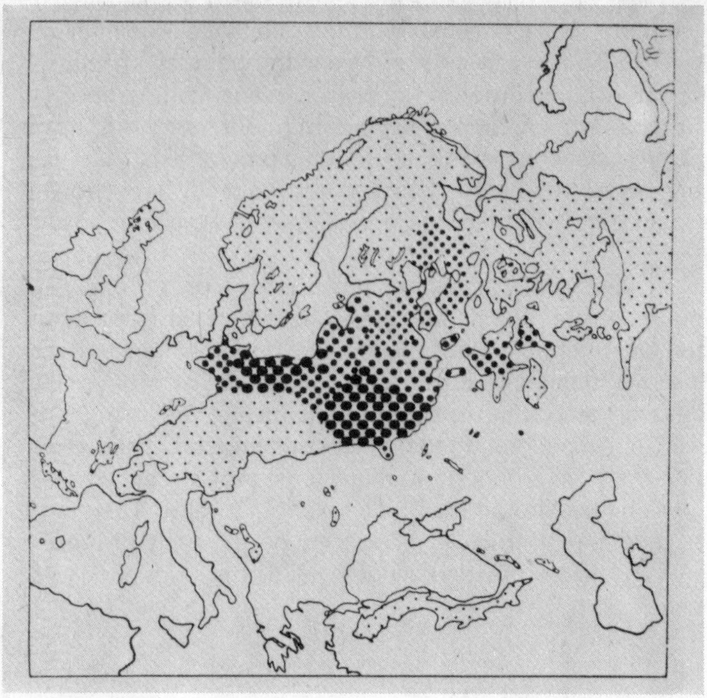

Fig. 6.1. Location of the most adaptable races of Scots pine recommended for introduction to other regions when local races are inadequate. The larger the spots, the better are the Scots pine races for transfer (Oleksyn et al., 1986)

The domiciled races from Belgium and the Netherlands are generally very good, as are the provenances from NW Germany. The direction of transfer appears to be of little significance. The provenances from western Poland and eastern part of (former) GDR are generally good, usually well above the average for a given site. Provenances from the Baltic countries are superior on most sites. The easternmost Griva provenance and the North-Ukrainian ones, Susk and Luboml, are less reliable.

The montane provenances from the Pyrenees, through the Swiss, Italian and Hungarian Alps to the Carpathians display generally average or below average tree heights and, therefore, are not recommendable anywhere. Even on the most elevated planting site (Arboretum Royat in France at 850 m) the montane provenances are not better than the continental lowland ones.

In the USA where the elevations are generally higher and latitudes lower than at European planting sites, the same general conclusions hold.

In Finowtal (former GDR), the experiment was planted out at two spacings. There is no provenance x spacing interaction. Also there is little difference between the two Italian planting sites located close to each other. It appears therefore that provenance x site interactions are of little importance. The interactions in Scandinavia appear to be primarily of provenance x climate character.

The latest review of the study (Giertych, 1979), while indicating the best areas of pine races suitable for seed transfer, also shows that local races are usually not substantially poorer than the best introduced ones. It is important to remember this.

There are not many reports on volume production per unit area from this study and no general review of this information, but with the increasing availability of such data some changes of conclusions may result. For example, in the Hungarian trial the Italian provenance proved most productive and the Dutch and Belgian ones proved slightly below average. The performance of other provenances was more or less as discussed above (Harkai and Mátyás, 1981).

On the basis of the IUFRO experiments and the Ogievskij's series in Russia, which appear to complement each other well, a synthetic map of the location of most adaptable races has been proposed (Oleksyn et al., 1986). This is shown in Fig. 6.1.

Provenance study in the North Central States (NC-51)

In 1959 another major provenance study (NC-51) was launched in the north-central states of the USA (31 locations) based on 122 seed lots collected since 1951 throughout Europe and also in some parts of Asia (Wright and Bull, 1963). The main use of Scots pine in North America is for Christmas trees and it is with this utilization in mind that most of the reports are given. However, the study produced much information on racial variation in general, including information about growth performance. J. W. Wright has co-ordinated the study and J. L. Ruby has investigated the intraspecific taxonomy on the basis of an analysis of the variation of 2-year-old seedlings and proposed to divide the provenances into 21 geographic races which were given varietal status (Wright et al., 1966). This was later altered to 20 varieties and 3 races (Ruby and Wright, 1976). Most papers from the study provide information on these "varieties" rather than on individual provenances.

On the basis of 5- to 7-year height at 16 test sites it was found that the fastest growing were the populations of var. *haguenensis* (Belgium, N France, FRG), race 'East Anglia' and var. *hercynica* (Germany, Czechoslovakia). The varieties *lapponica* (N Scandinavia), *mongolica* (Mongolia, E Siberia) and *altaica* (Altai Mountains of Siberia) grew 1/3 as fast as the above-mentioned populations. Central-European provenances were intermediate (Wright et al., 1966). In Maryland the Central-European ones grew best together with the Belgian ones (Genys, 1970). Mortality was high in var. *iberica* (Khalil, 1968) and also in French populations (Genys, 1970).

In N Dakota and Nebraska after 10 years best growth and survival were reported for provenances originating between 50° and 55° lat. N and from 20° to 40° long. E, i.e. from eastern Poland, N Ukraine, Bielorussia and the Baltic countries (Cunningham, 1973). This opinion coincides with the results summarized in Fig. 6.1.

In Michigan, at age 13 the best height growth was obtained with Central-European varieties *(carpatica, polonica, hercynica, haguenensis)* and with the race 'East Anglia', average and below average with fringe populations from Scotland through Spain, France and all the European ones south of the Alps and the Carpathians. The Scandinavian, N Russian and Siberian populations were the poorest. Among the latter var. *altaica* and *uralensis* were among the average (Ruby and Wright, 1976).

The results of this series of trials are usually discussed in the context of Christmas tree growing and intraspecific taxonomy (Wright, 1976), and since the trees are now well beyond any utility for these purposes the interest in the series is declining.

Siberian populations

Most of the major studies reviewed primarily the European part of the range of the species, yet half of the range and possibly also of timber productivity comes from beyond the Ural. It seems necessary therefore to try to collect any information available on the races of Scots pine from Siberia.

Ogievskij's experiment included two Siberian populations, from Semipalatinsk which perished and from Akmolinsk which grows very poorly in the Ukraine (Giertych and Oleksyn, 1981).

In the NC-51 series there are some Siberian provenances. The race 'NE Siberia' (Yakutsk) grows very slowly, like var. *lapponica* with Swedish and Finnish populations. Var. *mongolica* (from the watershed of Lake Baikal and the Angara river) is also slow growing. The populations from race 'Krasnoyarsk' coming from the more disjunct parts of the range between lat. 50 to 55° and long. 75 to 100° are still below average. On the other hand, var. *altaica* (Altai Mountains) and var. *uralensis* (Ural Mountains) are close to average in growth observed in Michigan (Ruby and Wright, 1976).

A major effort is in progress now in the USSR to test all the pine populations including those from Siberia. However, the material is still mostly young and few results are available. A planting in Hungary in 1978/79 has shown the provenances from around Angara river and Lake Baikal (var. *mongolica* of the NC-51 series) at age 11 to have tree heights well below average. Sources from the Sayan Mountains were a little better and those from the Altai Mountains still better, but all below the mean for the whole experiment. Populations from the Ural Mountains and the steppes of Kazakhstan are close to average (Mátyás, 1987).

Similarly in Czechoslovakia (near Plzeń) from among the material of Prokazin's 1961 collection the populations from the steppes of the southern foothills of the Ural Mountains and the outliers in the vicinity of Altai Mountains were among the best, together with those from the European central-west region of the USSR (see Fig. 6.1; Kaňák, 1980).

In a trial near Voronezh, on the river Don, where currently the work with USSR provenances of Scots pine is concentrated, it was found that while the best growing populations were European (Lvov, Sumy, Brest, Minsk) including the local one from Voronezh, the worst ones include not only the extreme northern ones (Karelia, Arkhangelsk) but also those from Siberia, Tomsk (on the river Ob) and Irkutsk (Lake Baikal) (Shutyaev, 1983).

A trial near Moscow gave a similar result with the local material among the best as defined in Fig. 6.1. Out of the Siberian sources the best were pines from the Krasnoyarsk and Altai regions, though the latter suffered from heavy mortality (Pravdin and Vakurov, 1968; Naryshkin et al., 1983).

At a planting site in the southwestern part of Siberia, in the forest steppe region near Novosibirsk at age 8 the highest plants were of local origin and a few other provenances were equivalent, mostly European ones from the region of central Russia referred to as optimal before (Demidenko et al., 1984).

In southern Siberia, at Minusinsk (Sayan Mountains), North-European races (Estonia, Leningrad, Arkhangelsk) as well as southern outliers (Rostov) grow poorly, while pines from Voronezh have a height growth akin to the local ones (Cherepin, 1977). Further north, near Krasnoyarsk, the Vyatka, Perm and Ural provenances grow somewhat below the local average, however, the best European provenances were not tested there (Iroshnikov, 1977).

Further east, on the Amur river, near Khabarovsk a 28-year-old trial testing —more than 30 eastern fringe populations—has shown that best height and diameters were obtained from local Amur origins, followed by Chita and Buryat (east of Lake Baikal) provenances with the Yakutsk (NE limit of the range) provenances growing slowest. An older, 30-year study which included also an Irkutsk (Lake Baikal) and a Kuibishev (SE European Russia) provenance gave a similar result but the latter two provenances had superior growth (Tagil'tseva, 1986).

To summarize, it appears that, for continental conditions, provenances of special interest are to be found among the Siberian populations in the Krasnoyarsk and Altai regions and the region directly southeast of the Ural. For more maritime conditions the local or best European provenances are preferable.

National studies with Poland as the example

Apart from the international studies or transcontinental trials as in the USSR, in most countries where Scots pine is an important forest species there are numerous provenance experiments established locally, often in several places, which test seed origins from within the country, sometimes including also populations from promising foreign sources. As an example I present a summary of results from Poland which is perhaps of more than just local interest due to the value of Polish races demonstrated in other countries.

In Fig. 6.2, the average growth performance of various Scots pine populations in numerous provenance trials is indicated in the form of histograms on the map of Poland (Cierniewski, 1981; Cierniewski and Giertych, 1982). Obviously montane (southern Poland) provenances are very poor. Sources from South-east and

North-west Poland are generally average and below average. Good provenances come from central, from northeastern (Tabórz 13, Krutyń 16, Ruciane 19, Supraśl 21) and particularly from lowland southwestern Poland (Rychtal 29, Prószków 28, Kubryk 27, Wołczyn 30). These same populations were indicated superior also in later reports (Korczyk, 1983; Giertych, 1988).

Prior to World War II some parts of Poland belonged to Germany and Scots pine populations from these were on several occasions included in German provenance experiments. Whenever the above-mentioned superior provenances were included they also proved to be among the best (Troeger, 1960; Schober, 1975).

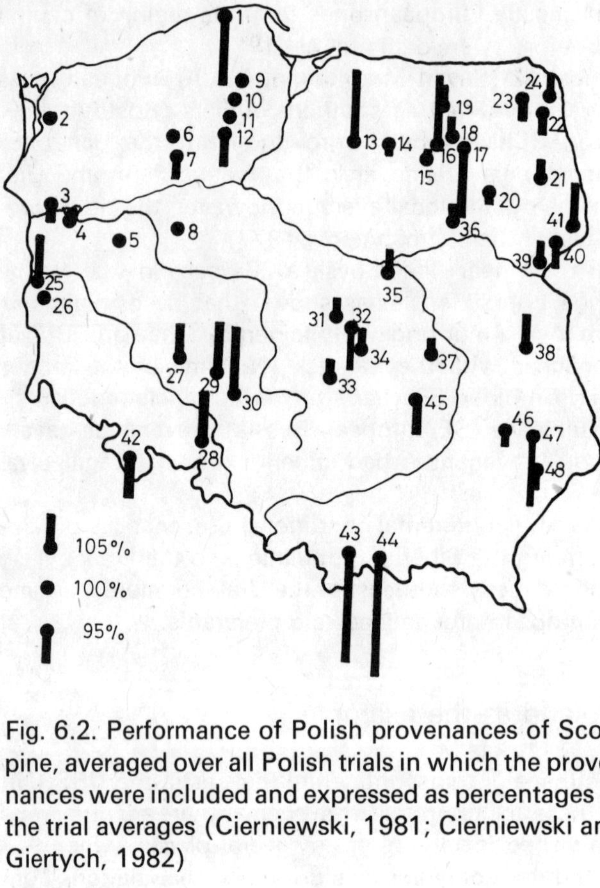

Fig. 6.2. Performance of Polish provenances of Scots pine, averaged over all Polish trials in which the provenances were included and expressed as percentages of the trial averages (Cierniewski, 1981; Cierniewski and Giertych, 1982)

Value of local provenances

The primary purpose of provenance experiments is to find populations better than the local ones. Usually such populations can be found and as I have indicated above there are certain regions that yield populations of sufficient adaptability to be useful in many places. However, having studied so many reports from provenance experiments of various species I am struck by the persistence of the conclusion for Scots pine, as distinct from other species, that the local populations are among the better ones. I have indicated this in the review of the major experiments. The common conclusion of national experiments is that introduction should be employed only if local populations of good quality are unavailable (Lines and Mitchell, 1964; Pravdin and Vakurov, 1968; Patlaj, 1973, 1984; Timofeev, 1974; Barniskis, 1978; Zhurova, 1983) or that transfers should be limited to only few degrees of latitude (usually southwards) or longitude (Stefanson and Sinko, 1967; Johnsson, 1971; Veresin, 1972; Remröd, 1973; Eiche and Anderson, 1974; Ulissova, 1984; Presher, 1986).

Mátyás (1987) summarized the topic as follows: "Outstanding performance of local populations was typical for two types of experimental locations: *(a)* experiments situated in the mentioned genetically valuable zone of the distribution area, and *(b)* experiments with extreme site conditions (higher altitudes, maritime or boreal locations).

The performance of local sources was significantly inferior at locations where isolation and drift caused a degeneration of the genetic pool of the population."

I would question the last sentence here. One must remember that there are very few trials located within outlier zones. We know about the performance of pines from such areas from experiments located elsewhere.

The information summarized in Fig. 6.1 and in the review given above refers to the continuous range of the species. That is where the most studied populations come from. The outliers have been studied much less. It is perhaps useful therefore to review the information about Scots pine from these regions.

Performance of outlier populations

Scots pine from Scotland is the only outlier population dealt with in most international studies. Locally it may be useful (Lines and Mitchell, 1964) but elsewhere it is usually well below the average in growth (Giertych, 1979).

Spanish populations are of interest in North America as Christmas trees due to their winter needle colour (Wright, 1976) but they suffer from considerable seedling mortality and poor and crooked growth (Wright et al., 1966; Khalil, 1968; Wright 1976). In New Zealand several Spanish populations were tested on the assumption that coming from similar latitudes they would prove most suitable. They performed very poorly discrediting the species in general (Sweet, 1964). In Korea, a test including 4 Spanish provenances showed them to be inferior in survival, form and height growth to populations form North France and Central Germany (Hwang and Kwon, 1969; Han et al., 1982).

The populations from South France (e.g. Millau and Les Angles of the IUFRO 1938 series, but other ones also) grew very poorly in various tests compared to populations from North France which belong to the continuous range (Thor, 1976; Giertych, 1979; Han et al., 1982).

Balkan populations have been seldom tested but, in general, perform very poorly in all studies. For example the population from Bulgaria (Creweno) used in the IUFRO 1907 series and from Romania (Talmacel and Tinoava) used in the IUFRO 1938 series yielded unsatisfactory results where tested (Giertych, 1979). In Tennessee (USA) populations from Turkey, Greece and Yugoslav Macedonia grew much slower than those from North France (Thor, 1976). A Hungarian study showed a South-Yugoslavian population (Kupresko Pl.) to grow much less than a North-Yugoslavian (Kranj), the former being an outlier and the latter belonging to the continuous range of the species (Mátyás, 1987). However, when Romanian provenances were compared with foreign ones in a trial in Romania, the local ones proved similar to the best foreign ones (Lazarescu et al., 1961). In Bulgaria, a 9-year study outside the range of the species with 12 Bulgarian and 3 Romanian provenances of Scots pine, the Romanian ones had better height growth and their diameters were equivalent to the best three Bulgarian sources (Tsanov and Kosmov, 1985). This study did not include other than Balkan seed sources.

In Turkey, a local Turkish provenance (Catacik) ranged among the better growing ones in the IUFRO 1939 series (Giertych, 1979).

In Russian experiments Caucasian provenances are always well below the average in growth (Giertych and Oleksyn, 1980; Patlaj, 1984), but in a Georgian trial the local population (referred to as *P. sosnovskyi*) proved superior to the introduced ones (Mandzhavidze, 1972).

Near Kharkov in a forest steppe region of the Ukraine at age 4 local sources yielded the highest biomass production, greatest seedling heights and stem diameters. Among the other sources tested the Ukrainian populations of the southern fringe were usually better than those from the continuous range. Among the latter Baltic ones were the best (Zhurova, 1983). At 12–13 years on a poor site the local population maintained a closed canopy as did the Lithuanian ones, while others were much too open (Patlaj and Zhurova, 1987).

Near Semipalatinsk, eastern Kazakhstan, in a study including only Kazakhstan outlier seed sources at 8 years the local provenance had grown the least. The best height was shown by a northern provenance followed by an eastern one and a southern one from Alma Ata. Another northern provenance (Pavlogorsk) and a western (Aktiubinsk) one were intermediate. In diameter the local population again was the poorest, northern and western provenances were the best while eastern and southern ones where intermediate. Thus, the preference here is decidedly against the local seed source and for the northern ones, towards the continuous range of the species (Gurina, 1984).

There are some straight-stemmed and wide-crowned outlier populations of Scots pine extensively used locally in Manchuria (China) and of possible utility in Hungary (Mátyás, personal communication). Other Far Eastern (Soviet) populations in Hungary are well below the average in survival and growth (Mátyás, 1987).

As mentioned above, local outlier populations along the Amur river in the Far East (Tagil'tseva, 1986) are substantially above the average in growth but are surpassed by several introduced, more western sources.

To summarize I would say that outlier populations are truly not of much use when transferred, but locally they have to be considered of interest because they are adapted and their performance is not drastically below any possible introductions. This is not so obvious in young trials where height is only measured but becomes increasingly valid when diameter and survival are considered.

It can be said for almost any place within the range of Scots pine that some populations, growing faster than the local one, can be found but also the latter is of interest and should be included in any tree improvement programme.

Phenology

Scots pine having evergreen leaves starts photosynthesis early in spring, as soon as temperature conditions permit, and continues late into autumn. However, most phenophases are of much shorter duration. In Jödras (Sweden) needle elongation and shoot elongation start together in late May, two months after the onset of photosynthesis. Shoot elongation terminates soon, in early July, while needle elongation persists till late August. In early June stem diameter increment starts and continues till late August. In mid-June bud primordia initiation begins and continues till early October (Rook, 1985 from a Ph. D. thesis of Troeng, 1981).

The actual date for a given phenophase depends, among other factors, on the inherited properties as is evidenced by many provenance studies. Most reports concern the onset of stem growth (bud growth or flushing as it is sometimes erroneously referred to) and its termination, sometimes defined as bud set.

At a given planting location the onset is earlier in more northern populations than in more southern ones (Kalela, 1937/38; Dietrichson, 1961; Gowin, 1972; Steiner, 1979; Vidyakin, 1979; Mátyás, 1981; Baranov and Manchaevich, 1983). Only Gatti (1971) who compared German and Italian provenances found the latter to be earlier starters in spring, but these were from much higher elevations and therefore from colder climates. Some authors note also a longitudinal cline with western, more maritime, provenances starting later (Kalela, 1937/38; Steiner, 1979; Vidyakin, 1979; Mátyás, 1981). When altitude was considered the provenances from higher elevations started earlier (Hagner, 1970; Gatti, 1971; Mátyás, 1981) though in a Polish study the latitudinal cline appears to have overruled the altitudinal one (Gowin, 1972).

On the basis of an analysis of many provenances, Mátyás (1981) suggests that the latitudinal and longitudinal clines combine into a pattern giving N–S isolines in western Europe and W–E isolines in that part of the range east of longitude 30°E, with the curve of the isolines hinged on the Mazury region of North-east Poland. This basically agrees with the January isotherms, and more especially with curves showing the number of frost-free days.

It appears that spring growth does not start until a specific temperature (or degree days) is attained. The colder the climate of origin the lower is the temperature that is sufficient to induce growth (Vidyakin, 1979; see also Chapter 4).

Termination of leader growth or bud set has a similar pattern. Northern (Kalela, 1937/38; Dietrichson, 1961; Patlaj, 1974; Vidyakin, 1979; Mátyás, 1981; Mikola, 1982; Baranov and Manchaevich, 1983; Koski and Sievänen, 1985) and more continental (Kalela, 1937/38; Dietrichson, 1961; Mátyás, 1981) provenances terminate earlier. Among Finnish origins continentality and altitudinal gradients are of lesser importance (Mikola, 1982).

Here again, the colder the climate of origin, the lower is the temperature needed to induce growth termination and bud set (Mikola, 1982; Koski and Sievänen, 1985). Photoperiod appears important in the first year (indeterminate) growth termination but not later (Koski and Sievänen, 1985).

Root hardiness for the winter is attained more slowly by southern provenances than by northern ones as evidenced by growth chamber studies where nursery grown seedlings were frozen to −20 °C or −25 °C at various dates from September to April (Lindström and Nyström, 1987).

The longer growing cycle of southern provenances, particularly in the case of diameter growth, constitutes a limiting factor in northward transfers of Scots pine because such populations tend to get damaged by early frosts (Dietrichson, 1961, 1964). Similarly, late frosts limit southward transfers because the early starting northern populations tend to get damaged in warmer regions. This may be one of the reasons why local populations have an advantage over introduced ones.

Literature

Baranov, M. I. and Manchaevich, B. D. (1983): Sezonnoe razvitie sosny obyknovennoj raznogo geograficheskogo proiskhozhdeniya (Seasonal development of Scots pine of various provenances). *Lesoved. i les. khozya.* **18**:54–56. *Ref. Zhur.* 3.56.141 (1984).

Barniskis, E. K. (1978): Geograficheskie kul'tury sosny obyknovennoï v Litovskoï SSR (Scots pine provenances in the Lithuanian SSR). In: Ramanauskas, V. (Ed.) *Selektsiya drevesnykh porod v Litovskoï SSR.* Raide Publ. Kaunas, pp. 142–153.

Cherepin, V. L. (1977): Geograficheskie kul'tury sosny obyknovennoj v yuzhnoj lesostepi Krasnoyarskogo kraya (Scots pine provenance experiments in the southern forest-steppe zone of the Krasnoyarsk region). In: Minina, E. G. (Ed.) *Geograficheskie kul'tury i plantacii khvoinykh v Sibiri.* Nauka Publ. Siberian Branch, pp. 111–123.

Cierniewski, M. (1981): Wyniki badań proweniencyjnych nad sosną zwyczajną (Results of provenance studies on Scots pine). *Las Polski* **24**:7–9.

Cierniewski, M. and Giertych, M. (1982): Pol'skie rasy sosny obyknoviennoj (*Pinus silvestris* L.) v geograficheskikh opytakh (Polish races of Scots pine in provenance research). In: Pirags, D. M. (Ed.) *Geograficheskie opyty v lesnoj selekcii Pribaltiki.* Zinatne Publ., Riga, pp. 105–119.

Cunningham, R. A. (1973): Scotch pine for the Northern Great plains. *USDA Forest Service Research Paper* **RM—114**:1–12.

Demidenko, V. P., Alekseev, Yu. B. and Urusov, V. M. (1984): Geograficheskie kul'tury sosny i eli na yuge zapadnoj Sibiri (Pine and spruce provenance studies in southwestern Siberia). *Lesnoe Khozyaistvo* **(3)**:40–42.

Dietrichson, J. (1961): Breeding for frost resistance. *Silvae Genet.* **10**:172–179.

Dietrichson, J. (1964): Proveniensproblemet belyst ved studier av vekstryme og klima (The provenance problem illustrated by studies of growth-rhythm and climate). *Medd. Norsk. Skogførsoksvesen* **71**:499–656.

Eiche, V. and Andersson, E. (1974): Survival and growth in Scots pine (*Pinus silvestris* L.) provenance experiments in Northern Sweden. *Theor. Appl. Genet.* **44**:49–57.

Gatti, F. (1971): Osservazioni sulla fenologia di alcune provenienze di *Pseudotsuga menziesii* e di *Pinus silvestris. Ann. dell'Instituto Sperimentale per la Selvicoltura* (Arezzo) **2**:217–243.

Genys, J. B. (1970): Intraspecific diversities among Scotch pines *(Pinus sylvestris)* of different origin studied in Maryland. *Chesapeake Sci.* **11**:1–15.

Giertych, M. (1979): Summary of results on Scots pine (*Pinus sylvestris* L.) height growth in IUFRO provenance experiments. *Silvae Genet.* **28**:136–152.

Giertych, M. (1988): Interakcja genotypu ze srodowiskiem oraz z wiekiem polskich proweniencji sosny zwyczajnej (*Pinus sylvestris* L.) (Interaction of genotype with environment and age for Polish provenances of Scots pine (*Pinus sylvestris* L.). *Arboretum Kórnickie* **33** (in print).

Giertych, M. and Oleksyn, J. (1981): Summary of results on Scots pine (*Pinus sylvestris* L.) volume production in Ogievskij's pre-revolutionary Russian provenance experiments. *Silvae Genet.* **30**:56–74.

Gowin, T. (1972): Development of apical buds of Scots pine (*Pinus silvestris* L.) seedlings of different provenience. *Ekologia Polska* **20 (48)**:771–779.

Gurina, T. F. (1984): Introduktsiya sosny obyknovennoj na poluostrove Mangyshlak (Introduction of Scots pine on the Mangyshlak peninsula). *Bull. Glavnogo Bot. Sada.* **131**:29–33.

Hagner, M. (1970): A genecological investigation of the annual rhythm of *Pinus silvestris* L. *Stud. For. Suec.* **80**:1–40.

Han, Ch. Y., Lim, S. Ch. and Lee, Y. K. (1982): Provenance test of Scotch pine (*Pinus sylvestris* L.) growth performance and some characteristics at age 15 years. *Res. Rep. Inst. For. Gen. Korea* **18**:47–53.

Harkai, L. and Mátyás, C. (1981): A hazai erdeifenyő-állományok minőségének megjavítása idegen származások alkalmazásával (Improvement of local Scots pine stands through introduction of foreign provenances). *Agrártud. Közl.* **40**:345–347.

Hwang, J. W. and Kwon, H. M. (1969): Provenance test of *Pinus sylvestris* in Korea. I. Survival and growth performance in juvenile age. *Res. Rep. Inst. For. Gen. Korea* **7**:91–99.

Iroshnikov, A. I. (1977): Geograficheskie kul'tury khvoinykh v yuzhnoj Sibiri. 1. Rost sosny (*Pinus silvestris* L.) raznogo geograficheskogo proishozkhdeniya v lesostepnykh i gornotaezhnykh rajonakh (Conifer provenance studies in southern Siberia. 1. Growth of Scots pine of various provenances in forest steppe and mountain taiga regions). In: *Geograficheskie kul'tury is plantatsii khvoinykh v Sibiri.* Nauka Publ., Siberian Branch, pp. 4–62.

Johnsson, H. (1971): Proveniensförsök med tall i Sydsverige (Provenance trials with Scots pine in Southern Sweden). *Arsb. Fören. Skogsträdsförädl.* **1970**: 87–122.

Kalela, A. (1937/38): Zur Synthese der experimentellen Untersuchungen über Klimarassen der Holzarten. *Comm. Inst. For. Fenn.* **26**:1–435.

Kanák, K. (1980): Proveniencni studie s borovici lesni v ceskych zemich (Provenance studies on Scots pine in Czech regions). *Proceedings of conf. Proveniencni vyzkum lesnich drevin, Praha, 9–11 Oct. 1979,* pp. 83–104.

Khalil, M. A. K. (1968): Provenance trials of Scotch pine in Minnesota. *Min. For. Res. Notes* **193**:1–4.

Korczyk, A. (1983): Badania nad wartością genetyczną drzew doborowych sosny wykorzystywanych do zakładania plantacji nasiennych. Etap II. Ocena 7-letnich potomstw z wolnego zapylenia drzew doborowych i porównawczych sosny zwyczajnej 8 polskich pochodzeń w uprawach doświadczalnych (Studies on the genetic value of pine plus-trees used in seed orchards. Stage II. Evaluation of 7-year-old half-sib progenies of plus and comparison trees of 8 provenances). Forest Research Inst. Warsaw. (unpublished MS)

Koski, V. and Sievänen, R. (1985): Timing of growth cessation in relation to the variations in the growing season. In: Tigerstedt, P. M. A., Puttonen, P. and Koski, V. (Eds) *Crop physiology of forest trees.* University of Helsinki, Department of Plant Breeding, Helsinki.

Lazarescu, C., Benea, V. and Carniatchi, A. (1961): Experimentări cu diferiţe provenienţe de pin silvestru efectuate la Predeal (Experiments on different provenances of Scots pine in Predeal). *Revista Pădurilor* **76**:585–587.

Lines, R. and Mitchell, A. F. (1964): Results of some older Scots pine provenance experiments. *Rep. For. Res.* yr. ended March 1964, 172–194.

Lindström, A. and Nyström, L. (1987): Seasonal variation in root hardiness of container-grown Scots pine, Norway spruce and lodgepole pine seedlings. *Can. J. For. Res.* **17 (8)**:787–793.

Mandzhavidze, D. V. (1972): Nekotorye rezultaty opyta geograficheskoj posadki sosny obyknovennoj i sosny sosnovskogo (Some results from a provenance trial of Scots pine and *Pinus sosnovskyi*). *Vopr. introd. rast. zelen. stroit. (Tbilisi)* **6 (75)**:100–105.

Mátyás, C. (1981): Kelet-európai erdeifenyő-származások fenológiai változékonysága (Phenological variability of East-European Scots pine provenances). *Erdészeti Kutatások* **74**:71–80.

Mátyás, C. (1987): Adaptation processes in forest tree populations. Abstract of habilitation thesis. University of Sopron, Faculty of Forestry.

Mikola, J. (1982): Bud-set phenology as an indicator of climatic adaptation of Scots pine in Finland. *Silvae Fenn.* **16**:178–184.

Naryshkin, M. A., Vakurov, A. D. and Peterson, Yu. V. (1983): Geograficheskie kultury sosny obyknoviennoï pod Moskvoï (Scots pine provenance experiments in the vicinity of Moscow). *Lesovedenie* **(2)**:50–57. *Ref. Zh.* 6.56.127 (1983).

Oleksyn, J. and Giertych, M. (1984): Results of a 70-year-old Scots pine (*Pinus sylvestris* L.) provenance experiment in Pulawy, Poland. *Silvae Genet.* **33**:22–27.

Ołeksyn, J., Giertych, M. and Red'ko, G. I. (1986): Novyi vglyad na geograficheskie kul'tury sosny obyknovennoj V. D. Ogievskogo (A new look at the Scots pine provenance experiments of V. D. Ogievskij). *Lesnoï Zh.* **(6)**:20–24.

Patlaj, I. N. (1973): Geograficheskie kul'tury sosny obyknovennoï v Ukrainskoï SSR (Scots pine provenance studies in the Ukrainian SSR). *Proceedings of IUFRO symposium on genetics of Scots pine.* Kórnik, pp. 1–25.

Patlaj, I. N. (1974): Sezonnyï prirost sosny obyknovennoï v vysotu v geograficheskikh kul'turakh i plantatsiyakh (Seasonal growth of Scots pine in provenance trials and plantations). *Lesovod. agrolesomeliorat. (Kiev)* **36**:87–95.

Patlaj, I. N. (1984): Selektsionno-ekologicheskie osnovy semenovodstva i vyrashchivanya vysokoproduktivnykh kul'tur sosny obyknovennoj, duba chereshchatogo i yasenya obyknovennogo v ravninnoï chasti Ukrainskoï SSR (Selection and ecological principles of seed management and nursery growing of highly productive Scots pine, Turkey oak and European ash in the lowland Ukrainian plains). Avtoreferat, Ukr. Selskokhozyaistvennaya Akademiya, Kiev, pp. 1–45.

Patlaj, I. N. and Zhurova, P. T. (1987): Svetovoi rezhim i mikroklimaticheskie osobennosti geograficheskikh kul'tur sosny obyknovennoj (Light regime and microclimate of Scots pine provenance trials). *Lesovod. Agrolesomeliorat.* **74**:37–45.

Pravdin, L. F. (1964): Sosna obyknovennaya (Scots pine). Nauka, Moscow,

Presher, F. (1986): Transfer effects on volume production of *Pinus sylvestris* L.: a response to surface model. *Scan. J. For. Res.* **1**:285–292.

Remröd, J. (1973): Variation in surviving ability and growth in northern populations of Scots pine (*Pinus sylvestris* L.). *Proceedings of IUFRO symp. genetics of Scots pine,* Kórnik, pp. 1–12.

Remröd, J. (1976): Val av tallprovenienser i Norra Sverige—analys av överlevnad, tillväxt och kvalitet i 1951 ars tallproveniensförsök (Choosing Scots pine (*Pinus sylvestris* L.) provenances in Northern Sweden—analysis of survival, growth and quality in provenance experiments planted in 1951). Inst. f. Skogsgenetik, Stockholm. *Research Notes* **19**:1–131.

Rook, D. A. (1985): Physiological constraints on yield. In: Tigerstedt, P. M. A., Puttonen, P. and Koski, V. (Eds) *Crop physiology of forest trees.* University of Helsinki, Department of Plant Breeding, Helsinki, pp. 1–19.

Ruby, J. L. and Wright, J. W. (1976): A revised classification of geographic varieties in Scots pine. *Silvae Genet.* **25**:169–175.

Schober, R. (1975): Beobachtungen über den Kiefern-Provenienzversuch im Forstamt Wolfgang. Abt. Pechhütte 50b. *Allg. Forest Jgd. Ztg.* **146 (5)**:81–87.

Shutyaev, A. M. (1983): Osobennosti klimatipov sosny obyknovennoj v geograficheskikh kulturakh vtoroi generatsii (Characteristics of Scots pine climatypes in second generation provenance trials). *Lesovedenie* **1**:69–71.

Stefanson, E. and Sinko, M. (1967): Försök med tallprovenienser med särskild hänsyn till norrländska höjdlägen (Experiments with provenances of Scots pine with special regard to high-lying forests in Northern Sweden). *Stud. For. Suec,* **47**:1–108.

Steiner, K. C. (1979): Patterns of variation in bud-burst timing among populations in several Pinus species. *Silvae Genet.* **28**:185–194.

Sweet, G. B. (1964): Growth of five Spanish provenances of *Pinus sylvestris* at age six years in New Zealand. *Res. Leaflet For. Res. Inst. Rotorua, NZ.* **4**:1–2.

Tagil'tseva, V. M. (1986): Sosna obyknovennaya v khabarovskom krae (Scots pine in the Khabarovsk territory). *Lesnoe Khozyaistvo* **1**:44–46.

Thor, E. (1976): Tree breeding at the University of Tennessee 1959–1975. *Univ. Tenn. Bull.* **554**:1–48.

Troeger, R. (1960): Kiefernprovenienzversuche. I. Teil. Der große Kiefernprovenienzversuch im südwürttembergischen Forstbezirk Schussenried. *Allg. Forst. Jgd. Ztg.* **131**:49–59.

Tsanov, G. and Kosmov, K. D. (1985): V'rkhu ustoichivostta k'm naskemni vrediteli i rastezha na nyakoi proizhodi na beliya bor (*Pinus silvestris* L.) izv'i estestveniya im areal v raione na s. Rabisha, belogradchishko gorecko stopanstvo (Pest resistance and growth of Scots pine provenances outside their natural range at Rabisha, Belogradchik forest). *Gorskostopanska Nauka* **22 (1)**:23–30.

Ulissova, N. V. (1984): Geograficheskie kul'tury sosny obyknovennoj v Vologodskoi oblasti (Scots pine provenance studies in the Vologda region). *Lesnoe Khozyaistvo* **3**:42–44.

Veresin, M. M. (1972): Vliyanie proiskhozhdeniya semyan sosny obyknovennoj na rost kul'tur (Influence of Scots pine seed origin on stand growth). *Dokl. Symp. Genetika, selektsia i semenovodstvo khvoinykh porod.* Pushkino, pp. 38–48.

Vidyakin, A. I. (1979): O sezonnom razvitii sosny obyknovennoj raznogo geograficheskogo proiskhozhdeniya (On the seasonal development of Scots pine of various geographic origins). *Lesnoj Zh.* **6**:107–108.

Wiedemann, E. (1930): Die Versuche über den Einfluß der Herkunft des Kiefernsames aus der preußischen forstlichen Versuchsanstalt. *Ztschr. Forst. Jagdwesen* **62**:498–522, 809–836.

Wright, J. W. (1976): *Introduction to forest genetics.* Academic Press, New York, p. 463.

Wright, J. W. and Bull, W. I. (1963): Geographic variation in Scots pine. *Silvae Genet.* **12**:1–40.

Wright, J. W., Pauley, S. S., Polk, R. B., Jokela, J. J. and Read, R. A. (1966): Performance of Scots pine varieties in the North Central Region. *Silvae Genet.* **15**:101–110.

Zhurova, P. T. (1983): Fitomassa nadzemnoj chasti geograficheskikh kultur v pristepnykh gorakh Ukrainy (Phytomass of aerial parts of provenance cultures in the steppe mountains of the Ukraine). *Lesovod. agrolesomelior.* **66**:34–38.

CHAPTER 7

Breeding strategies for timber production

H. H. Hattemer

Introduction

A *breeding goal* is a desirable expression of phenotypic traits of a population in a given environment. The trait expressions are desirable because they possess *economic value,* although the present average expression of these traits may differ from this goal. Selecting those portions of the population approaching the breeding goal most closely is one approach to improving the economy of harvesting the population. This makes clear, that breeding has a biological basis, but that it is aimed at an economic objective.

Breeding continued over generations

The sequence of decisions and operations in a long-term breeding project devoted to increased timber production is outlined following those of Namkoong (1976), Namkoong et al. (1988) and other authors. Two issues arise at the outset: *(i)* The *decision to embark* on a breeding project should be made after evaluating competitive approaches to an improvement of yield. More adequate methods of growing the tree species, and more intense forest protection or site improvement suggest themselves although these measures must be applied anew in every tree generation. This can make breeding so attractive that projects have been initiated without a thought being given to whether or not economic benefit can be expected from them. *(ii)* The *base populations* must meet certain requirements. Their choice is usually based on comparisons among several candidate populations possessing a high level of productivity under given conditions. These populations are eligible under the premises that the respective traits vary, that they are at least under partial genetic control, and that the populations are adapted to the environments of future use of the breeding products.

Several issues arise anew in every generation of the project: *(i)* In advanced generations, there is an increasing tendency for trees in populations to be related. This has multiple implications for the breeding strategy, as discussed by Lindgren and Gregorius (1976). *(ii)* In every generation, the parent trees for the next generation are chosen on the basis of the level and stability of their relative merit. Average merit and phenotypic stability are observed in *tests conducted in several environments,* preferably in the field. *(iii)* In every generation, some *constraints on selection intensity* are imposed to maintain minimal population sizes. Maintaining a certain population size helps to avoid major losses in genetic variation. The populations, or replicates exposed to different environmental conditions, may respond to heterogeneous conditions of natural selection or diverge genetically for other reasons. This is important in conserving genetic variation (see section on Conserving genetic variation). *(iv) Multiple populations* are maintained. Populations may differ in the combination of trait expressions; therefore, the breeder might select populations in different ways depending on their distance from the optimum combination of trait expression. Alternatively, the breeder might direct his attention toward a single objective and apply some system of reciprocal recurrent selection. In the latter, multiple replicates of a single population are advisable. *(v)* Volume alone is rarely assessed directly because the rotation period is long and results are desired before it ends. This leads to prediction of yield from observations of several traits which are assessed at juvenile stages. The selection is, therefore, of the *indirect* type. In deciduous trees, all parts above the ground may be observed by a video camera and measured after digitalization. The number of predictors of timber production in a non-deciduous conifer like Scots pine must not be too small to ensure reliability. The formation of stem volume of trees is controlled by growth in both the primary and secondary meristems. The earlier the traits are assessed, the lower the effeciency of prediction. *(vi)* The traits are entered into some *index*. Index selection has become the standard procedure since the advent of electronic data processing in breeding. Little information on test performance during a particular generation may be utilized in the subsequent generation, since *environments* vary among tree generations. This condition reduces the correlation between the results of testing in subsequent generations. In addition, the entries of test experiments in predominantly cross-pollinating species such as Scots pine are not *effectively fixed,* and the intermittent sexual process adds greatly to the diminishing of correlations. The prospects of stepwise selection are improved in domestic animals under artificial environments or in clones. The integration of stepwise testing is advantageous if *the same* seedlings or seedlings raised from the same seed lot enter into the various testing steps. There are several types of indices. An index on relatives integrates the merit of relatives into that of an individual to be selected (Namkoong 1981, Burdon 1982*b*). Namkoong (1981) explained that a complete evaluation of relatives requires an index on traits. Burdon (1979) derived an index which utilizes the expression of one trait at several sites. Baradat's (1976) index may be used to combine the merits of relatives and the merits in multiple traits for selection. Namkoong (1976) discussed the advantage of having indices of multiple populations. *(vii)* A breeding goal may rarely be defined in terms of volume alone. An additional determinant of yield is the *quality of timber,* since it has an impact on price. Simultaneous selection for more traits reduces selection intensity in the individual traits but serves economy. Tandem procedures of selecting for only one trait in a given generation usually means less pay-off in the initial generations. *(viii)* Indices making efficient use of economic weights of the various traits are difficult to construct, and their application leads to severe losses if these weights are not explored with adequate precision. The weights change over generations according to the situation at the market. *(ix)* One particular component trait of yield is *resistance,* since it has an effect on the safety of production. Scots pine hosts many pathogens.

Many types of populations and sublines have to be considered, as discussed by Lindgren, and Gregorius (1976), Kang (1982), Burdon and Namkoong (1983), and Namkoong et al. (1988).

Attempts to penetrate the complexity of a long-continued breeding process have encouraged model experiments with animals selected for body weight in the laboratory. Something which can be learned from these experiments is the fact that the variation at the gene loci controlling the traits under selection is unlikely to be exhausted after few generation cycles. However, selection limits become manifest only at later stages and are, therefore, likely to be neglected at the beginning.

Besides exhaustion of the genetic variation at the controlling gene loci, another condition can imply a selection limit; that is, natural selection becomes opposed to artificial selection if the lines with the potential of highest yields possess lower field resistance.

In reviewing new developments in the application of quantitative genetics in tree breeding, Van Buijtenen (1987) listed other items of breeding strategies related to mating and environmental designs which may not be empirically explored by means of laboratory experiments. Hood (1982) used simulation for the construction and analysis of breeding strategies.

Genetic basis of trait variation

The genetic basis of a complex metric trait, such as timber production, is generally unknown. The breeder only infers gene actions with the help of quantitative genetic models. Little progress is to be expected if *heterozygosity at many loci* is advantageous for trait expression in the sense of the breeding goal. It is difficult to find out whether increased volume is accompanied by heterozygosity at many gene loci. However, evidence of the advantage of heterozygosity for viability has been compiled by Ledig (1986); according to these data, heterozygosity is high also in Scots pine. Information on the frequency and evolutionary importance of heterozygosity has been reviewed by Namkoong (1985). Heterosis requires a different type of breeding strategy (Schnell, 1982); possible gains from the use of inbreeding in tree breeding have been discussed by Lindgren and Gregorius (1976).

The response of the population to artificial selection may be small if the number of genes controlling the traits is large; the same is true if the environment exerts strong influences on the expression of the traits and genotypic and environmental effects are non-separable. Both conditions decrease the selection coefficients acting on allele frequencies at the controlling gene loci. Environmental effects on volume are larger than those on component traits and presumably not all gene effects on volume are of the so-called additive type (Burdon, 1982*a*). These conditions not only encumber breeding progress (Namkoong, 1982), they require consideration in setting up a breeding strategy.

Recruitment

Several authors have stressed the need for using genetically variable, locally adapted base populations for breeding. The structure of the population can respond to a selection pressure exerted by the breeder only at polymorphic gene loci. Namkoong et al. (1988) emphasized the start from provenance experiments. In a case study, Burdon (1987) described procedures applied in recruiting the base population of an exotic tree species.

Unfortunately, the testing of provenances in the past has provided many examples for a non-optimal sequence of operations: A few provenances were first tested in large plots before a larger number of provenances were tested in small plots. The results of both existing types of experiments must be utilized in view of the investments made.

Periodic controls

The most important source of information for devising a selection step is to anticipate the progress to be expected from alternative measures of selection. The most promising measure is then chosen. In the subsequent generation, it can be verified empirically whether the prediction was actually realized. The comparisons of *expected and realized responses* to selection, therefore, must be monitored. In every generation, a re-evaluation of the breeding progress serves as a source of justification to continue the project.

Present empirical progress for Scots pine evaluated at an early age is large (Ericsson, 1986). This is presumably due to enormous intensities of selection applied in the initial plus tree selection.

Monitoring concomitant effects

Breeding for a group of traits may bring about two types of involuntary side-effects. These must be identified if measures of avoidance are to be designed.

Those of the first type are commonly called *correlated phenotypic responses;* two examples may illustrate this. *(i)* Due to physiological correlations with the traits under selection, other traits may react to breeding although they were not selected themselves. An example is the decrease in cell wall thickness and hence wood-specific gravity, which implies corresponding losses of timber quality with increasing growth speed. Negative correlations of this type may be compensated for by including the correlated trait in the selection index. They nevertheless represent biological obstacles consuming part of the breeding progress. *(ii)* Genes controlling the traits under selection influence pleiotropic control over other traits. These may change their average expression in an undesirable way. This occurs if the subsequent generation is initiated before the majority of trees have entered the reproductive phase. Selection for early and possibly prolific reproduction brings about a correlated change in timber production for physiological reasons. This side-effect can be kept at a minimum by lowering the speed of progression.

In outlining a breeding strategy for Scots pine, Müller (1937) advocated a generation length of 10 years. He found indications of inheritance of various components of reproductive behaviour, but no differences in growth rate between cone-bearing trees and the others. This might be different in pines beyond the age of 10 years. In other tree species, the investment in female reproduction evidently decreases timber production.

Burdons theory

Side-effects of the second type pertain to the genetic variation and are not necessarily expressed phenotypically. They will be dealt with in the following section.

Conserving genetic variation

The size of the populations which may be handled by the breeder over generations is finite, and perhaps even small. The risk of random loss of genes is determined by the drift-effective population size. This is strongly influenced by the harmonic mean of the population sizes during a sequence of generations, so that one drastic reduction has a pronounced and long-lasting effect (Crow and Kimura, 1970, p. 360, quoting earlier results of S. Wright). Conservation of genetic variation is motivated by two considerations. *(i)* The variation at the controlling loci is exhausted when all of these loci are fixed. This is not likely to happen soon in trees. *(ii)* The gene pool of a population represents its *adaptive potential.* The variable gene pool is the only instrument with which the population is able to react to environmental change. Every reduction in the number of genes in the gene pool means a loss in potential genotypic multiplicity, and, thus, adaptive potential (Gregorius et al., 1979). This can be remedied only by later additions from a genetically differentiated population.

The individuals carrying the genetic information worth conserving are called a *gene resource.* A conservation deme is declared a gene resource if it can be expected to carry specific genetic information or because the information it carries can be expected to be particularly variable. This criterion has been defined by Ziehe et al. (1989), who also distinguished between three different goals of conservation: *(i)* Conservation of the *potential* of a deme to produce a *desirable expression of certain traits.* Because only the potential may be conserved, pursuing this goal has good prospects under the pre-condition that trait expression is largely independent of the environment. *(ii)* Conservation of the *genetic adaptability* of a deme. Various parts of the distribution range of Scots pine are affected by environmental pollution, which exerts an unanticipated selection pressure. *(iii)*

Conservation of as much variation as possible regardless of whether it is identified or not.

These three goals require different conservation procedures. The demes most eligible for pursuing the two latter goals are identified by a *genetic inventory* of populations. In the evaluation of such inventories, and in decision-making, it is useful to have appropriate population genetic parameters allowing quantification of variation and differentiation. It is possible to measure the fraction of genes that a deme has in common with the others and, thereby, identify the demes *most representative* of the population. It is also applicable at the level of populations (see section on Integration of genetic principles). This has to be combined with experimental findings on the phenotype of that deme in making a decision.

Consequences for application

Integration of genetic principles

As was outlined above, breeding is an application of population genetics. These foundations of breeding were not widely known before the empirical study of marker loci in genetic inventories became possible. The marker loci are identified using certain gene markers such as isozymes or the relative composition of the terpene fraction of the resin. Adams (1983), Cheliak et al. (1987), and Adams et al. (1988) described further applications of genetics in tree breeding (cf. also Hattemer, 1985). Only a very brief review will be given below in order to reduce present scepticism. *(i)* The size of the gene pool reacts most drastically to reductions in population size. Such a reduction has a most unspecific impact since alleles at *all* loci are liable to be lost according to their present frequency and to population size. Losses in adaptive potential of a population can therefore be monitored at marker loci.

Hence, if marker genes are representative of the totality of the genome in this respect, the genetic structure at marker loci yields a relative measure of *genetic variation,* even though a generally valid relationship between the condition of markers and the phenotypic merit of their carriers may be difficult to establish. Markers allow inference on the presumed number of progenitors of a certain deme and enable decisions to be made whether the deme is eligible for inclusion as part of the base population or declared a gene resource. The components of a seed orchard cannot be expected to carry all of the alleles of marker loci which were originally encountered in their base population. It is thus possible to verify shrinkage of the adaptive potential due to shrinkage of the population size. *(ii)* The impact of inbreeding on heterozygosity is also unspecific and can thus be monitored at any marker loci under the precondition of codominance. *(iii)* The artificial selection scheme devised by the breeder of forest trees is not the only selection pressure to which a population is exposed. Numerous reports exist on *natural* genetic selection acting on tree populations. Carriers of certain genotypes differ in viability or even in their capacity to submit their genetic information to their progeny. These processes of natural selection are monitored at marker loci even though the selective forces do not necessarily act on the marker loci themselves. However, there exists experimental evidence that some of these loci participate in the genetic control of viability. A difference in genotypic structures at marker genes between the orchard clones and the seed collected from them indicate processes impairing the precision of realizing a breeding progress. In his contribution to this volume, Müller-Starck reviewed selective and other genetic processes active in seed orhards of Scots pine. *(iv)* In maintaining conservation demes, in collecting seed from stands to produce the subsequent generation, or in collecting seed from seed orchards, it is important to know distances and effective quantities of gene flow and the homogeneity of the pollen sources which become effective in different individuals. *(v)* It is crucial to know whether two demes differ genetically or not before a decision can be made as to whether both are to be used in designing a breeding population or whether they are to be declared gene resources. It is advantageous to be able to estimate the proportion of genes which they have in common, and the degree to which either of them is representative of the totality of the others. Gregorius, (1985, 1988) and Gregorius and Roberds (1986) developed such a concept and presented examples. *(vi)* Marker genes enable the breeder to infer on the descent of certain trees from others. This is important in checking the validity of crosses and the correct establishment of field tests. It is relatively easy to check the homogeneity of clones and the layout of seed orchards.

It has been questioned recently whether conventional quantitative genetic models are adequate if applied as widely as is being done in breeding practice (Feldman and Lewontin, 1975, Gregorius, 1977). The most crucial utilization of quantitative genetics is gain prediction under varied selection schemes for choosing selection methods. Hence, it might be interesting to find out whether discrepancies between predicted and realized breeding progresses can be attributed to incorrect prediction. Unfortunately, the offspring of existing seed orchards are not eligible for this analysis. Alternatives to the evaluation of linear models in the analysis of variance have clear advantages, even though more operational methods need to be developed (Gregorius and Namkoong, 1986, 1987).

Gene conservation

It should be clear that gene conservation must accompany breeding operations. However, very little effort has been devoted to developing a methodology of gene conservation when compared to the existing literature on the development of mating designs and selection methods. This is all the more ironical since one generation of a long-term project is the crucial determinant of the genetic variation transferred to all subsequent generations. Lindgren and Gregorius (1976) discussed criteria for allocating conservation efforts.

The greater the *conversion rate* of present populations by breeding products, the more urgent is the need for conservation. A high conversion rate means that breeders in the near future will have to rely on the introduction of genetic variation from forests which were planted with trees produced in breeding projects. The conversion rate will be high if intermediate results of breeding projects are utilized on a large scale. Some breeders stress the early release of short-term outputs because they are then able to raise additional funds for more extensive testing.

In some parts of the world awareness for the need for gene conservation has increased due to forest decline resulting from man-made environmental changes.

Keeping records

Records should be duplicated and stored in several locations to reduce the risk of accidental loss. Modern technology of data storage greatly facilitates the maintenance of *detailed* data. This includes information on pedigrees and coancestry (Lindgren and Gregorius, 1976), both essential for designing a strategy avoiding inbreeding during many generations.

Methods of evaluating relative breeding merit evolve more rapidly than tree generations emerge. This adds to the importance of storing as much original data as possible. Any condensations may impede later re-evaluation with more efficient or more adequate methods.

Maintaining breeding material

Multiple demes must be conserved as insurance against accidental loss. This also provides increased numbers of individuals for exposure to different environments. Gene flow between demes, selected under different regimes of natural selection, will help to maintain genetic polymorphisms (Gregorius and Namkoong, 1983).

Spreading the risk of loss is advised for the selection demes. This has gained importance since parts of the range of Scots pine are presently menaced by environmental pollution.

Implementation of a strategy

The elements of a strategy are normally presented in a flow chart; illustrative examples are presented by Namkoong et al. (1988). These elements receive differential weights according to the local conditions: *(i)* Biological conditions, such as the existing variation, greatly influence the efforts required for attaining a certain breeding progress. The existence of indigenous populations urgently calls for their conservation and for their utilization as a source of adapted genetic information. *(ii)* Ecological conditions, such as the severity and heterogeneity of the physical environment or the population dynamics of parasites, require locally different strategies. The efforts necessary for attaining a certain efficiency of selection largely depend on ecological parameters. *(iii)* Economic conditions, such as uses of timber and cost structure of labor, land, and equipment, vary in different parts of the distribution range of Scots pine. These conditions exert pressure for financial expenditure. *(iv)* Breeders are almost exclusively members of public organizations. These have certain policies regarding the thinning of stands and the production of planting stock. These forestry operations also have genetic implications. Wilusz and Giertych (1974) reported on breeding implications of silviculture for Scots pine. These may be considered as alternatives to more aggressive breeding. In contrast with agronomic crops, only a small portion of the trees are harvested in timber-producing stands. This allows for reductions in the efforts of the breeder, because he can select more mildly and needs to produce less improved planting stock.

Although few of the above points apply exclusively to Scots pine, all is relevant because this species is propagated only generatively. If large-scale vegetative propagation could be practised, it would have strong impacts on breeding strategy (Burdon, 1982*c*).

like what?. Pointless open-ended statement

Literature

Adams, W. T. (1983): Application of isozymes in tree breeding. In: Tanksley, S. D. and Orton, T. J. (Eds) *Isozymes in plant genetics and breeding.* Part A. Elsevier, Amsterdam, pp. 381–400.

Adams, W. T., Neale, D. B. and Loopstra, C. A. (1988): Verifying controlled crosses in conifer tree-improvement programs. *Silvae Genet.* **37**:147–152.

Baradat, Ph. (1976): Use of juvenile–mature relationships and information from relatives in combined multitrait selection. In: *Proceedings of IUFRO joint meeting on advanced generation breeding.* INRA, Cestas, pp. 121–138.

Burdon, R. D. (1979): Generalisation of multiple-trait selection indices using information from several sites. *N. Z. J. For. Sci.* **9**:145–152.

Burdon, R. D. (1982*a*): Breeding for productivity—jackpot or will-o'-the-wisp? In: *Proceedings of North American forest biology workshop.* Lexington, Ky. pp. 35–51.

Burdon, R. D. (1982*b*): Selection indices using information from multiple sources for the single-trait case. *Silvae Genet.* **31**:81–85.

Burdon, R. D. (1982*c*): The roles and optimal place of vegetative propagation in tree breeding strategies. In: Spethmann, W. (Ed.) *Proceedings of IUFRO joint meeting of working parties on genetics about breeding strategies including multiclonal varieties.* Lower Saxony Forest Research Institute, Staufenberg–Escherode, pp. 66–83.

Burdon, R. D. (1987): Recruitment for breeding populations: Objectives, genetics, and implementation. In: Weir, B. S., Eisen, E. J., Goodman, M. M. and Namkoong, G. (Eds) *Proceedings of second international conference of quantitative genetics.* Sinauer, Sunderland, pp. 555–572.

Burdon, R. D. and Namkoong, G. (1983): Multiple populations and sublines. *Silvae Genet.* **32**:221–222.

Cheliak, W. M., Yeh, F. C. H. and Pitel, J. A. (1987): Use of electrophoresis in tree improvement programs. *For. Chron.* **63**:89–96.

Crow, J. F. and Kimura, M. (1970): *Introduction to population genetics theory.* Harper and Row, New York–Evanston–London.

Ericsson, T. (1986): Pine seed orchard tests in Northern Sweden. Preliminary results from an assessment in summer 1984 of a series of field trials planted 1973–75. In: Lindgren, D. (Ed.) *Provenances and forest tree breeding for high latitudes.* SLU, Dept. For. Genet. Plant Physiol. Rep. No. 6, Umeå, pp. 113–126.

Feldman, M. W. and Lewontin, R. C. (1975): The heritability hang-up. *Science* **190**:1163–1168.

Gregorius, H.-R. (1977): The genotype x environment-to-phenotype relationship. *Theoret. Appl. Genet.* **49**:165–176.

Gregorius, H.-R. (1985): Measurement of genetic differentiation in plant populations. In: Gregorius, H.-R. (Ed.) *Population genetics in forestry.* Springer, Berlin–Heidelberg–New York–Tokyo, pp. 276–285.

Gregorius, H.-R. (1988): The meaning of genetic variation within and between subpopulations. *Theoret. Appl. Genet.* **76**:947–951.

Gregorius, H. R., Bergmann, F., Müller-Starck, G. and Hattemer, H. H. (1979): Genetische Implikationen waldbaulicher und züchterischer Massnahmen. *Allgem. Forst-Jagdztg.* **150**:30–41.

Gregorius, H.-R. and Namkoong, G. (1983): Conditions for protected polymorphism in subdivided populations. 1. Uniform pollen dispersal. *Theor. Pop. Biol.* **24**:252–267.

Gregorius, H.-R. and Namkoong, G. (1986): Joint analysis of genotypic and environmental effects. *Theor. Appl. Genet.* **72**: 413–422.

Gregorius, H.-R. and Namkoong, G. (1987): Resolving the dilemmas of interaction, separability and additivity. *Math. Biosci.* **85**:51–69.

Gregorius, H.-R. and Roberds, J. H. (1986): Measurement of genetical differentiation among subpopulations. *Theor. Appl. Genet.* **71**:826–834.

Hattemer, H. H. (1985): Genetics, breeding strategies and healthy forests. In: Oswald, M. (Ed.) *Human impacts on forests.* Institut National de la Recherche Agronomique, Paris, pp. 355–373.

Hood, J. V. (1982): Evaluating tree breeding strategies by computer simulation. In: Spethmann, W. (Ed.) *Proceedings of IUFRO joint meeting of working parties on genetics about breeding strategies including multiclonal varieties.* Lower Saxony Forest Research Institute, Staufenberg–Escherode, p. 163.

Kang, H. (1982): Components of a tree breeding plan. In: Spethmann, W. (Ed.) *Proceedings of IUFRO joint meeting of working parties on genetics about breeding strategies including multiclonal varieties.* Lower Saxony Forest Research Institute, Staufenberg–Escherode, pp. 119–135.

Ledig, F. Th. (1986): Heterozygosity, heterosis, and fitness in outbreeding plants. In: Soulé, M. E. (Ed.) *Conservation biology.* The Science of Scarcity and Diversity. Sinauer, Sunderland, pp. 77–104.

Lindgren, D. and Gregorius, H.-R. (1976): Inbreeding and coancestry. In: *Proceedings of IUFRO joint meeting on advanced generation breeding.* INRA, Cestas, pp. 49–69.

Müller, K. O. (1937): Über die Geschlechtsverteilung und den Eintritt der Geschlechtsreife bei der Waldkiefer (*Pinus silvestris* L.). *Zeitschr. Forst. Jagdw.* **69**:177–201.

Namkoong, G. (1976): Multiple-index selection strategy. *Silvae Genet.* **25**:199–201.

Namkoong, G. (1981): *Introduction to quantitative genetics in forestry.* Castle House, Tunbridge Wells.

Namkoong, G. (1982): Challenging tree breeding theory. In: Spethmann, W. (Ed.) *Proceedings of IUFRO joint meeting of working parties on genetics about breeding strategies including multiclonal varieties.* Lower Saxony Forest Research Institute, Staufenberg–Escherode, pp. 155–161.

Namkoong, G. (1985): The population genetic basis of breeding theory. In: Gregorius, H.-R. (Ed.) *Population genetics in forestry.* Springer, Berlin–Heidelberg–New York–Tokyo, pp. 2–15.

Namkoong, G., Kang, H. C. and Brouard, J. S. (1988): *Tree breeding: principles and strategies.* Monographs on Theoretical and Applied Genetics No. 11. Springer, New York–Berlin–Heidelberg–London–Paris–Tokyo.

Schnell, F. W. (1982): A synoptic study of the methods and categories of plant breeding. *Z. Pflanzenzüchtg.* **89**:1–18.

Van Buijtenen, J. P. (1987): Quantitative genetics in forestry. In: Weir, B. S., Eisen, E. J., Goodman, M. M. and Namkoong, G. (Eds) *Proceedings of second international conference of quantitative genetics.* Sinauer, Sunderland, pp. 549–554.

Wilusz, W. and Giertych, M. (1974): Effects of classical silviculture on the genetic quality of the progeny. *Silvae Genet.* **23**:127–130.

Ziehe, M., Gregorius, H.-R., Glock, H., Hattemer, H. H. and Herzog, S. (1989): Gene resources and gene conservation in forest trees. In: Scholz, F., Gregorius, H.-R. and Rudin, D. (Eds) *Genetic effects of air pollutants in forest tree populations.* Springer, Berlin–Heidelberg–New York–Tokyo, pp. 173–186.

CHAPTER 8

Breeding strategies for Christmas trees*

D. F. Van Haverbeke and H. D. Gerhold

Introduction

Scots pine is the leading pine species grown for the commercial production of Christmas trees in the northern United States and Canada. Because of its extensive natural distribution throughout much of Eurasia, the species has great genetic variability and, thus, adaptability to diverse environments. This variability offers excellent opportunities for genetic improvement and refinement for specialized uses, such as Christmas trees.

As the planting of Scots pine for Christmas trees expanded in North America, growers expressed an urgent need for genetic improvement in the quality of planting stock: faster growth, more desirable winter foliage colour, straighter trunks and smaller butts, increased cold-hardiness, longer needle retention, and greater disease resistance and insect tolerance.

Precocious flowering in Scots pine permits the rapid development of successive generations, making it a desirable species for improvement through breeding and progeny testing. Selection and breeding can begin in Scots pine by age 7, the approximate rotation age for Christmas trees (Gerhold, 1966).

Christmas tree growers in the United States generally attempt to purchase the best stock available from commercial nurseries, whose material may or may not be of known or certified origin. Some large growers obtain seed from selected stands, generally of observed, but unproven, genetic quality.

Few independent growers, however, have the resources or technical expertise to independently embark on an extensive and expensive breeding programme. Genetic improvements have been developed mainly by researchers, and the resulting improved seeds or seedlings have been distributed to growers through government and commercial nurseries.

Interim steps toward improvement

Planted stands can be rogued, retaining individual trees exhibiting one or more of the previously described Christmas tree characteristics, as a quick initial step toward an improved source of seed. Open-pollinated progenies of these residual

* Research on this subject has been supported through USDA Regional Research Projects NC–99 and NE–27.

trees should produce progenies of somewhat higher genetic quality than that obtained from random collections in nature.

An alternative approach, resulting in a somewhat higher degree of genetic improvement, would consist of establishing a clonal seed orchard with ramets from the best selected trees (Van Haverbeke, 1974; 1981*a,b*).

Because many, widespread provenances of Scots pine are commonly grown in the United States, inter-racial hybridization is an option. Until recent decades, experience with racial hybridization was confined to Europe. Results of these experiments suggested the possible attainment of heterosis, the combining of desirable traits of different populations, and the creation of greater variability within populations for recurrent selection. Such provenance material, consisting of many geographically diverse and genetically variable races, is available in the United States (Ruby and Wright, 1976).

Breeding strategies

Genetic gain above that which may be realized through roguing, provenance trials, and subsequent seed orchard establishment with this unproven material, may be attained through open- or control-pollinated progeny testing of selected trees, followed by recurrent selection.

Breeding programmes in the United States, designed specifically to improve the genetic quality of Scots pine for Christmas trees, have utilized the factorial, a variation of the factorial (factorial population polycross), and a variation of the nested and diallel (nested population diallel) mating designs. Breeding methods have included recurrent selection and synthetic varieties.

Factorial design and recurrent selection

The factorial design and recurrent selection are being used in Nebraska. Forty-two ovulate trees with desirable Christmas tree characteristics were selected from a 7-year-old, 36-origin provenance plantation of Scots pine (Van Haverbeke, 1976). Each of the ovulate plus-trees was crossed with seven staminate testers, omitting selfs. A clonal seed orchard of a randomized complete block design, but with the testers structured systematically within each block to ensure adequate pollination, was established concurrently with the initiation of the breeding programme. Following evaluation of the control-pollinated progeny tests, the seed orchard will be rogued of undesirable clones, and the open-pollinated seedlings derived from the orchard will be distributed to Christmas tree growers as the "Nebraska" strain of Scots pine. The factorial gives a good assessment of parental breeding value, and provides unrelated families of known parentage from which selections for advanced generations can be effectively made. However, unless the number of testers is large, or material can be exchanged among programmes using unrelated testers, the potential for recurrent selection is quickly exhausted.

Synthetic varieties

The Pennspanish Scots pine variety released by Pennsylvania State University in 1984 (Gerhold, 1985) was developed by a combination of recurrent selection (mainly in the Spanish Guadarrama Mountains variety), and hybridization with trees selected from several other provenances. Following incomplete factorial matings in which 9 ovulate Spanish trees were crossed with 11 Spanish and 10 non-Spanish staminate trees, a progeny test of the resulting 59 families was converted to a seedling seed orchard. The 362 trees remaining after roguing are the best 23% selected genotypically and phenotypically for dense and symmetrical crown form, fast growth, and bluish-green foliage. If random mating occurs, the composition of seed harvests would be 73% pure Spanish, 22% backcrosses to Spanish, and 5% simple or complex hybrids. Five successive seed crops starting at age 10 have totaled 66 kg; at 86 000 seeds per kilo, this could produce 4 ½ million seedlings. The estimated value of these seedlings is $678 600, and they could be grown into Christmas trees worth about $86 million. Compared with ordinary Spanish Scots pines, the "Pennspanish" variety is estimated to be 20% better in crown form and density, 7% in growth rate, and 15% in foliage colour, based on progeny test results.

Provenance crosses

During the time when the "Pennspanish" variety was being developed, hybridization among provenances was also explored as a means of improvement. The only barrier to hybridizing trees of diverse provenances was the phenological difference in time of flowering among some races (Karrfalt et al., 1975). Genetic variability in height among Scots pine individuals was larger than that among populations; genetic variations and combining abilities of individuals and populations in needle length were about equal. General combining ability of individuals was the most important source of genetic variance (Gerhold and Park, 1986; Park and Gerhold, 1986). Several sizeable positive, and negative, heterosis values were observed, but in general hybrids were not superior. The authors believe that population hybridization will likely be important in combining characters and in producing populations of high diversity in future generations.

Provenance hybrids in the F_2 generation are now being used to form synthetic varieties at Pennsylvania State University. Progeny tests planted in 1986 contain mainly three-way and four-way hybrids, and back-crosses involving provenances from Belgium, England, France, Poland, Scotland, Spain, Sweden, Turkey, and southern USSR. The gene pool of the families produced from 77 parent trees is 39% Spanish germplasm; 9% of the families are pure Spanish, and 27% are entirely non-Spanish. The goals are to improve winter hardiness and disease resistance, along with the form, growth, and colour traits mentioned previously. Thus better combinations of traits are being sought, rather than heterosis.

Literature

Gerhold, H. D. (1966): Selection for precocious flowering in *Pinus sylvestris.* In: *Joint proceedings of the 2nd genetics workshop of Society of American Foresters and 7th Lake States forest tree improvement conference. Oct. 21–23, 1965.* Michigan State University, East Lansing, pp. 4–7.

Gerhold, H. D. (1985): 'Pennspanish' Scotch pine released for production. *Am. Christ. Tree J.* **29 (2)**:30, 31, 33.

Gerhold, H. D. and Park, Y. S. (1986): Population hybridization in Scotch pine (*Pinus sylvestris* L.): II. Combining ability comparisons. *Silvae Genet.* **35**:195–201.

Karrfalt, R. P., Gerhold, H. D. and Palpant, E. H. (1975): Inter-racial hybridization in Scotch pine: geographic flowering patterns and crossability. *Silvae Genet.* **24**:107–110.

Park, Y. S. and Gerhold, H. D. (1986): Population hybridization in Scotch pine (*Pinus sylvestris* L.): I. Genetic variance components and heterosis. *Silvae Genet.* **35**:159–165.

Ruby, J. L. and Wright, J. W. (1976): A revised classification of geographic varieties in Scotch pine. *Silvae Genet.* **25**:169–175.

Van Haverbeke, D. F. (1974): A Scotch pine clonal seed orchard of provenance origin. In: *Proceedings of the 9th Central States forest tree improvement conference, Oct. 10–11,* Ames, Iowa, pp. 62–70.

Van Haverbeke, D. F. (1976): Interprovenance breeding in Scots pine. In: *Proceedings of the 10th Central States forest tree improvement conference. Sept. 21–23.* Purdue University, Lafayette, Indiana, pp. 171–182.

Van Haverbeke, D. F. (1981*a*): Seed orchards – a way to better Christmas trees. Part I. Establishment. *Am. Christ. Tree J.* **25 (2)**:22–25.

Van Haverbeke, D. F. (1981*b*): Seed orchards – a way to better Christmas trees. Part II. Management. *Am. Christ. Tree J.* **25 (3)**:22–25.

CHAPTER 9

Seed stands and plus trees

E. Pihelgas

Seed stands

Introduction

The selection of seed stands from the existing natural and cultivated stands is of great importance as long as the supply of seed orchard seed is inadequate. The need will not completely cease even after the attainment of full seed production in seed orchards. The maintenance of seed stands permits the use of the best populations for regeneration and breeding and the more perfect preservation of the local gene pool, which is the result of a long-lasting natural selection process.

Most countries with commercial plantations of Scots pine have a programme of selection of seed stands and plus trees. Following earlier discussion about the concept of seed stands (Andersson, 1963; Matthews, 1963) the procedures used in most countries are nowadays similar, therefore, it will be perhaps sufficient to give a detailed account of the programme in my country (Estonia) as an example of the selection work performed throughout Europe.

Selection and management procedures

The establishment of seed stands must be preceded by a genetical inventory of stands. The selection of seed stands has to be planned with consideration of the limits of geographical regions or seed zones. As Scots pine grows on a wide range of sites, differences in ecological adaptation must also be taken into account: the crop of seed stands should be utilized for regeneration under more or less similar site conditions (Pihelgas, 1975).

Seed stands are selected from among plus and above average stands. Depending on the genetic value of the stand and on the planned duration of their maintenance three types (temporary, permanent and selectional seed stands) are distinguished (Pihelgas, 1970; Molotkov et al., 1982; Veresin et al., 1985).

Temporary seed stands

Temporary seed stands are mainly chosen from above-average stands approaching rotation age (60 to 80 years, sometimes up to 100 years depending on the felling age) as well as from the mature stands ready for felling. Stands from better site classes (III and better) are preferred. In mixed stands the percentage of Scots pine must however surpass 70% in immature age classes and 80% at felling age. (Occasionally heath pine stands of low productivity are selected as seed stands, but the seed harvested is utilized only on similar site types.) A seed stand may be autochthonous or cultivated, but previously regularly tended and resistant to storms. It should be bordered by stands of at least average quality. The stand should be isolated by narrow felled strips and marked by corner-posts.

To favour seed production, other tree species present, as well as malformed and suppressed Scots pine trees are removed. This is to be carried out 5 to 8 years before the final cut. Admixed tree species, which do not shade the crowns of pines and the felling of which would create a gap greater than the average stand height, are not cut. The density and standing volume must remain above 50% of the normal. If necessary, cutting should be done in two steps to allow for some recovery of the crowns. When marking out trees for felling it is advisable to mark also the plus trees. These are destined to grow until the harvest cut.

To increase the seed yield it is of importance to fertilize the stand after thinning. The fertilizer rate depends on the growing conditions. In Estonia mostly nitrogen fertilizers are recommended in the *Pinetum oxalidosum* and *Pinetum myrtillosum* forest types. The application of N fertilizer at a rate of 200 to 300 kg/ha may bring about an increase in the seed yield of 2 to 4 times. At the same time the volume yield will increase as well. The fertilizer should be applied to the surface during the vegetative period. As the effect of fertilizer diminishes with time, it is expedient to refertilize in every 3 to 5 years.

The cone yield is monitored and the stand has to be designated for the final cut when the cone yield exceeds the average. The trees are cut during winter and the cones are picked from the felled trees. Cones harvested from different stands are treated separately. If necessary, cones from the felled plus trees may be separately collected to be used for breeding and the establishment of progeny tests.

Seed stands envisaged for harvesting must meet the seed demand until the next seed year. The annual demand for seeds, the expected seed yield and the frequency of seed years has to be considered.

Permanent seed stands

Where sufficient temporary seed stands are lacking it is expedient to make use of permanent ones. Permanent seed stands are natural or cultivated stands selected for seed harvesting from standing trees during a longer period. Although several proposals have been made and devices constructed for collecting of cones from full-size trees, they cannot be applied on a large scale in Scots pine forests at present. It is expedient, therefore, to establish permanent seed stands in young forests, preventing intensive height growth and favouring their fertility.

The proper selection of suitable young stands is of primary importance. At a young age not all quality traits of the trees are visible, and it is almost impossible or very difficult to judge them, therefore indirect characteristics must be taken into account. The selection of permanent seed stands may be based on their origin. Natural regenerations on sites where previously high quality stands were growing and where the neighbouring stands are at least above average, are suitable for that purpose. A planted stand can be taken into consideration only if its origin form a local superior or above-average stand is documented. An age between 5 to 10 years is the most suitable for stand selection because at this age the lower branches are still alive and simple techniques can be applied for cone harvesting over a long period.

The measures to be taken in the designated seed stands are aimed at

— the genetical improvement of the stand;
— increasing the seed crop;
— creation of favourable conditions for cone harvesting.

At first large-scale thinnings are undertaken and later repeated 3–4 times. The rogueing procedure has to be adjusted to local conditions. For example, in planted stands two adjacent rows are left to grow during the first rogueing and depending on the distance between the rows, 3–4 rows are cut to obtain a cleared strip of 6–8 m wide. The strips may be cleared gradually, by cutting one or two strips at a time. The strips ensure the spatial order of the remaining seed trees which has a favourable effect on the insolation of crowns and allow the movement of machines. In the remaining rows a selection thinning is carried out based on the phenotype. In the taiga forest zone the rogueing is repeated every 3 to 5 years, in more southern regions of the pine area every 2–4 years. The last rogueing should leave 200 to 300 trees per hectare (Veresin et al., 1985). Trees of superior growth, with a straight stem, a narrow crown, and of high vitality and resistance are selected to grow as seed trees. Trees with several leaders, curved stems, damaged by diseases and insects, with too long and too thick branches as well as the natural regeneration of other tree species are cut first.

The subsequent rogueings are carried out based upon the main phenotypic characters such as height growth and crown diameter, leaving trees of more rapid growth and narrower crowns. In case of similar growth traits, trees with a higher number of female strobili are favoured. The so-called indirect traits (darker and longer needles, lower number of branches per whorl, longer life-time of needles, 3-needle bundles on leading shoots, etc.) which suggest a better growth in the future, are also considered. If desired, resin yield may be included in the selection criteria. For the assessment several methods have been worked out (Prokazin, 1963; Pihelgas, 1970).

In naturally grown stands rogueing is carried out evenly over the whole area or strips of 6–8 m width are cut, leaving tree belts up to 4 m wide in between. The belts are, in turn, rogued further based on the phenotype of the trees.

To increase the seed crop and to facilitate the collecting of cones, various measures can be applied after the rogueings: crown formation, soil cultivation and herbicide application, sowing of perennial lupine, mineral fertilization, and control of insects and diseases of cones and seeds.

Crown forming should inhibit height growth and enlarge the size of the crown. After the rogueings are finished, the top growth of 2–3 years is cut off (Veresin et al., 1985). If necessary, the topping should be repeated during subsequent years. N or NPK fertilizers are used depending on soil conditions.

Selection seed stands

Exceptionally good pine stands are managed as selection seed stands (Pihelgas, 1970). The purpose of their maintenance is not so much the seed growing but the conservation of the valuable gene pool and its utilization in breeding. The stands must be of high productivity with a volume yield of at least 60% above average. The volume yield should be maintained after the removal of inferior tress. The majority of trees must be of high quality, straight stemmed, well self-pruned, with a relatively narrow crown, thin branches and resistant to diseases and insects. Selection seed stands are excluded from forest management for timber. In course of the selection plus, normal and minus trees have to be marked and the latter are removed together with all other tree species. Grafts are taken from selected plus trees to establish seed orchards. If light conditions are sufficient, the soil is cultivated in the stand gaps to achieve natural regeneration. The seedlings of the natural regeneration and the seeds collected from standing trees are utilized for the establishment of permanent seed stands and experiments. To reduce the influx of undesirable pollen the surrounding minus stands and minus trees are eliminated.

Plus trees

Introduction

Virtually all countries cultivating Scots pine on a larger scale have been engaged in the selection of plus trees and their vegetative propagation to establish seed orchards (Lindquist, 1948; Larsen, 1956; Rohmeder and Schönbach, 1959; Pihelgas, 1961; 1962; 1963; 1968; 1977; Kärki, 1963; Prokazin, 1963; Oskarsson, 1972).

Tree selection is based on various classification criteria. According to the grouping most widely used, the phenotypes are qualified as plus, average and minus trees. Such rating of trees is rather subjective in spite of numerous selection instructions and criteria (Lücke, 1963). Owing to the extensive range of Scots pine and the varying growing conditions the selection criteria greatly differ in various regions. Requirements for concrete characters of plus trees can be set only for smaller regions taking into account the local conditions. Stands of high quality and of high productivity at an age of 60–80 years are best suited for the selection of plus trees. A whole complex of criteria is considered in the course of selection: growth rate, stem and wood quality, self-pruning, shape and arrangement of branches, size of crown, resistance to diseases and insects, hardiness, etc. The rating of trees is relative and in most cases average trees abound. Minus trees include first of all the slow growers belonging to classes III–V (according to Kraft) and regardless of the growth rate the trees with the following defects: crooked or forked stem, wide crowns, heavy, thick branches, poor self-pruning and infection with diseases (Pihelgas, 1986).

Growth traits

In Estonia a plus pine must exceed the average stand height at least by 10%, and in case of even-aged stands by 15%. Analogous criteria have been set up in various other regions. In general, the height of a plus tree should exceed the average by twice the standard deviation (H + 2s; Kotov, 1982). Apart from a comparison with the stand average, in order to limit the effect of the microenvironment, plus trees must also be compared with the neighbouring trees. The height of a plus tree must exceed that of neighbouring ones at least by 10%.

According to several criteria and proposals in the USSR, the DBH of a plus tree must exceed the stand average and that of neighbouring trees by 20 to 30% (Girgidov et al., 1977; Veresin et al., 1985). We are of the opinion that there is no need for such a requirement as diameter is strongly affected by light conditions. Very thick trees have often a conical stem, thick branches, a broad crown and often inferior wood quality.

Stem quality

The most valuable part of a plus tree is of course the stem. Stem quality is determined by the form factor, straightness, self-pruning, thickness of branches, etc. No absolute criteria can be given for the mentioned traits, since they greatly depend on the regional growing conditions and on the age of the stand (Bolland, 1965). In any case the stem of a selected plus tree must be straight. In mature stands, leaning of trees in the direction of prevailing winds is not an obstacle to their selection since this defect is evidently not connected with inherited characters. Trees which spiral growth indicated by bark cracks must be avoided.

From the point of view of wood quality self-pruning is of essential importance. Self-pruning is greatly influenced by the site quality: on fertile soils self-pruning is always considerably slower. However, well-pruned trees and stands can also be found on fertile soils and, in reverse, poorly self-pruned ones on sites of low fertility. This indicates that self-pruning does not only depend exclusively on growing conditions but also on inheritance.

The height of the self-pruned stem section is increasing with age. It is followed by the medium section from the first dry branch to the beginning of the living crown. The living crown part itself makes up 25–30% of tree height in older trees and up to 50% at medium age. The longer the clean bole and the shorter the unpruned part with dead branches, the more valuable is the stem.

Wood quality

According to the studies carried out in Estonia, trees with relatively long crowns have lower wood density. The same can also be said for trees with a higher extension of coarse bark on the bole (Pihelgas, 1986). Thus, to some extent, it is possible to assess wood quality of pines by the phenotype. For the more detailed analysis of wood properties increment cores are used to determine ring widths, precentage of latewood and approximate wood density.

Percentage of latewood is important: the higher, the better the wood quality. In Estonian plus pines an average of 44% was measured (Pihelgas, 1986).

Crown shape and size

One of the most essential indices of the growth energy and quality of a tree is its crown; its diameter, length, density, thickness and length of branches, branching angle, colour and persistence of needles. The shape and the traits of the crown are greatly influenced by the geographical location, exposure and site conditions. Crown forms, however, show a great variability also under uniform environmental conditions and the basic crown shape of races persists after transfers from one geographical region to another, indicating that the crown form is genetically determined.

Cone-shaped, pointed crowns indicate intensive, continuous height growth, therefore these genotypes are of great value (Pravdin, 1963; Pihelgas, 1970). The crown diameter is of essential importance for the evaluation of a tree: the narrower the crowns, the more trees can be cultivated per unit area and the stand volume depends after all on the number of trees. Trees with narrow crowns utilize the area more effectively. In Estonia the requirement for the ratio of crown diamater and DBH is set at 10/1 for better plus trees, the acceptable minimum being 15/1. Compared to height, crown diameter must not exceed 15% of tree height (at 60 to 80 years of age), preferably not more than 10% (Pihelgas, 1970). In trees with narrow crowns the branches are also thinner and their number per whorl is lower. As a result crowns are lighter and stems are quicker to self-prune.

The branching angle is also of great importance. The branches set at an acute angle are usually longer and are exposed to snowbreak damage. According to several studies branch angle is strongly hereditary. In most cases its heritability is higher than for height or diameter (Pöykkö, 1982; Velling, 1982; Velling and Tigerstedt, 1984).

The selection of superior phenotypes and the determination of phenological qualities in situ must be followed by testing their clones and generative progenies (Fröhlich, 1980).

Literature

Andersson, E. (1963): Seed stands and seed orchards in the breeding of conifers. *FAO/FORGEN Conf.* **63–8/1**:1–16.

Bolland, G. (1965): Züchterische Möglichkeiten zur Verbesserung der Schaftform der Kiefer. *Tagungsber. Dtsch. Akad. Landwirtschaftswiss. (Berlin)* **75**:161–165.

Fröhlich, H.-I. (1980): Forstpflanzenzüchtung. Wege und Ziele 1955 bis 1980. *Allg. Forstzeitschr.* **35 (26)**:667–673.

Girgidov, D. Y. and Tolstopyatenko, A. I. (1977): *Sozdanie lesosemennykh plantatsij khvoinykh (Establishment of seed orchards of conifers).* Gosleskhoz, Leningrad.

Kärki, L. (1963): Metsäpuiden jalostuksen nykyvaihe (State of forest tree breeding). *Metsätalondellinen aikakauslehti* **1**:1–2.

Kotov, M. M. (1982): *Organizatsiya lesosemmennoj bazy (Organization of forest seed basis).* Gosleskhoz, Moscow.

Larsen, C. S. (1956): *Genetics in silviculture.* Oliver and Boyd, Edinburgh.

Lindquist, B. (1948): *Genetics in Swedish forestry practice.* Svenska Skogsvardföreningens Förlag, Stockholm.

Lücke, H. (1963): Sind gute Phaenotypen bessere Vererber als schlechte? *Silvae Genet.* **12 (2)**:62–63.

Matthews, L. (1963): Seed production and seed certification *FAO/FORGEN Conf.* **63–8**:1–31.

Molotkov, P. I., Patlaj, I. N. and Davydova, N. I. (1982): *Selektsiya lesnykh porod (Selection of forest trees).* Gosleskhoz, Moscow.

Oskarsson, O. (1972): Suomelaiset plusmännyt ja plusskuuset (Finnish plus pines and plus spruces). *Folia Forest.* **150**:138.

Pihelgas, E. (1961): Tähelepanu männi eliitpuude valikule (Attention should be paid to the selection of plus-trees of Scotch pine). *Sotsialistlik Pollumajandus* **16 (10)**:471–472.

Pihelgas, E. (1962): Hariliku männi plusspuude valik ja nende paljundamine pookimise teel (Selection of plus-trees and vegetative propagation of Scotch pine). *EPA tead. toode kogumik* **23**:159–169.

Pihelgas, E. (1963): *Hariliku männi selektsioon (Selection of the Scots pine).* Tallin.

Pihelgas, E. (1968): Plussmändide valikust Eestis (Selecting plus trees of Scots pine in Estonia). Metsamajandus. Tallin, pp. 24–30.

Pihelgas, E. (1970): *Geneetika ja selektsioon metsamajanduses (Genetics and selection in forestry).* Tallin.

Pihelgas, E. (1975): Dependence of seed and culture properties of Scots pines on site types of mother stands. *Estonian contributions to the international biological programme, VIII.* Tartu, pp. 25–61.

Pihelgas, E. (1977): Kuidas parandada metsapuude omadusi (Ways and means of improving the quality of forest trees). *Eesti Loodus:* **7**:427–435.

Pihelgas, E. (1986): Plussmändide fenotuupilistest tunnustest ja nendevahelistest seostest (About phenotypical properties of plus pines and their interrelations). Metsamajandus, Tallin, pp. 43–75.

Pöykkö, T. (1982): Genetic variation of quality characters of Scots pine. *Silva Fenn.* **16 (2)**:135–140.

Pravdin, L. (1963): *Zadachi i metody sovremennogo lesnogo semenovodstva (Tasks and methods in modern seed growing).* Gosleskhoz, Moscow.

Prokazin, E. (1963): *Uusi meetode id männiseemne kasvatamisel (Some new methods in growing of pine seeds).* Tallin.

Rohmeder, E. and Schönbach, H. (1959): *Genetik und Züchtung der Waldbäume.* Parey, Hamburg und Berlin.

Velling, P. (1982): Genetic variation in quality characteristics of Scots pine. *Silva Fenn.* **16 (2)**:129–134.

Velling, P. and Tigerstedt, P. M. A. (1984): Harvest index in a progeny test of Scots pine with reference to the model of selection. *Silva Fenn.* **18 (1)**:21–32.

Veresin, M. M., Yefimov, Y. P. and Arefyev, Y. F. (1985): *Spravochnik po lesnomu selektsionnomu semenovodstvu (Reference book of forest seed management on genetical basis).* Gosleskhoz, Moscow.

CHAPTER 10

Seed orchards

Cs. Mátyás

Introduction

From among European tree species Scots pine is the one where the seed orchard idea of Syrach Larsen (1936; 1972) proved to be the most successful. At present most countries which grow Scots pine commercially have a seed orchard programme (Table 10.1).

In spite of the relative success with the species, criticism of the orchard concept has not ceased (Kiellander, 1963; Stern, 1972; Schlenker, 1979; Giertych, 1987). Apart from concerns about the reduction of genetic diversity the main objections against seed orchards are their costly operation, irregular seed production and largely uncontrollable genetic contribution of the different clones.

To some extent these shortcomings may be corrected by proper establishment and management practices. Any shortcomings must, however, be compared with other silvicultural options, for example, natural regeneration, seed collections from seed stands, or vegetative propagation. When so doing, expected genetic and financial gains have to be set against the costs of planting stock production and regeneration within the general concept of the silvicultural strategy.

Flowering and yield of orchards

The genetic efficiency of seed orchards depends upon the genetic value and gene contribution of the selected clones, the frequency of selfed embryos and the frequency of embryos fertilized by external pollen. The unequal genetic contribution from clones decreases the effective clone number. Therefore, keeping records of male and female flower production and the phenology of the clones is essential, if it is to be implemented by later selection or rogueing. Similarly, the local pollination conditions should be monitored to indicate the amount of contamination, selfing, etc.

Flowering and yield as clonal traits

Many authors have stated that infraspecific variation in flowering is extremely strong and clonally fixed (Sarvas, 1962; Bánó, 1963; 1969; Baumeister, 1975; Jonsson et al., 1976), though the relation between male and female flowering may change with age (Bhumibhamon, 1978).

Table 10.1. Area of Scots pine seed orchards in some countries (grafted clonal orchards if not otherwise indicated)

Country, source	Area
Belgium (Nanson, 1987**)	1 orchard, area unknown
Canada (Rauter and Beaudoin, 1987**)	15 ha (in Ontario and Quebec)
China (Hong, 1987**)	477 ha
Denmark (Dietrichson et al., 1982)	9 ha
Finland (Dietrichson et al., 1982)	3100 ha
France (Roman-Amat, 1987*)	17 ha
GDR (Schütze, 1988*)	182 ha, incl. 55 ha tested
FRG (1987**)	56 ha, incl. 12 ha tested
Great Britain (Samuel, 1987**)	23 ha, 29 ha abandoned
Hungary (1988***)	140 ha, incl. 54 ha tested
Ireland (O'Driscoll, 1987*)	4 ha, incl. 2 ha tested
The Netherlands (de Vries, 1987*)	10 ha
Norway (Dietrichson et al., 1982)	12 ha
Poland (Kocięcki, 1987*)	210 ha grafted 75 ha seedling
Sweden (Hadders and Samuelson, 1987*)	477 ha, 94 ha being established
Turkey (Boydak, 1987**)	64 ha
USA	several orchards, seedling and grafted, total area unknown
USSR (Novoseltzeva, 1980), *Total:*	5171 ha, 18 500 ha planned
— Chernozem region (Efimov et al., 1987)	80 ha
— Latvia (Pirags, 1987*)	700 ha
— Lithuania (Dreimanis, 1987*)	325 ha
— Ukraine (Molotkov, 1988*)	400 ha

* Personal communication;
** Personal communication through a survey by J. Mikola;
*** The author's unpublished data.

Considerable variation exists in seed and cone characters which are also genetically determined and stable traits (Table 10.2), and may be utilized also for clonal identification (Mátyás, 1983). Similarly the percentage of full seed (Bánó, 1963; Kiellander, 1963) and the abortion rate of flowers and conelets are clonal traits (Zviedre, 1985). The undoubted importance of phenologic differences between clones is somewhat reduced due to differences in anthesis and gynesis between different parts of the crown which are similar in extent to the between-clone variation (Jonsson et al., 1976).

Table 10.2. Correlation between traits of 10- and 30-year-old grafts (calculated from clonal averages of 30 West-Hungarian clones)

Trait	Unit	r	F	Level of sign.
Cone crop	kg/graft	0.834	38.8	xxx
Cone size	g	0.771	25.0	xxx
Seed yield	seeds/graft g/graft	0.845 0.763	42.3 23.7	xxx xxx
Empty seed cont.	%	0.251	1.14	–
Thousand grain wt.	g	0.860	48.1	xxx
Graft diameter	cm	0.361	—	–
Graft height	m	0.784	—	xxx

xxx: P=0.1%

Periodicity and climatic influences

Numerous fruit tree crops are biennial due to the reduction of nutrients in developing buds; for Scots pine the prolonged development cycle of the seed suggests a 3-year cycle. In spite of a broad variation, on favourable sites the 3-year cycle is observable (Fig. 10.1).

Both flowering and cone yields are influenced by weather conditions (Efimov, 1980; Ilstedt and Ericson, 1982). Warm and dry early summers seem to encourage early and profuse flowering (Hadders, 1981). Recent data show that conditions during primordia differentiation, rather than their initiation, have an effect on flowering. According to Efimov (1980) in the year of primordia initiation female flowering is stimulated by warm and dry early summers, while cool weather with higher humidity is favourable for male flower initiation. McLemore (1977) suggests that variations in the size of flower- and cone-destroying insect populations may also contribute to periodicity.

Abscission of flowers and conelets

Abscission of plant parts is generally observable in trees. The abscission of flowers and conelets is a serious limiting factor to the seed yield for Scots pine. The percentage of dropped conelets may reach or even surpass 40% (Dreimanis, 1975; Jonsson et al., 1976; Zviedre, 1985). Similar observations are known for flower and fruit losses in fruit trees (Luckwill, 1970).

Three main periods of abortion are distinguished: immediately after flowering, in early summer, and before seed ripening. In Scots pine the abortion of ovules, flowers, conelets and embryos is mainly attributed to poor pollination conditions

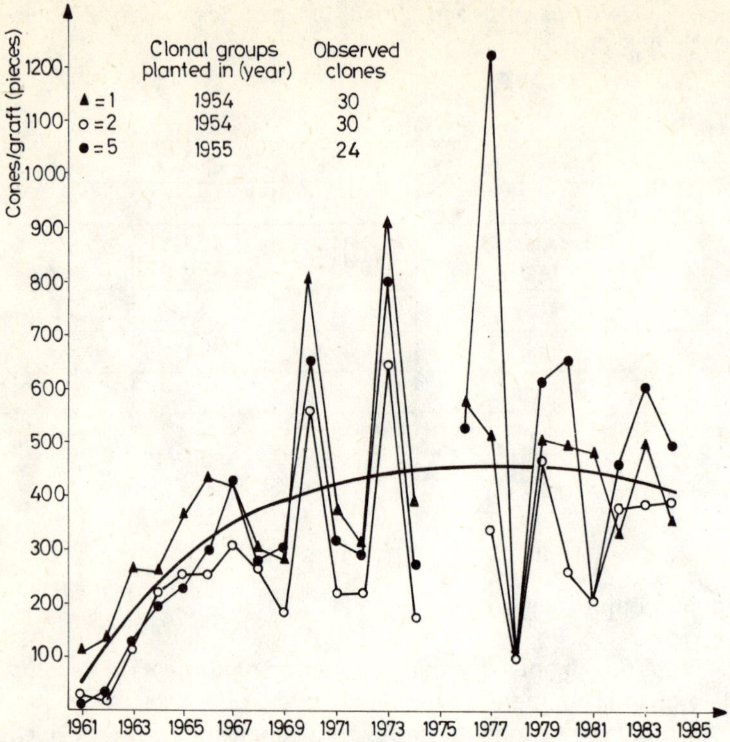

Fig. 10.1. Cone yield of grafts over three decades in a clonal archive in Hungary (averages of clonal groups; 3 ramets per clone) (Mátyás, unpublished data)

and lethality due to selfing (Sarvas, 1962; Koski, 1973, 1975). Insects may cause severe conelet losses, too. A certain level of abortion remains, however, even when pollination conditions are favourable and damages to flowers are excluded. This is especially true of flower losses occurring soon after pollination. It is believed that competition between generative and vegetative organs for nutrients, assimilates and water may be the reason (Priestley, 1970; Dreimanis, 1975; Sweet, 1975). Kurm (1987) observed the least flower losses in the best-nourished upper crown (although the effect of pollination was not investigated). Similar results were observed in an experiment where vegetative buds were removed from branches bearing second-year conelets. The abortion rate of conelets was less than the control, but higher than those which received supplemental pollination (Mátyás, unpublished paper; Fig. 10.2). From this it may be assumed that an improvement of nutrient and carbohydrate supplies to generative organs will limit abscission. Fertilization, the retardation of vegetative growth and partial interception of sap flow by light stem girdling may also improve flower and conelet retention.

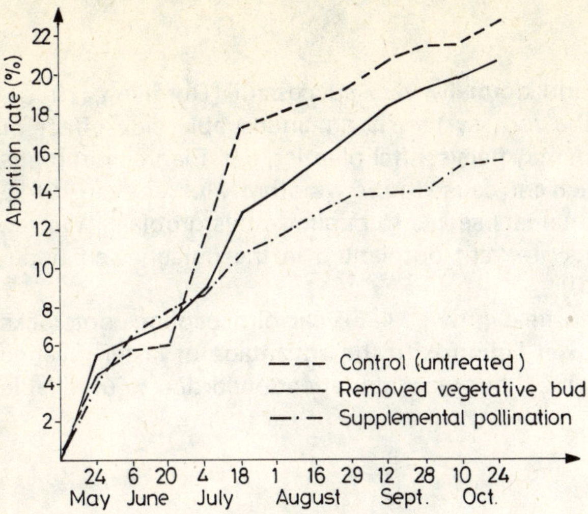

Fig. 10.2. Abortion rate of second-year conelets (Mátyás, unpublished data)

Seed yields

Under climatically optimal conditions of Central Europe, Scots pine seed crops per hectare may surpass 30 kg per annum without any stimulative or protective measures. Further north and east yields are lower. In northern Sweden yields around 15 kg/ha occur in good seed years (Kiellander, 1963). If germinable seed number per area unit is considered, the yield differences are less.

Establishment

Seedling versus clonal orchards

From the genetic point of view, establishment of seedling seed orchards is justified if heritability is low (as for most quality traits) and if the difference in precocity of flowering between grafts and seedlings is negligible. Cost comparisons between clonal and seedling orchards for establishment and maintenance show that they are not markedly different (Mátyás and Rauter, 1987), and so under certain conditions the seedling option may be an unconventional but sensible solution for Scots pine.

Grafting

Grafting Scots pine is simple and normally very successful (Dormling, 1963), although the deformation of the root system is an undesirable side-effect of grafting potted rootstocks which may persist after planting out. Planted out grafts retain the spiral root growth which can cause future windthrow losses even 10–15 years later. Using large-size containers seems to minimize this problem (Weiser, 1985). In some countries grafts are kept outplanted in the nursery bed under polythene cover (Altonen, 1980).

Grafting at breast height onto field-grown, 4–6-year-old seedling rootstocks was widely practised in the Soviet Union with the advantage of faster juvenile height gain (Molotkov et al., 1982), but was later abandoned due to unreliable survival.

Location

Selecting a suitable planting site is one of the main factors which determines the success of an orchard. Slightly elevated locations, protected from cold winds, with a light, well-drained soil of at least medium fertility are preferred (Werner, 1975; Mikola, 1987).

For Scandinavian conditions the following climatic minimum requirements are defined (Ilstedt and Ericson, 1982): a heat sum in July of 50–55 degree days (at 13:00 h., above 20 °C) for the initiation year and 15–17 dry days in June in the pollination year.

Establishing seed orchards on more favourable southerly sites is an obvious way to circumvent some climatic difficulties. Selected sites need not be optimal for growth or within the area of natural distribution of the species. Southward transfer of Scots pine orchards was advocated by numerous authors (Wright et al., 1966; Sarvas, 1970; Molotkov et al., 1982). The expected advantages are:

— reduction of pollen contamination because of fewer external pollen sources of the same species,

— accelerated initial development and increased cone production (Wright et al., 1966; Werner, 1975; Schmidtling, 1984; 1987),

— phenological isolation from local populations of the same species as temperature sum requirements for flowering of different sources are not similar.

The question of reduced contamination was investigated by Koski (1987). His results showed that even long distance transfers, for example from Finland to Hungary (2 000 km), do not provide for complete isolation. In Sweden a cone yield decrease was observed if the grafts were transferred southward further than 1 200 km from the origin of the parent trees. Sarvas proposed a transfer southward of 200–300 degree days (above 5 °C) difference. Transfers too far south may affect the chilling requirements of the species and could cause physiological upsets as observed with southern pines in the USA (Schmidtling, 1987). Effects of changed photoperiod and temperature conditions on the seed ("Lamarck-effects") as reported by Björnstad (1981) are not fully understood but lasting genetic effects are improbable. Lindgren and Quiyu (1986) found that the locality of seed maturation is of low importance.

In spite of all difficulties the option of southward transfer has to be considered seriously.

Isolation

Contrary to earlier beliefs effective isolation of orchards is often a serious problem as Scots pine stands produce large amounts of background pollen. The arbitrarily recommended isolation distance varies, between 100 (Koski, 1980) and 1 500 m (Anon., 1987). In areas with extensive Scots pine forests even larger distances will not provide sufficient isolation. "Protective courtains" planted around the orchard with broadleaved species may help to filter out some of the contaminant pollen.

Based on accumulated information on pollen flight and mating conditions within orchards, requirements for isolation may be formulated more precisely, though not conclusively, as follows (Koski, 1980; Hadders, 1981):

— in order to achieve reasonably well-matched flowering, the heat sums of the original habitats of the clones should not differ by more than 140 degree days (above 5 °C). (In addition to securing panmixia, phenological uniformity is also important for any hormone treatment, supplemental pollination and pest control measures);

— self-pollination should not exceed 20%;

— pollen production should reach 10–20 kg/ha for commercial harvests. This means 500 or more male strobili per graft;

— background pollen influx should be less than 20%;

— the minimum size of the orchard should be 5 ha.

Planting designs

A wide variety of planting layouts exists (summarized by Giertych, 1975). More recently breeders have become increasingly aware that between-clonal differences in their contribution to the generative potential of the orchard renders the layout of lesser importance. To minimize the occurrence of selfing, a minimum distance of 25 m between ramets of the same clone is now recommended (see also Chapter 11).

Ideally, designs should reflect the variation in the genetic potential of each clone, and so the ratio of outplanted ramets per clone should be in inverse relation to its genetic contribution (both male and female) to the orchard seed.

The design should allow for at least one mechanical thinning at which every second plant is removed.

depends on whether its a tested or untested orchard

Spacing

Planting distance determines the age when crowns meet. In clonal orchards with square spacing, crown closure is reached at the following ages (in Central Europe):

Spacing: 3×3 4×4 5×5 6×6 8×10 m
Age: 9–11 13–15 16–18 20–22 25< years

Because of increasing costs of both grafts and thinning operations, current spacings are wider than previously. In Sweden, for example, the cost of mechanized removal of an orchard tree was around 3–5 US$, whereas a new graft costs approx. 10 US$/piece (Alfjorden, 1979). For similar reasons the interplanting of orchard rows with precocious clones was unsuccessful because of the high costs (Bánó et al., 1972). In a spacing study, both per hectare and per graft cone yields were optimal at a spacing of 8×8 m, already at the age of 10–12 years (Table 10.3).

Table 10.3. Kiln-dry average weight (grams) of cone crop per graft in a Hungarian spacing experiment (G. Varga, personal communication)

Growing space (m^2)	Age from outplanting (years)				
	6	7	8	9	10
4	75	197	159	479	999
16	128	416	492	1550	2896
36	171	723	974	2999	4067
64	185	828	971	3234	5289
100	119	444	582	2341	4869
144	96	342	540	2128	4125

Orchard maintenance

Maintenance activities should aim to secure optimal seed yields, for the lowest costs. In general, an abundant crop requires a good health condition and a favourable nutrition status of the trees which are also preconditions for vigorous vegetative growth. Applied measures have to exert their effect through
— slowing down vegetative growth,
— encouraging the nutrient supply to terminal shoots and buds bearing generative organs,
— increasing the initiation of flower primordia, and
— prevention of abortion.

Soil cultivation

Cultivation measures should maintain conditions favouring growth and flowering and reducing competition for water and nutrients. Especially in young orchards the elimination of weed competition (either by cultivating, herbicides or mulching) is important. Rosvall and Andersson (1982) observed better growth and significantly higher cone crops in herbicide-treated or harrowed plots, while grass mowing and no treatment at all had the same effect. Systemic herbicides (e.g., atrazine, amitrol) may be used in Scots pine orchards with necessary care. It has to be remembered that grafts have their roots closer to the surface than seedlings and are therefore more sensitive (Georgi et al., 1984). Soil cultivation has positive effects also in preventing the overwintering of insects in the soil and disturbing the habitat of voles. Also, intercrops attracting voles (e.g., alfalfa) should be avoided.

Pruning

Crown shaping of Scots pine seed orchard grafts is a controversial subject. Through topping many future flowering points are lost. Scots pine does not respond well to shaping and usually one of the branches of the remaining top whorl takes over the role of the leader. According to Toda and Akasi (cit. by Werner, 1975), a lanoline paste application of 8% IAA inhibits the upward development of branch tips.

Other side-effects of pruning are: more difficult collection by traditional climbing, higher risk of snowbreak and faster crown closure.

It follows that both too frequent and too vigorous pruning leads to unsatisfactory results. The removal of the top 3 whorls seems to be the most appropriate measure. Limiting height growth should start after the onset of female flowering at around age 10. Topping should be repeated every 3–5 years. Experiments have proved that cone yields of pruned trees recover after 2–3 years to 80–90% of the crop of unpruned trees (Hellström and Werner, 1984). In addition to topping it is advisable to cut back any vigorous side branches (especially in the upper crown) to retain a more regular crown form and to prevent an early dieback of lower branches (Baumanis and Ronis, 1985).

To avoid a sudden slump in cone yields, pruning can be restricted to a part of the orchard (one half, one third each year). Provided the next pruning is 0.5–1 m higher than the previous one, starting at 5 m at age 10, trees can be kept below 10 m height during the life-span of the orchard.

Thinning

Genetic thinnings in Scots pine clonal seed orchards are less frequent than anticipated. The main reasons are: relatively low initial clone numbers in many older orchards and restrictions imposed by spatial arrangement. The latter problem may be solved by transplanting. Werner (1982) reported successful transplantation of Scots pine grafts up to 5 m high using tree transplanting machines at acceptable costs.

Reductions in cone yields after thinning generally last 3–4 years. There is a temporary decline in male flowering too because of the improved lighting of the lower branches.

In areas prone to root rot *(Heterobasidion annosus)* the prophylactic treatment of stumps with a culture of the antagonistic fungus *Peniophora* should be applied immediately after felling!

Supplemental mass pollination (SMP)

For Scots pine this method is both technically and biologically possible on a practical scale. Male strobili are collected shortly before pollen is shed. One male-flowering branch may yield 50–60 mg of pollen. For the SMP of one hectare of orchard approximately 1 kg of pollen is required and this quantity may be collected by 5–7 workers in one day. Prior to use the pollen may be diluted with talcum powder in the ratio 1:1 (Hadders, 1984). Because fertilization occurs on a "first come, first served" basis, SMP should precede general anthesis in the orchard.

SMP is most profitable when pollen from good general combiners is used in order to produce seed of high genetic value. The construction of isolating foil houses where SMP may be applied was initiated in Sweden, where Hadders (1984) calculated the additional cost of SMP per plant to be only 0.12 ore (= 2 cents US).

Fertilization

The conclusion from numerous, partially contradictory experimental results is that mineral fertilization is only effective if precipitation is sufficient, and weed competition is eliminated. Although the amount of nutrient uptake is moderate in Scots pine orchards, significant stimulation of flowering has been achieved even on sites where the soil nutrient supply seems to be more than sufficient (Mátyás, 1979).

Application of fertilizers should be preceded by an analysis of the topsoil (approx. 60 cm deep). The nutrient content (mg/g) may be converted into kg/ha data. Fertilization is required if soil nutrient content falls below the following minimum levels (Efimov et al., 1987):

N: 100 t/ha humus
P: 130 kg/ha
K: 200 kg/ha

Outward signs of nutrient deficiency on Scots pine needles are seldom seen in orchards. The normal nutrient content in foliage (on vigorously growing branches, collected in winter) should reach values of 1.6–2.0% for N, 0.15–0.18% for P and

0.4–0.5% for K. If both foliar and soil nutrients are near minimum, fertilization is justified.

The N content of graft needles is relatively higher than that of seedlings (Rosvall and Untinen, 1982). Even so, N fertilization is recommended as a first priority. Combined (NPK) fertilizers did not result in convincingly better flowering than N alone (Mátyás, 1979, Wesoły et al., 1987). Mátyás found, that P and K application improved the general condition of the grafts and reduced conelet abortion. It is believed that ammonium nitrate is the best N source for pines.

Timing of fertilization seems to be unimportant (Wesoły et al., 1987). Herbicide applications or soil cultivation are, however, essential for positive results. The response of grafts is clone-specific; poor flowering clones are less responsive (Mátyás, 1979).

Flower stimulation with biologically active substances

Unreliable response and comparatively high costs still limit the large-scale application of gibberellins or other hormones in seed orchard management. According to calculations of Luukkanen (1980) the net cost of treating 12-year-old Scots pine grafts (800 mg GA/year/graft) amounts to 40 FMk = 10 US$ per graft. Compared to the value of seed, this is still unacceptably high for operational use.

An economically feasible stimulation method was worked out in the (former) GDR which is in use on an operational level. The patented procedure brings yield increases up to 100% (Matschke et al. 1981; Sachert et al. 1981). During successful application of the method in Hungarian seed orchards the clonal response was found to depend on the initial flowering ability and phenology. An increased abortion rate of flowers and cones was observed on the treated trees (Fodor and Mátyás, 1982) (see Chapter 13).

Establishment and maintenance of seedling seed orchards

Initial spacing distances depend on the planned rogueing (selection) age. Data on selection efficiency support the view that optimum selection age is not much above 10 years (Mátyás and Rauter, 1987).

Scots pine seedling orchards are most widespread in Poland, where 75 ha were recently planted at 2.5 × 2.5 m. First thinnings are carried out at age 6–7; by age 25 stem numbers are reduced to 300/ha and 200/ha, respectively, after age 35 (Anon., 1987).

In Middle Russia 1 × 5 m initial spacing is recommended; this is reduced to 5 × 5 m by age 15 to 18; around age 20 every second row is removed. Subsequent thinnings are carried out only in favour of well-formed, well-flowering trees. The planting of clusters of 5 trees at 1 × 1 m and at a cluster spacing of 5 × 5 m was not successful; thinning is too early for reliable selection and costs are relatively high (Efimov et al., 1987).

Protection

Insects cause most damage in Scots pine seed orchards. Damage by mammals (particularly by voles) are more pronounced in northern Europe; showbreak threatens many orchards in milder areas.

Concept of insect control

The entomofauna of seed orchards is influenced by specific conditions. Higher cone crops, better developed and insolated crowns, the finer bark of grafts attract insects, but better health conditions make orchard trees more resistant than forest-grown crops. Good management, e.g. soil cultivation, removal of diseased trees, etc. discourages insect attacks.

Specialized cone and seed insects have relatively stable populations, and so in poor crop years damage is greater. In barren years a diapause is observed (Vilks, 1985). Constant pest control is therefore unnecessary and measures should be aimed at reducing population size of the main pests and preventing the restoration of the population.

Effective control of insects presupposes a monitoring system of the entomofauna and of the potential cone crop. "Life tables" showing the development of flower primordia into cones with all the losses occurring are necessary to assess the feasibility of action to be taken.

Table 10.4. Effect of stimulative and control measures on the cone crop of a 50-ha seed orchard (Falco Timber Co. Hungary) (Fodor and Mátyás, 1982)

Year	Average age (yr)	Cone yield (tons)		Remark
		estimated*	collected	
1977	10	200	211	No measures
1978	11	220	284	Pest control
1979	12	250	510	Control + hormonal stimulation
1980	13	380	915	As above

* Without control and stimulation measures.

Crops achieved without protection are only a small fraction of the biological potential of orchards. Prevention of primary and secondary insect damage contributes to the faster development of a fertile crown. Combined with stimulation treatments insect control may bring significant crop increases and savings in collection costs (Table 10.4).

Primary damage on flowers, conelets and cones by specialized insects

Flower and conelet damage is very difficult to detect; a number of less conspicuous insects are probably still unrecorded.

Polydrosus sp.: Fodor and Mátyás (1982) observed the previously undescribed damage by beetles feeding on female strobili. A loss of up to 10% was recorded.

Dioryctria abietella Schiff.: Hatching larvae feed on conelets and scales of second-year cones, seeds are destroyed later.

Dioryctria mutatella Fuchs: In Hungary damage was registered on conelets and seldom on buds. Attacked conelets abort. Damage has also been reported from the USSR.

Pissodes validirostris Gyll. causes minor external damage on buds and shoots, imagos also feed on cone scales and destroy developing seeds. Larvae feed inside the cone, in later stages destroying the axis, causing brown discolouration of the cones. One larva can destroy up to 60% of seeds in a cone. Infested cones do not open when kilned. Vilks (1985) reported up to 30% damage in Latvia.

Main secondary pests

A number of other non-specific insects often common in Scots pine stands may cause losses of generative organs and strobili-bearing branches through direct or indirect damage. Their presence depends on local infestation conditions and on the level of management.

Petrova resinella L. was originally described as a minor pest; recently serious infestations were reported in Baltic and Central-European orchards.

Rhyacionia sp. are sporadically reported in orchards. In Hungary *Rh. buoliana* is one of the main pests, feeding not only on shoots and buds but also on conelets and cones (Fodor, 1976).

For controlling the two main pests, *Pissodes* and *Dioryctria* sp. spraying with dimethoate (Annila, 1973), trichlorfon or formothion (Fodor and Mátyás, 1982) are recommended. It has to be pointed out that favourable site conditions diminish the necessity of chemical control to a significant extent. Table 10.5 and Fig. 10.3 give a summary of the damage types and periods when insect pests occur.

Cone collection

The method and cost of collection are determined by the height of trees, and the quantity and distribution of the cone crop in the crown. The fertile crown sector (which is clonally determined) shifts steadily upwards, making collecting increasingly expensive with age (Fig. 10.4).

Up to a height of 4–5 m, collecting can be done from the ground; at heights between 6–9 m, 50 to 74% of the crop has to be collected from ladders or platforms. Collecting times increase by 30–40% for every metre increment.

Table 10.5. Grouping of seed orchard insects by damage types (Mátyás, 1979)

Species	Indirect damage				Direct damage			
	first-	second-	shoots, buds	female flowers	female flowers	first-year conelets	second-year cones	buds, shoots
	year shoots							
Polydrosus sp.					+			+
Pissodes validirostris				+			+	+
Magdalis sp.	+	+		+				+
Blastaphagus piniperda	+	+	+	+				+
Petrova resinella	+	+	+	+		+		
Rhyacionia duplana				+				+
Blastethia turionana				+				+
Rhyacionia buoliana				+		+	+	+
Dioryctria mutatella				+		+		+
Dioryctria abietella				+		+	+	+

of disproportionate importance!

Seed is physiologically mature in early autumn, but if collecting starts in the second half of October, a storage or drying period of at least a month is necessary before kilning. Beginning from late November, cone moisture content falls sharply making handling and extraction simpler (Cram and Lindquist, 1979).

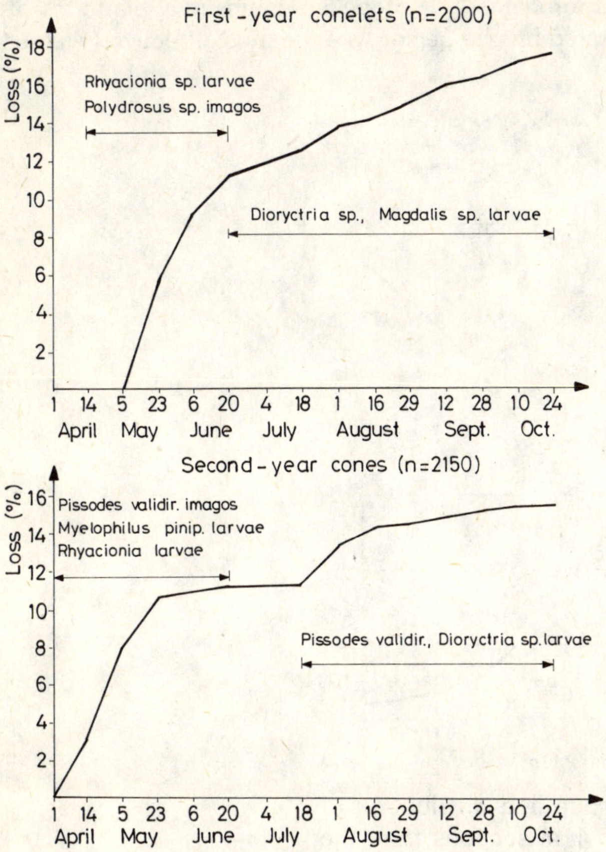

Fig. 10.3. Conelet and cone losses caused by insects in a 10-year-old seed orchard (Fodor and Mátyás, 1982)

Technical answers

Machines for collecting seeds or cones from the ground are out of question, since cones or seeds cannot be shaken from Scots pine trees (Kellison, 1975). Furthermore the natural seed release is spread over several months (80% of the crop falls during one month in spring), and quantities per unit area are low; the seeds are small, difficult to recover and are easily swept away by wind. In a study only 14%

of a seed crop was recovered from the ground; bird predation contributed to this (Bánó and Mátyás, 1976).

To harvest cones from higher branches, unadjustable platforms operated between rows are inefficient since only a small proportion of cones can be reached (see Fig. 10.4). Hydraulically adjustable platforms provide the best solution. According to tests in Swedish orchards, self-propelled commercial platforms of medium-lifting capacity, operated by one person, are the most effective (Werner and Petterson, 1980).

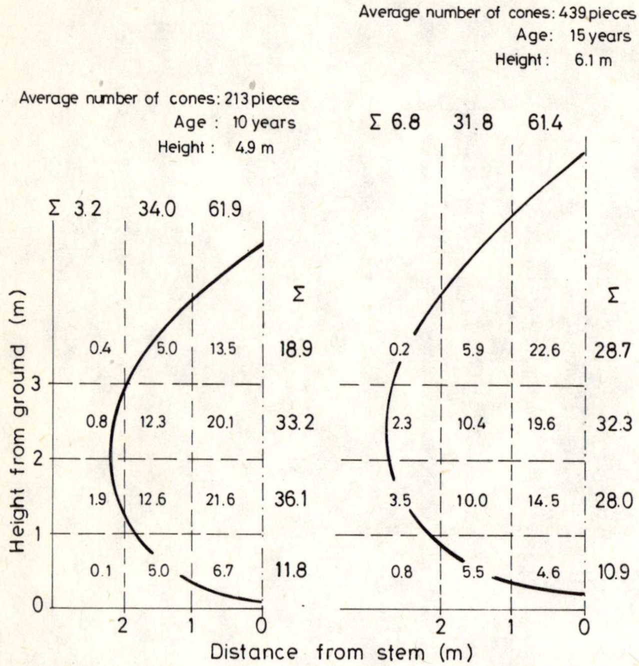

Fig. 10.4. Average crown shape and distribution of cones on unpruned grafts in 1 m sections from the ground (in per cent) (Mátyás, 1979)

Effectivity greatly depends on cone crop quantity and the skill of the picker. Average collected quantities per person per hour are given in Table 10.6.

Specialized machinery for cone collections can only be efficiently used provided alternative uses are available outside the collecting season.

Economics

Seed orchards are costly in terms of time, space and money. Although production orchard systems have been subjected to economic analysis in horticulture, forest seed orchards have rarely been subjected to such treatment.

Table 10.6. Average collected cone quantities per hour and person (in kg of cones)

	Orchard age (years)			
	10*	15*	20*	20**
From the ground	7.7	8.7	8.7	4.5
From ladders	2.8	4.6	4.9	
From vertically adjustable platform	2.9	8.8	8.2	4.8
From vertically and horizontally adjusted platform	3.6	10.8	11.2	8.5

* Data from Inst. f. Sogksförb., Sweden (unpublished);
** Bánó and Mátyás, 1976, Hungary.

Accurate economic analysis is difficult due to the long-term nature of cost and benefit effects and especially when realized genetic gains are considered. Feasibility of seed orchard activities should, however, be investigated without calculating genetic gains. Once an orchard programme has been launched, its profitability is directly influenced by management decisions on its life-span and the degree of maintenance.

Lack of economic analyses is also a result of the widespread opinion that orchard costs must be accepted as an integral part of silvicultural activities, and because they represent only a small fraction of the forest regeneration costs. This is, however, only true if the genetic improvement of the orchard seed justifies the investment. If genetic gains are low, or the site conditions of the potential regeneration areas are not favourable enough, the use of seed stand seed is a valid option.

Apart from the capital investment itself, decisions on management practices cannot be judged without some kind of accountability. Management costs, compared to investment and collecting costs, are more easily adjusted to current conditions. Knowledge is now accumulating to enable sound decisions to be made as to which intensity of management will give a sufficient increase in seed yield to justify the additional costs. Obviously similar quantities of seed may be produced on smaller, more intensively managed areas, or on larger, less well-cared-for ones. There is, however, also a lower limit to the size for a seed orchard to secure effective pollination conditions. From this point of view larger, less-intensive orchards might be favoured.

The economic turnover period (from establishment to liquidation) is a function of the unit production cost of the seed and of the expected financial return from the stands grown from the seed. The first is relatively easy to assess, the second is not. Production costs include establishment (investment) and seed collection costs. Maintenance costs may be disregarded in this context, because they are relatively independent of orchard age and have to be incurred in orchards of any generation.

Orchard investment costs per unit weight of produced seed decrease as an orchard ages until at a certain point they start to soar again due to the compound-

Table 10.7. Non-maintenance (investment + harvesting) costs for various life-spans in Scots pine seed orchards (in CDN $) (Mátyás and Rauter, 1987)

Life-span (years)	Investment + interest ($/ha)*	Total harvested cones (kg/ha)	Capital	Collection	Total
			costs per kg of cones		
1–10	32 578	1 500	21.72	2.20	23.92
1–15	41 579	6 050	6.87	4.05	10.92
1–20	53 066	14 250	3.72	5.10	8.82
1–25	67 727	23 700	2.86	6.25	9.11
1–30	86 439	29 350	2.95	9.40	12.35
1–35	110 313	33 850	3.26	10.00	13.26

* CDN $ 20 000/ha initial investment, 5% interest.

ed interest burden coupled with diminishing cone yields. Collection costs rise also with the age of the orchard (Table 10.7).

By combining the direct costs of collection and the proportion of capital costs per unit weight of collected cones, some indication may be obtained as to the most effective life-span period of the orchard. In the example presented in Table 10.7, the economically effective life-span is rather short and falls between 20 and 25 years. With higher interest rates, the economic life-span would be even shorter. Another factor influencing the turnover period of the orchard is the progress of the improvement work, which may render a productive orchard obsolete earlier than the planned life-span.

Literature

Alfjorden, G. (1979): *Maskinell gallring i tallplantager (Machine thinning in Scots pine seed orchards).* Inst. f. Skogsförb. Inf. 1979/80 No. 1, p. 2.

Altonen, M. (1980): *The techniques applied in Finland in mass production of grafts for pine seed orchards.* Found. For. Tree Breed., Inf. 2/1980, p. 4.

Annila, E. (1973): The biology of *Pissodes validirostris* Gyll. and its harmfulness, especially in Scots pine seed orchards. *Comm. Inst. For. Fenn. (Helsinki)* **85**:6.

Anon. (1987): Wytyczne w sprawie zakładania i prowadzenia plantacyjnych upraw (Guidelines for the establishment and management of seedling seed orchards). IBL Warsaw, p. 16.

Bánó, I. (1963): Ermittlungen über Fruktifikation und Samenertragsfähigkeit an Kiefernpfropflingen. *Tag. Ber. D. A. L. (Berlin)* **69**:103–128.

Bánó, I. (1969): Scots pine clone tests for determining seed growing value. *Erdészeti Kutatások (Budapest)* **65 (1)**:118–129.

Bánó, I. and Mátyás, Cs. (1976): Nemesített magtermesztési rendszer (Improved seed production system). Res. Rep. No. 1, NYFK Timber Co., Szombathely, (unpublished report).

Bánó, I., Mátyás, Cs., Retkes, J. and Szőnyi, L. (1972): Planning and establishment of Scots pine seed orchards in Hungary. Part II. *Erdészeti Kutatások (Budapest)* **68 (2)**:129–145.

Baumanis, I. and Ronis, E. (1985): Formirovanie kron privitykh derev'ev na semennykh plantacijakh khvoinykh (Crown pruning of grafted seed orchard trees). In: *Semennie plantacii v lesnom semenovodstve (Seed orchards in forest seed production).* Zinatne, Riga, pp. 36–42.

Baumeister, G. (1975): Möglichkeiten der Früherkennung quantitativer Saatgutertragsleistungen bei Klonen von *Pinus sylvestris* L. in Samenplantagen. *Silvae Gen.* **24**:5–6, 177–187.

Bhumibhamon, S. (1978): Studies on Scots pine seed orchards in Finland with special emphasis on the genetic composition of the seed. *Comn. Inst. For. Fenn.* **94 (4)**:118.

Bjornstad, A. (1981): Photoperiodical after-effects of parent–plant environment in Norway spruce seedlings. *Medd. Norsk. Inst. Skogsforsves.* **36**:1–30.

Cram, W. H. and Lindquist, C. H. (1979): Maturity of Scots pine cones. *For. Chron., Oct.* :170–174.

Dietrichson, J., Feilberg, L., Hadders, G. and Koski, V. (1982): Fröplantageverksamheten i de nordiska landerna (Situation of seed orchards in Nordic countries). In: *Fröproduktion in nordiska fröplantager (Seed production in Nordic seed orchards).* Part II. Rapp. och Upps., Uppsala, No. 34, p. 32.

Dormling, I. (1963): Anatomical and histological examinations of the union of stock and scion in grafts of Scots pine and Norway spruce. *Stud. For. Suec.* **13**:132.

Dreimanis, A. (1975): Izmenchivost' opada zhenskikh strobilov sosny obykhnovennoi (Variability in female strobilus abortion of Scots pine). In: Pirags, D. M. (Ed.) *Geneticheskoe issledovanie drevesnykh v Latvijskoi SSR (Genetic investigation of forest trees in Latvia).* Zinatne, Riga, pp. 89–93.

Efimov, J. P. (1980): Some results on the regularities of seed crops in Scots pine seed orchards. *Silva Fenn.* **14**:1, 63–70.

Efimov, Yu. P., Beloborodov, V. M. and Stebakova, V. N. (1987): Rekomendacii po sozdaniyu, expluatacii i povysheniyu urozhainosti semennykh plantacii sosny obyknovennoi v CChO (Recommendations for the establishment, utilization and yield increase in Scots pine seed orchards of the Central Chernozem Zone). N. O. T. Minleskhoz RSFSR, Moscow, p. 24.

Fodor, S. (1976): Biologia i fenologia pobegovyuna zimuyushchego (Biology and phenology of the pine shoot borer *Rhyacionia buoliana* Schiff). *Erdészeti Kutatások (Budapest)* **72 (2)**: 55–61.

Fodor, S. and Mátyás, Cs. (1982). Magtermésfokozás erdeifenyő magtermesztő ültetvényekben (Increasing seed yield in Scots pine seed orchards). Hung. Ac. Sci. Veszprém (unpublished rep., p. 94).

Georgi, E., Bolland, G. and Wenzel, W. (1984): Rationalisierung der Samenplantagenbewirtschaftung durch kombinierte Verfahren der Bodenpflege. *Beitr. Forstwiss. (Berlin)* **18 (2)**:55–58.

Giertych, M. (1975): Seed orchard designs. In: Faulkner, R. (Ed.) *Seed orchards.* For. Comm. Bull., London, No. **54**, pp. 25–37.

Giertych, M. (1987): Seed orchards in crisis. *Forest Ecol. Manag.* **19**:1–7.

Hadders, G. (1981): Isolering av fröplantager (Isolation of seed orchards). *Inst. Skogsförb. Inf.* 1980/81, **3**:4.

Hadders, G. (1984): Supplemental and controlled mass pollination in seed orchards of Scots pine. *Inst. Skogsförb. Arsb.* 47–64.

Hellström, C. and Werner, M. (1984): Beskarningsförsök i twa sydsvenska tallfröplantager (Pruning experiments in two south Swedish Scots pine seed orchards). *Inst. Skogsförb. Inf.* 1984/85, No. **6**., p. 3.

Ilstedt, B. and Erikson G. (1982): Orsaker till skillnader i fröproduktion mellan plantager och ar (Reasons for seed crop irregularities between years and seed orchards). *Inst. Skogsgen. Res. Notes,* No. **33**, p. 83.

Jonsson, A., Ekberg, I. and Ericson, G. (1976): Flowering in a seed orchard of *Pinus sylvestris* L. *Stud. For. Suec.* **135**:38.

Kellison, R. C. (1975): Cone and seed harvesting from seed orchards. In: Faulkner, R. (Ed.) *Seed orchards.* For. Comm. Bull., London, No. **54**, pp. 101–107.

Kiellander, C. L. (1963): Ertragsverhältnisse schwedischer Kiefern- und Fichtensamenplatagen. *Soz. Forstw., Sond. Heft Samenplantagen* **III**:1–4.

Koski, V. (1973): On self-pollination, genetic load and subsequent inbreeding in some conifers. *Metsatiet. Tutkimusl. Julk.* **78**:10, 42.

Koski, V. (1975): Natural pollination in seed orchards with special reference to pines. In: Faulkner, R. (Ed.) *Seed orchards.* For. Com. Bull., London, No. **54**, pp. 83–91.

Koski, V. (1980): Minimivaatimukset mannyn siemenviljelyksille Suomessa (Minimum requirements for seed orchards of Scots pine in Finland). *Silva Fenn.* **14 (2)**:136–149.

Koski, V. (1987): Long geographic transfers, a possible way of eliminating pollen contamination in advanced generation seed orchards of Pinus sylvestris. *Forest Ecol. Manag.* **19**:267–271.

Kurm, M. (1987): Stimulation of Scots pine seed crop with bioactive substances in Estonia. *Forest Ecol. Manag.* **19**:173–176.

Lindgren, D. and Quiyu, W. (1986): Are genetic results influenced by the environment during seed maturation? *Proceedings of IUFRO conference on breeding theory, progeny testing and seed orchards.* Williamsburg, pp. 192–199.

Luckwill. L. C. (1970): The control of growth and fruitfulness of apple trees. In: Luckwill, L. C. and Cutting, C. V. (Eds) *Physiology of tree crops.* Academic Press, London, pp. 237–253.

Luukkanen, O. (1980): Flower induction by exogenous plant hormones in Scots pine and Norway spruce grafts. *Silvae Fenn.* **14 (1)**:95–105.

Matschke, J., Schachler, G. and Schönborn, H.–J. (1981): Enzymscreening zur Optimierung von Bioregulatoren zur Blühinduktion bei Koniferen. *Beitr. Forstw. (Berlin)* **15 (2)**:86–90.

Mátyás, Cs. (1979): Szaporítóanyagtermesztés (Propagation material production). Res. Rep. No. 3, NYFK Timber Co., Szombathely, (unpublished report).

Mátyás, Cs. (1983): *An introduction to forest tree improvement.* Sopron, University Multipliers, p. 175.

Mátyás, Cs. and Rauter, R. M. (1987): Feasibility of seed orchard establishment, rogueing and lifespan. *Forest Ecol. Manag.* **19**:247–256.

McLemore, B. F. (1977): Strobili and conelet losses in four species of southern pines. *Res. Note, South. For. Expt. Sta.,* New Orleans, No. SO–226, p. 5.

Mikola, J. (1987): Effects of fertilizer and herbicide application on the growth and cone production of Scots pine seed orchards in Finland. *Forest Ecol. Manag.* **19**:183–188.

Molotkov, P. I., Patlaj, I. N. and Davidova, N. I. (1982): *Selekcia lesnykh porod (Selection of forest trees).* Lesnaya Promyshl., Moscow, p. 224.

Novoseltzeva, A. I. (1980): On the application of forest genetics and forest tree breeding research in practical forest seed production. *Silva Fenn.* **14 (1)**:28–31.

Priestley, C. (1970): Carbohydrate storage and utilization. In: Luckwill, L. C. and Cutting, C. V. (Eds) *Physiology of tree crops.* Academic Press, London, pp. 113–125.

Rosvall, O. and Andersson, B. (1982): Slatter, harvning och herbicidbehandling i tallfröplantager (Soil treatment in Scots pine seed orchards). *Inst. Skogsförb. Inf.* 1982/83, **7**:4.

Rosvall, O. and Untinen, P. (1982): Fröplantagernas naringsstatus (Nutrient status of seed orchards). *Inst. Skogsförb. Inf.* 1981/82, **4**:8.

Sachert, E., Eifler, I. and Matschke, J. (1981): Blühstimulierung in Kiefernsamenplantagen. *Beitr. Forstw. (Berlin)* **15 (3/4)**:129–131.

Sarvas, R. (1962): Investigations on the flowering and seed crop of Pinus silvestris. *Metsatiet. Tutkimusl. Julk.* **53**:4, 198.

Sarvas, R. (1970): Establishment and registration of seed orchards. *Folia For.* **89**:24.

Schlenker, G. (1979): Sind Langner's Warnungen vor dem Plantagen-Saatgut gerechtfertigt? *Allg. Forstzschr.* **41**:1123–1124.

Schmidtling, R. C. (1984): Planting south of origin increases flowering in shortleaf and Virginia pines. *Silvae Gen.* **33**:140–144.

Schmidtling, R. C. (1987): Locating pine seed orchards in warmer climates: benefits and risks. *Forest Ecol. Manag.* **19**:263–271.

Stern, K. (1972): Nature's diversifying selection and its impacts on orchard breeding. *For. Tree Impr. (Copenhagen)* **4**:41–51.

Sweet, G. B. (1975): Flowering and seed production. In: Faulkner, R. (Ed.) *Seed orchards.* For. Comm. Bull. London No. **54**, pp. 79–82.

Syrach-Larsen, C. (1936): The importance of vegetative propagation in respect of forest improvement plans. Proc. 11e Congrès Intern. de IUFRO, Budapest, Vol. 3, pp. 44–49.

Syrach-Larsen, C. (1972): Seed orchards in forestry. *Forest Tree Impr. (Copenhagen)* **4**:7–27.

Vilks, M. (1985): Vrediteli lesosemennykh plantacii sosny i borba s nimi (Scots pine seed orchard pests and their control). In: *Semennie plantacii v lesnom semenovodstve (Seed orchards in forest seed production).* Zinatne, Riga, pp. 47–51.

Weiser, F. (1985): Untersuchungen zur vergleichsweisen Anzucht von Kiefernpfröpflingen durch Freilandverschulung und in Containern. *Beitr. Forstw. (Berlin)* **19 (4)**:148–153.

Werner, M. (1975): Location, establishment and management of seed orchards. In: Faulkner, R. (Ed.) *Seed orchards.* For. Comm. Bull., London, No. **54**, pp. 49–57.

Werner, M. (1982): Flyttning av uppvuxna ympar (Transplanting grown-up grafts). *Inst. Skogsförb. Inf.* 1981/82, **5**:5.

Werner, M. and Petterson, H. (1980): Kottskord i uppvuxna tallplantager (Cone harvest in grown-up Scots pine seed orchards). *Inst. Skogsförb. Inf.* 1979/80, **8**:4.

Wesoły, W., Urbański, K. and Barzdajn, W. (1987): Effect of mineral fertilization on flowering of Scots pine. *Forest Ecol. Manag.* **19**:191–198.

Wright, J. W., Lemmien, W. A. and Bright, J. (1966): Early flowering patterns in Scots pine. *Michigan Agric. Expt. Sta. Bull.* **49**:189–199.

Zviedre, A. (1985): Proizvoditel'nost lesosemennikh plantacij sosny obikhnovennoj v Latvijskoi SSR i kachestvo urozhaya (Productivity and quality of crop of Scots pine seed orchards in the Latvian SSR). In: Pirags, D. M. (Ed.) *Semennie plantacii v lesnom semenovodstve (Seed orchards in forest seed production).* Zinatne, Riga, pp. 52–62.

CHAPTER 11

Genetic processes in seed orchards

G. Müller-Starck

Introduction

Seed orchards are defined here in a very broad sense as populations which are established for the purpose of producing offspring by means of sexual reproduction. The members of the parental population are derived from selected trees which appear to be superior with respect to certain phenotypic characters. In this sense seed orchards can be considered to be forest breeding populations (for classification of populations and for breeding concepts see e.g., Namkoong et al., 1989).

In clonal seed orchards, the parental population consists of vegetatively propagated trees (clones), each of which is represented by a number of genotypically identical individuals (ramets). Orchards which are established by generatively propagated plants ("seedling seed orchards") are rare in Scots pine and have not been subjected to detailed genetic studies. In the following, the term "seed orchard" will refer exclusively to the clonal seed orchard.

In contrast to natural populations, seed orchards are reproductive communities which contain ramets and consist of only small numbers of different genotypes (parental clones). In most cases seed orchards are better isolated spatially from external pollen immigration of the same species than forest stands.

Objectives of seed orchard establishment

Contrary to many agricultural breeding populations, forest seed orchards usually aim at a maximum phenotypic similarity between the set of the selected clones and their offspring from open pollination. The greatest similarity with respect to certain genetically controlled characters can be expected if the genetic structures among the parental population and their offspring are identical at the same life stage (age class). This condition is fulfilled in the case of the genetic equilibrium.

Consequently, two aspects are involved, firstly the mode of reproduction resulting in zygote formation and secondly the viability of the offspring until the life stage of the previous generation at census, i.e. the age class of the selected phenotypically superior trees. During both stages, conditions which favour the approach to the genetic equilibrium appear to be the exception rather than the rule, especially in the case of stationary, longlived organisms which are exposed to extremely heterogeneous environments.

Studies on the characterization of reproduction in seed orchards can help to reveal predominant trends in the changes of the genetic structures between a breeding population and its offspring. In this chapter, processes will be taken into consideration which operate until the formation of fully developed seeds or, in some cases, the seedling stage. By this, it is intended to contribute to a settlement of the question of whether the reproductive conditions in Scots pine seed orchards can allow realization of the breeding objectives.

Components of the reproductive system

Following the classification described by Ziehe (1982), three levels of reproduction will be distinguished, all of which are related to one reproductive period:

1. Fertility of the parental trees: This part of the reproductive system is defined here to include the spatial distribution of sexual types and the production of female and male gametes.

2. The mating system: It describes the mode of association of female and male gametes to form diploid zygotes. This mode can be related to the given fertilities (mating among the available gametes) or to the members of the parental population (mating among genotypes, i.e. individuals in the diplophase). Deviations from random mating will be classified as preferential or non-preferential mating with respect to certain characters (positive or negative assortative mating), and preferential mating among relatives (inbreeding systems).

3. Viability of the offspring: As the final stage of the reproductive system, it is defined here to comprise all postzygotic processes until the formation of fully developed seeds.

Results of studies on reproductive processes

As in many other species, the majority of the studies on Scots pine seed orchards were performed using enzyme gene markers. For a recent survey of applications of biochemical gene markers in population genetic studies on various tree species see Rudin (1986). (Further information is given by Prus-Głowacki in Chapter 5.)

It is intended to classify the results of experimental studies in Scots pine seed orchards and demonstrate deviations from randomness during reproduction by examples rather than to supply a complete list of all known studies. Furthermore, the aim is to point out what kind of information is lacking and to propose methods which might be suitable to reveal more detailed information on reproductive processes in forest tree populations.

Non-random segregation

In several coniferous species other than Scots pine, alleles were found to deviate from the 1:1 segregation expected among the macrogametophytes of single trees (e.g. Cheliak et al., 1984, for *Pinus banksiana;* Boyle and Morgenstern, 1985, for *Picea mariana;* Müller-Starck, 1977*a,* for *Picea abies*).

Non-random segregation among two-locus types was observed in Scots pine and in many other tree species: Rudin and Ekberg (1978) found one linkage group among 12 studied loci, Szmidt et al. (1984) reported four linkage groups among 27 loci and Niebling et al. (1987) six groups among 30 loci.

Differential fertilities

Differences between clones in flowering intensity and asymmetries within the clonal production of female and male flowers are well known from various inventories in Scots pine seed orchards (e.g. Baumeister, 1975; Andersson and Hattemer, 1978; Ross, 1984).

This phenomenon has genetic consequences if it induces changes in the genetic structures among the bulked offspring as compared to the parental population. This is the case if fertility is genotype-dependent, i.e. if differential ovule and pollen production depends on the respective genotype (sexually asymmetrical fertility selection; Ziehe and Gregorius, 1981). If Mendelian segregation following meiosis, random gametic fusion and equal zygotic viabilities can be expected, selection will result in the "multiplicative structure" among the offspring (Ziehe, 1982). Observed and expected genetic structures among the offspring should then be identical. Non-random fusion of gametes (see subsequent topic) can cause characteristic deviations from the multiplicative structure which allow its detection in combination with fertility selection.

For example, in Table 11.1 the observed and the expected genotypic structures under panmixia are given for the bulked offspring of a Scots pine seed orchard which consists of 36 clones with an average of 25 ramets per clone.

The deviations between the panmictic structure and each of the observed offspring structures are highly significant ($P<0.001$, with genotypes 33, 13, and 31 pooled). In contrast with this, the calculated multiplicative structure (genotype frequency $P_{kl}=p_k^f p_l^m$, where p_k^f and p_l^m denote allele frequencies among ovules and pollen, respectively) is nearly identical to the observed one in the flowering period 1974. This means that the deviations from the panmictic reproductive conditions can be fully explained by the assumption of genotypically differential pollen and ovule production. Consequently, random processes are suggested with respect to the fusion of gametes and subsequent stages.

In the case of the period 1975, the surplus of homozygotes in the observed structure indicates that partial self-fertilization may play a role in addition to the genotypically differential fertilities. It cannot be ruled out that a larger surplus of

Table 11.1. Comparison between two observed offspring structures at the LAP-B gene locus and the corresponding reference structures (panmixia and multiplicative). Genotypes are ordered (f = female, m = male); sample size is 2 × 640 seeds (selected data from Müller-Starck et al., 1983)

LAP-B genotype f	LAP-B genotype m	Expected under panmixia	Frequencies between two flowering periods: Observed 1974	Observed 1975	Multiplicative structure 1974	Multiplicative structure 1975
1	1	7.9	9	9	9.4	2.7
2	2	490.0	444	528	444.3	520.2
3	3	0.1	–	1	0.5	0.6
1	2	62.2	96	41	96.9	48.1
2	1	62.2	44	22	42.9	28.7
1	3	1.0	8	3	6.7	2.2
3	1	1.0	–	1	0.7	0.7
2	3	7.8	30	23	30.8	24.2
3	2	7.8	9	12	7.7	12.7

homozygotes due to inbreeding could have been counterbalanced by asymmetrical fertility selection (for proof see Ziehe and Gregorius, 1981; Ziehe, 1982).

These results are confirmed in a study which deals with the locus GOT-C (Müller-Starck, 1982*a*). Other investigations on fertility selection in Scots pine are not known.

Non-random fusion of gametes

Assortative mating

Positive assortative mating occurs, for instance, in the case of preferential mating among early or late flowering trees, providing that this character is under genetic control. Deviations up to several days between Scots pine clones in the initiation and the duration of the period of receptivity and of pollen shedding were reported by Jonsson et al. (1976) and Chung (1981).

In the former study, the hypothetical contributions of single clones to the bulked orchard offspring were calculated on the basis of the flowering phenology observed in three successive years. Strong deviations were noticeable among the hypothetical reproductive success of clones: Over the three years, the contribution of a certain clone was 10 times that of another clone. The results indicate that asynchronies in the flowering processes can fundamentally affect the genetic structures among the orchard offspring.

Negative assortative mating results from the non-preferential mating due to asynchronous flowering of trees (see above-cited studies). In conifers, it cannot be caused by prezygotic incompatibilities because the contact of pollen and egg cell cannot be prevented by sporophytic or gametophytic systems as in angiospermous trees.

Inbreeding systems

Inbreeding affects the whole genome and not only certain gene loci, as opposed to common forms of assortative mating. Inbreeding systems, i.e. systems of preferential mating among genealogically related individuals (in this sense inbreeding can also be considered as a genealogically assortative system), are the consequence of a constellation in which only a limited number of individuals are in mating contact. In the most intensive form, such limitations are the mating between gametes of one and the same individual (self-fertilization).

The resulting increase of homozygotes as compared to the corresponding Hardy–Weinberg proportions is well known as is the disadvantageous consequence with respect to the expression of metric characters and survival (inbreeding depression). As a representative of many other studies in the family of *Pinaceae* and other families of plants see Eriksson et al. (1973), Andersson et al. (1974), and Lundkvist et al. (1987).

(a) Self-fertilization

The occurrence of viable offspring from self-fertilization in Scots pine is proven by controlled crosses, although the percentages of empty seed were considerably greater than after cross-fertilization or open pollination (e.g. Plym Forshell, 1974; Baumeister, 1975). Results from experimental studies on self-fertilization following open pollination in Scots pine seed orchards are summarized in Table 11.2 and compared with the results obtained in forest stands.

Table 11.2 exhibits a large variability in the estimates of self-fertilization within both stands and seed orchards. This should not be a question of the methods applied: Even two entirely different markers like chlorophyll deficiency and allelic enzymes revealed very similar results in a study in which both methods were used (Müller-Starck, 1977*a*).

The variation of results among the stands should primarily be a consequence of deviations in the sizes of the effectively reproducing populations: Some of the studies suggest that deviations are due to different flowering intensities and/or stand densities (e.g. Koski and Malmivaara, 1974; Yazdani et al., 1985*a*). The stand density in particular shows tremendous differences for the number of reproducing adult trees per hectare can range between only 10–20 in northern Sweden (Rudin et al., 1977) and 250–300 trees in Central Europe.

The harvesting mode can also account for the variation: The study of Rudin and Ekberg (1982) demonstrates that the percentage of offspring from self-fertilization in cones from the upper part of the crown can substantially exceed the corresponding values from the lower part (on average 22.0% *vs.* 11.0%). An

Table 11.2. Estimates of the proportion of offspring from self-fertilization in Scots pine stands (ST) and seed orchards (OR). (CM = chlorophyll mutants, EGM = enzyme gene marker. The total number of orchard clones is added in brackets.)

Marker trees/clones	Gene marker (CM, EGM)	Studied population (ST, OR)	Self-fertilization (%)		References
			Range	Mean	
20 6	CM CM	ST ST	2.8– 5.2 0.4– 3.6	<5.0*[1] <3.0*[2]	Koski and Malmivaara, 1974
4 6	CM EGM	ST ST	1.8– 8.2 2.9– 9.6	5.2* 6.2	Müller-Starck, 1977*a*
3 5	EGM EGM	ST ST	7.0–24.0 17.0–48.0	17.0* 24.3	Rudin et al., 1977
13	EGM	ST	0.0–24.0	11.8*	Yazdani et al., 1985*a*
5	EGM	ST	1.0–29.0	11.9	Rudin et al., 1986
2 (25)	EGM	OR	0,0– 5.7	2.3*	Rudin and Lindgren, 1977
1 (36)	EGM	OR	11.5–13.9[3]	12.6	Müller-Starck, 1979
–[4] –[4]	EGM EGM	OR[5] OR[6]		6.1 0.5	Chung, 1981
1 (51)	EGM	OR		6.0	Shen et al., 1981
3 (36)	EGM	OR	0.0–19.2[7]	9.0[7]	Müller-Starck, 1982*a*
5 (36)	EGM	OR	12.0–30.0	16.0	Rudin and Ekberg, 1982
5 (51)	EGM	OR	0.0–18.0	8.3	Rudin et al., 1986

* Seedling stage (otherwise embryo)
[1] Northern marginal population;
[2] Central population;
[3] Different flowering periods;
[4] Estimates on clonal groups;
[5] Young clone bank (9 years);
[6] Older clone bank (19 years);
[7] Averaged female and male selfing rates derived from bulked orchard offspring.

opposite trend was found by Shen et al. (1981) with 2.9% for the upper and 8.5% for the bottom part of the crown. The latter values confirm earlier estimates which were based on empty seed percentages (see Hadders and Koski, 1975). An additional limitation on the size of the effectively reproducing population can arise from the age of the harvested seed orchard (Chung, 1981; see also Table 11.2). Furthermore, great variation in the number of ramets can be expected among seed orchards. Unfortunately, such information is missing in most of the studies.

Comparing both categories of populations, two trends can be observed: The studies in middle or northern Sweden seem to reveal greater probabilities of self-fertilization in forest stands than in seed orchards (with the exception of Rudin and Ekberg, 1982) while the opposite tendency was obtained in the studies of the author with populations in the central part of Germany with presumably larger reproductive efficiencies. The trend under the nordic conditions is confirmed by Rudin et al. (1986) who studied the same trees in both population types (see Table 11.2). This could be explained by the extremely small number of reproducing trees in nordic stands.

Generally, it must be kept in mind that self-fertilization at the clonal level comprises both individual self-fertilization (each ramet) and the additional cross-fertilization between genotypically identical individuals. This surplus in self-fertilization compared to stands becomes larger with increased ramet number and decreased distance among ramets of the same clone. Consequently, the amount of self-fertilization in seed orchards seems to be primarily a function of the number of clones and the seed orchard design.

(b) Cross-fertilization

In Table 11.3 the present information on cross-fertilization (i.e. the mode of the spatial dispersal of gametes after zygote formation) in Scots pine seed orchards is surveyed briefly. As in Table 11.2, the results from forest stands are given for comparison.

The information on cross-fertilization in Scots pine stands are interpreted for the present as an indication of preferential mating within limited neighbourhoods due to restricted pollen transmission. The study of Yazdani et al. (1985*a*) supports this assumption only within a range of 25 metres around the marker trees; the contradictory tendency for larger distances cannot be explained satisfactorily and may also reflect the particular experimental conditions (sampling after seed dispersal and 10 to 20 years of viability selection).

The results demonstrate that the large hypothetical range of pollen dispersal (e.g. Koski, 1970) is no longer evident if postzygotic stages are considered.

In seed orchards, a large variation in the results can be observed. For instance, the proportion of seeds of a tree containing the gametic contribution of its nearest neighbour varies between 1% (Rudin and Ekberg, 1982) and 31% (Shen et al., 1981). This is interpreted for the present as an indication of limited pollen transmission which, in contrast to Scots pine stands, is compensated to a varying extent by the successful pollen of other individuals of the same clonal genotype. This masking effect should depend strongly on the density of the representatives of the same clone within the seed orchard population.

(c) Estimates of the amount of inbreeding

The observed deviations from random fusion of gametes with respect to both self- and cross-fertilizations should result in increasing proportions of homozygotes among the offspring as compared to the Hardy–Weinberg proportions. In the past, attempts were made by several authors to predict the amount of inbreeding in seed orchards (mostly Scots pine or Norway spruce) without having the possibility to apply gene marker studies (e.g. Stern, 1959; Hadders and Koski, 1975; Gregorius and Müller-Starck, 1975; Lindgren and Gregorius, 1976).

Table 11.3. Estimates on the mode of cross-fertilization in Scots pine stands (ST) and seed orchards (OR) derived from the offspring of the given number of sample trees. (The total number of studied trees in stands or clones in orchards is given in brackets.) All studies refer to enzyme gene markers

Sample trees	Studied population (ST, OR)	Comments on the mode of cross-fertilization in relation to distance (m)	References
15 (105)	ST	Strong decrease within studied range of 80 m ($f(x) = 0.052e^{-0.038x}$)	Müller-Starck, 1977*b*
8[1] (121)	ST	Decrease until 25 m, indifferent or increase from 40 – 90 m*[2]	Yazdani et al., 1985*a*
14 (25)	OR	Random-like mode of cross-fertilization*	Rudin and Lindgren, 1977
10 (51)	OR	Decrease until 40 m; up to 31% of fertilizations by nearest neighbour	Shen et al., 1981
14 (36)	OR	Random-like mode or[3] decrease until 21 m ($f(x) = 19.01\ e^{-0.025x}$)	Müller-Starck, 1979 Müller-Starck, 1982*b*
8 (36)	OR	Random-like mode within 14 m; 1 – 3% of fertilization by next neighbour	Rudin and Ekberg, 1982
23 (>12)	OR	Restricted pollen transmission; estimated size of neighbourhood is 21 trees	Cheliak, 1985

* Seedling stage (otherwise embryo)
[1] Offspring are 10 – 20-years-old plants around the adult trees;
[2] Migration of seeds included;
[3] Two flowering periods.

In Scots pine, estimates on inbreeding on the basis of enzyme gene marker studies were first derived in forest stands: Rudin et al. (1977) calculated Wright's fixation indices for an adult stand ($F = 0.0$) and single tree progenies ($F = 0.1$). The assumption of viability selection against homozygotes during the entire life-span is supported under conditions of natural regeneration by Yazdani et al. (1985*b*), who found an excess of homozygotes at the embryo stage but conditions close to panmixia in young plants and in the adult stand. This trend was less pronounced in the studies of Tigerstedt et al. (1982), who still found indications for inbreeding at the age of 100 years.

In seed orchards, only a few estimates on inbreeding are available. Cheliak (1985) compared fixation indices for both parental population and offspring and found significant deviations for only one out of six gene loci. Various deviations from random mating between genotypes due to asynchronous flower phenology and restricted pollen transmission (see Table 11.3) are not appropriately reflected by the applied measurement of inbreeding. In such cases, it cannot be ruled out that inbreeding is masked by other phenomena in reproduction such as asymmetrical fertility selection (see explanations for Table 11.1).

Müller-Starck (1982*b*, see also 1979) calculated coefficients of inbreeding and kinship on the basis of information on the actual self- and cross-fertilization in a seed orchard with 36 clones (see Tables 11.2 and 11.3). Inbreeding was predicted for several generations and different hypothetical population types. The results obtained clearly show that already in the first generation the amount of inbreeding and kinship decreases with decreasing ramet number and increasing number of clones.

For instance, the coefficient of inbreeding is reduced from 0.096 to 0.060 (0.055) if the number of clones is increased from 10 to 50 (90). From this, a clone number of 50 seems to guarantee an efficient reduction in inbreeding if all clones have the same fertilities and no other restrictions of the reproductive efficient population size have to be taken into account.

Contamination by non-orchard pollen

Reproductive processes in seed orchards are particularly subject to interferences from outside pollen due to their small population size and the small pollen production per individual especially at early life stages.

Even in the case where detailed information is available on the frequencies of multilocus genotypes in seed orchards and in stands of the same species, estimates of the amount of contamination are biased: Generally, only the obligatory non-orchard pollen can be monitored but other pollen can originate from either population. Estimates on interferences by successful non-orchard pollen must therefore be considered as lower bounds.

Two studies in Scots pine seed orchards revealed very different results, which may be considered for the present to represent the range of pollen contamination:

One was conducted by Nagasaka and Szmidt (1985) in a 30-clone orchard (area of 6 hectares). By utilizing multilocus gene markers in an 18-locus study, the minimum percentage of embryos which contain non-orchard pollen contributions was estimated to be 38%. The orchard was 18 to 25 years old and can be considered to be at the stage of full seed production. It is important to know that the only isolation from external pollen immigration is a strip of birch trees measuring 10 m in width and 3 to 5 m in height.

An earlier study (Müller-Starck, 1982*c*) refers to a 36-clone orchard (area of 2 hectares) which can be considered to be well isolated because of its location in an area of deciduous trees with sporadic Scots pine stands. The minimum percentage of embryos which contain non-orchard pollen contributions was estimated for two flowering periods on the basis of only four loci to be 1.1% and 2.3%.

These two studies illustrate how much the involvement of non-orchard pollen depends on local conditions. More detailed information is required before general conclusions can be drawn. There are several methods to restrict external pollen contamination, the most effective of which appears to be supplemental mass pollination (e.g. Hadders, 1978; Yazdani, 1986).

The question should be raised as to whether the additional pollen contributions from surrounding stands are necessarily always disadvantageous. Particularly in small seed orchards with large numbers of ramets per clone, external pollen can reduce inbreeding and increase the genetic variation. A certain amount of successful non-orchard pollen from surrounding forest stands, the characters of which do not appear to counteract the aims of the respective tree breeding programme, should be tolerated.

Differential viabilities

Viability selection between the fusion of gametes and the stage of fully developed seeds cannot be studied explicitly because the genetic structures are not observable directly after zygote formation.

There is much evidence in support of the assumption that differential viabilities are primarily a consequence of the occurrence of sublethal or lethal factors: The probability for a coincidence of such factors at the same locus (loci) within the zygote increases with increasing kinship between the fused gametes and is at a maximum in the case of self-fertilization.

There are several studies on the estimation of frequencies of such sublethal or lethal factors in the genome of Scots pine (e.g. Sarvas, 1962; Koski, 1971; Plym Forshell, 1974). Furthermore, Lindgren (1975) demonstrated effects of competition during archegonial polyembryony on the viability of zygotes under various model assumptions.

All of these results strongly support the hypothesis of viability selection during the early embryonic stages but cannot quantify its impacts on the genetic structures of the offspring independently of other reproductive processes.

Inferences on reproductive systems as a whole

Deviations from the randomness of reproductive processes can be observed at the clonal level (or level of single trees or certain genotypes) by monitoring its reproductive success, or at the population level by means of a comparison between the genetic structures among parental population and bulked offspring.

Reproductive success

In three Scots pine seed orchards, Müller-Starck and Ziehe (1984) and Müller-Starck (1985) found substantial deviations between the reproductive success, i.e. the gametic contribution of single clones to the bulked seed orchard crop, and the expected panmixia values (for certain genotypes see Müller-Starck et al., 1983).

In the second study, the gametic contributions of the same clones could be monitored in two different seed orchards for two flowering periods. In these 2 × 2 environments the clonal reproductive success varied substantially within and among the studied environments. This concerned the female and the male contribution to the orchard seeds and also the pooled gametic contribution of each clone. Generally, the obtained results reveal a surprisingly large dependency of the clonal reproductive success on environmental conditions.

Deviations in the genetic structures

The comparison of parental and filial genetic structures allows a global characterization not only of the reproductive process as a whole but also of certain components. In a Finnish clone bank, Chung (1981) observed only small deviations within three different groups of parents and offspring: The comparison between the frequencies of single genetic types resulted only for one genotype in having statistically significant deviations and for none of the alleles. Deviations between the three groups of parental populations were not significant, those between the different offspring populations were.

Cheliak (1985) did not find any significant changes between the parental and the filial structures although pronounced deviations from random processes were observed with respect to single components of the reproductive system (see Table 11.3). In another Scots pine seed orchard, highly significant deviations were observed between parental and filial structures as well as between filial structures of different flowering periods (Müller-Starck, 1982*a;* Müller-Starck et al., 1983; see also "Differential fertilities").

Ziehe (1982) proposed a concept for a systematic comparison of the genotypic filial structures with reference structures which are derived from the parental and/or the filial (male and female) allele frequencies. By means of these methods, dominant components in reproduction can be revealed, such as fertility selection and partial self-fertilization. For applications see the above-cited studies of the present author. Reference structures other than Hardy–Weinberg provide an information potential which could be used in the study of population progeny to a much greater extent.

The need for new concepts

In many tree species the reproductive system is characterized by means of the "mixed mating model" (for survey see Brown et al., 1985). Unfortunately this concept has severe limitations such as:

— Restriction on random cross-fertilization.
— Population selfing rate without consideration of the proportionate contribution of each seed parent to the total contribution.

Table 11.4. Allelic preference patterns U for genotypes at three gene loci based on the observed parental and filial genetic structures (selected data from Müller-Starck and Gregorius, 1988)

Gene locus	Genotype	$U_{Mj:Fi}$	$U_{Fi:Mj}$	Genotype	$U^{*}_{Fi:Mj}$
GOT-B	3:3	.966	1.015	3:3	1.000
	5:3	1.013	.982	3:5	.991
	3:5	1.035	1.005	5:3	1.029
	5:5	1.005	.938	5:5	.960
GOT-C	2:2	1.152	.737	2:2	1.009
	4:2	.878	1.211	2:4	.991
	2:4	1.138	.724	4:2	.996
	4:4	.890	1.222	4:4	1.004
MDH-C	3:3	1.180	1.141	3:3	1.064
	4:3	.633	.721	3:4	.799
	3:4	.914	.857	4:3	.824
	4:4	1.229	1.357	4:4	1.551

— Frequencies among the cross-fertilizing pollen are assumed to be identical to the population frequencies.
— Assortative mating, mating among relatives and sexual selection cannot be considered.

The results from Scots pine stands and seed orchards surveyed above show most of these model assumptions to be unrealistic. In order to proceed in the analysis of reproductive systems, Müller-Starck and Gregorius (1988) proposed a set of methods which goes back to the concept of preference parameters (Gregorius and Hattemer, 1987). Mating preferences between genetic types were used to differentiate between the following classes of mating systems: assortative mating, sexual selection, random mating, and random preferential mating.

Preference parameters were calculated for two Scots pine seed orchards and applied for two modes of seed collections, namely single tree (clone) progeny and population progeny. In Table 11.4, an example is presented which demonstrates that each of three gene loci can reveal different trends. Data refer to the bulked seed orchard crop with known parental structures.

The parameter $U_{Mj:Fi}$ (or $U_{Fi:Mj}$) is equal to the ratio of the filial genotype frequency $P_{Fi,\,Mj}$ to the product of the parental and the female, F (or the male, M) gametic allele frequencies. The assortative mating preference $U^{*}_{Fi:Mj}$ refers to the same ratio, with parental allele frequencies replaced by the filial ones among the offspring. The values obtained reveal the following: the evenness of the U-values in the case of the GOT-B locus suggests random-mating conditions. The particular pattern of deviations at GOT-C is the precondition for random preferential mating (with respect to both sexes); the evenness of the corresponding U^{*} values suggests non-assortative mating conditions, finally, the symmetrical deviations at MDH-C indicate assortative mating for this gene locus. This conclusion could be based solely on the U* values, which are exclusively derived from the filial population.

The example given serves to demonstrate the information potential of the methods applied. It may also indicate that complex reproductive conditions are the rule rather than the exception.

Concluding remarks

1. The reviewed results indicate that a maintenance of the parental genetic structures among the offspring appears unrealistic. Although many proposals were made for more appropriate seed orchard layouts (Giertych, 1975), the realization of breeding objectives is therefore uncertain.

2. Inbreeding is an inherent problem in seed orchards, and it increases with decreasing numbers of clones. It cannot be predicted which portion of inbreds will be eliminated by viability selection in later life stages. At a minimum, clone numbers should exceed 50, particularly in the case of evidence for further restrictions of the reproductively effective population size.

very low!

3. More refined methods are required for a better quantification of reproductive processes and interpretation of the obtained results. Proposals are made for a new concept which can be further extended.

4. Although the genetic variation has not explicitly been considered, it is evident that most present seed orchards contain fewer different genotypes than forest stands (possible exceptions under nordic conditions). The reproductively efficient population size is further restricted by differential fertilities and non-random mating. Supplemental mass pollination and the tolerance of non-orchard pollen in certain cases may prevent severe losses of genetic variation.

5. Seed orchards should be more systematically compared with alternatives, such as other sexually-reproducing populations or multiclonal mixtures. A comparison based on genetic criteria is an essential requirement for any long-term strategy in tree breeding.

6. Finally, in a period of severe environmental stress and increasing demands for survival abilities of populations, the objective of a maximum attainable economic gain has to be considered relative to the risks caused by losses of genetic variation and the resulting reduction of the adaptive potential of tree populations. From this point of view, a relinguishing of parts of the economic gain in favour of considerably greater breeding populations appears necessary.

Literature

Andersson, E. and Hattemer, H. H. (1978): Variation among clones and ortet-ramet relationship in grafted Scots pine (*Pinus sylvestris* L.). *Stud. For. Suec.* **148**:31.

Andersson, E., Jansson, R. and Lindgren, D. (1974): Some results from second generation crossings involving inbreeding in Norway spruce *(Picea abies). Silvae Genet.* **23**:34–43.

Baumeister, G. (1975): Möglichkeiten der Früherkennung quantitativer Saatgutertragsleistungen bei Klonen von *Pinus sylvestris* L. in Samenplantagen. *Silvae Genet.* **24**:129–200.

Boyle, T. J. B. and Morgenstern, E. K. (1985): Inheritance and linkage relationships of some isozymes of black spruce. *New Brunswick Can. J. For. Research* **15**:992–996.

Brown, A. H. D., Barrett, S. C. H. and Moran, G. F. (1985): Mating system estimation in forest trees: Models, methods and meanings. In: Gregorius, H.-R. (Ed.) *Population genetics in forestry. Lecture notes in biomathematics.* Springer, Berlin–New York, 60, 32–49.

Cheliak, J. (1985): Mating system dynamics in a Scots pine seed orchard. In: Gregorius, H.-R. (Ed.) *Population genetics in forestry. Lecture notes in biomathematics.* Springer, Berlin–New York, 60, 107–117.

Cheliak, W. M., Morgan, K., Strobeck, C., Yeh, F. C. H. and Dancik, B. P. (1984): Segregation of allozymes in megagametophytes of viable seeds from a natural population of jack pine, *Pinus banksiana* Lamb. *Theor. Appl. Genet.* **69**:145–151.

Chung, M.-S. (1981): Biochemical methods for determining population structure and mating system in *Pinus sylvestris* L. *Acta For. Fenn.* **173**:28.

Eriksson, G., Schelander, B. and Akerbrand, V. (1973): Inbreeding depression in an old experimental plantation of *Picea abies. Hereditas* **73**:185–194.

Giertych, M. (1975): Seed orchard designs. In: Faulkner, R. (Ed.) *Seed orchards.* Forestry Comm. Bulletin **54**:21–37.

Gregorius, H.-R. and Hattemer, H. H. (1987): Characterisation of mating systems. *J. Theor. Biol.* **128**:407–422.

Gregorius, H.-R. and Müller-Starck, G. (1975): Genetic structures in finite open-pollinated plant populations. A model and its application to seed orchards. *Theor. Appl. Genet.* **46**:295–305.

Hadders, G. (1978): Experiments with supplemental mass pollination in seed orchards of Scots pine (*Pinus sylvestris* L.) 3. *World Cons. For. Tree Breeding, 21–26 March 1977, Canberra, Australia* 2, 967–979.

Hadders, G. and Koski, V. (1975): Probability of inbreeding in seed orchards. In: Faulkner, R. (Ed.) *Seed orchards.* Forestry Comm. Bull. No. **54,** 108–117.

Jonsson, A., Ekberg, I. and Eriksson, G. (1976): Flowering in a seed orchard of *Pinus sylvestris* L. *Stud. For. Suec.* **135**:38.

Koski, V. (1970): A study of pollen dispersal as a mechanism of gene flow in conifers. *Comm. Inst. Forest. Fenn.* **70**:5–78.

Koski, V. (1971): Embryonic lethals of *Picea abies* and *Pinus silvestris. Commun. Inst. For. Fenn.* **75, 3**:1–30.

Koski, V. and Malmivaara, E. (1974): The role of self-fertilization in a marginal population of *Picea abies* and *Pinus sylvestris.* IUFRO Joint Meeting of Work. Part. S2.04.1, S2.04.2, S2.04.3, 1–11.

Lindgren, D. (1975): The relationship between self-fertilization, empty seeds and seeds originating from selfing as a consequence of polyembryony. *Stud. For. Suec.* **126**:24.

Lindgren, D. and Gregorius, H.-R. (1976): Inbreeding and coancestry. Proc. IUFRO Joint meeting, S2.04.01, Bordeaux, 49–69.

Lundkvist, K., Eriksson, G., Norell, L. and Ekberg, I. (1987): Inbreeding depression in two field trials of young *Pinus sylvestris* (L). *Scand. J. For. Res.* **2**:281–290.

Müller-Starck, G. (1977*a*): Untersuchungen über die natürliche Selbstbefruchtung in Beständen der Fichte (*Picea abies* (L.) Karst) und Kiefer *(Pinus sylvestris). Silvae Genet.* **26**:207–217.

Müller-Starck, G. (1977*b*): Cross-fertilization in a conifer stand inferred from enzyme gene-markers in seeds. *Silvae Genet.* **26**:5–6.

Müller-Starck, G. (1979): Estimates of self- and cross-fertilization in a Scots pine seed orchard. In: Rudin, D. (Ed.) *Proceedings of the conference on biochemical genetics of forest trees, Umeå, Sweden, 1978.* Swedish University of Agricultural Sciences, *Rep.* **1**:170–181.

Müller-Starck, G. (1982*a*): Sexually asymmetric fertility selection and partial self-fertilization. 2. Clonal gametic contribution to the offspring of a Scots pine seed orchard. *Silva Fenn.* **16**:99–106.

Müller-Starck, G. (1982*b*): Reproductive systems in conifer seed orchards. I. Mating probabilities in a seed orchard of *Pinus sylvestris* L. *Silvae Genet.* **35**:5–6, 188–197.

Müller-Starck, G. (1982*c*): Tracing external pollen contribution to the offspring of a Scots pine seed orchard. In: *Proceedings of IUFRO joint meeting on breeding strategies including multiclonal varieties, Sensenstein 1982,* Lower Saxony Forest Research Institute, Department of Forest Tree Breeding, Escherode, p. 176.

Müller-Starck, G. (1985): Reproductive success of genotypes of *Pinus sylvestris* L. in different environments. In: Gregorius, H.-R. (Ed.) *Population genetics in forestry. Lecture notes in biomathematics.* Springer, Berlin–Heidelberg–New York, 60, 118–133.

Müller-Starck, G. and Gregorius, H.-R. (1988): Analysis of mating systems in forest trees. In: Weir, B. S., Eisen, E. J., Goodman, M. M. and Namkoong, G. (Eds) *Proceedings of the 2nd international conference on quantitative genetics.* Sinauer, Sunderland, Mass., pp. 573–595.

Müller-Starck, G. and Ziehe, M. (1984): Reproductive systems in conifer seed orchards. 3. Female and male fitness of individual clones realized in seeds of *Pinus sylvestris* L. *Theor. Appl. Genet.* **69**:173–177.

Müller-Starck, G., Ziehe, M. and Hattemer, H. H. (1983): Reproductive systems in conifer seed orchards. 2. Reproductive selection monitored at an LAP gene in *Pinus sylvestris* L. *Theor. Appl. Genet.* **65**:309–316.

Nagasaka, K. and Szmidt, A. E. (1985): Multilocus analysis of external pollen contamination of a Scots pine (*Pinus sylvestris* L.) seed orchard. In: Gregorius, H. R. (Ed.) *Population genetics in forestry. Lecture notes in biomathematics.* Springer, Berlin–Heidelberg–New York, 60, 134–138.

Namkoong, G., Kang, H. C. and Brouard, J. S. (1989): *Tree breeding: principles and strategies.* Monographs on Theor. Appl. Genet. 11, Springer, Berlin–Heidelberg–New York.

Niebling, C. R. (1987): *Isozyme analysis of linkage in Scots pine (Pinus sylvestris* L.). Thesis for Master of Science, Pennsylvanian State University, p. 115.

Niebling, C. R., Johnson, K. and Gerhold, H. D. (1987): Electrophoretic analysis of genetic linkage in Scots pine (*Pinus sylvestris* L.). *Biochem. Genet.* **25**:803–814.

Plym Forshell, C. (1974): Seed development after self-pollination and cross-pollination of Scots pine, *Pinus sylvestris* L. *Stud. For. Suec.* **118**:1–24.

Prus-Głowacki, W.: Biochemical polymorphism. (See this volume, Chapter 5.)

Ross, M. D. (1984): Frequency-dependent selection in hermaphrodites: the rule rather than the exception. *Bio. J. Linn. Soc.* **23**:145–155.

Rudin, D. (1986): Developmental trends in the field of biochemical genetics of forest trees. *Proceedings of the 18th IUFRO World Congress, Ljubljana. Division 2,* Vol. II, pp. 577–588.

Rudin, D. and Ekberg, I. (1978): Linkage studies in *Pinus sylvestris* L. using macrogametophyte allozymes. *Silvae Genet.* **27**:1–12.

Rudin, D. and Ekberg, I. (1982): Genetic structure of open-pollinated progenies from a seed orchard of *Pinus sylvestris. Silva Fenn.* **16**:87–93.

Rudin, D. and Lindgren, D. (1977): Isozyme studies in seed orchards. *Stud. For. Suec.* **139**:23.

Rudin, D., Eriksson, G. and Rasmuson, M. (1977): Inbreeding in a seed tree stand of *Pinus sylvestris* L. in Northern Sweden. A study by the aid of the isozyme technique. Royal College of Forest Genetics, Res. Note 24, p. 45.

Rudin, D., Muona, O. and Yazdani, R. (1986): Comparison of the mating system of *Pinus sylvestris* in natural stands and seed orchards. *Hereditas* **104**:15–19.

Sarvas, R. (1962): Investigations on the flowering and seed crop of *Pinus sylvestris. Commun. Inst. For. Fenn.* **53, 4**:1–198.

Shen, H. H., Rudin, D. and Lindgren, D. (1981): Study of the pollination pattern in a Scots pine seed orchard by means of isozyme analysis. *Silvae Genet.* **30, 1**:7–15.

Stern, K. (1959): Der Inzuchtgrad in Nachkommenschaften von Samenplantagen. *Silvae Genet.* **8**:37–42.

Szmidt, A. E., Muona, O. and Yazdani, R. (1984): Linkage relationships in Scots pine (*Pinus sylvestris* L.) In: Szmidt, A. E. (Ed.) *Genetic studies of Scots pine* (*Pinus sylvestris* L.). Domestication by means of isozyme analysis. *Swedish University of Agricultural Sciences, Umeå, Schweden,* Rep. *IV:1–24.*

Tigerstedt, P. M. A., Rudin, D., Niemelä, T. and Tammisola, J. (1982): Competition and neighbouring effects in a naturally regenerating population of Scots pine. *Silva Fenn.* **16**:122–129.

Yazdani, R. (1986): Efficiency of artificial mass pollination in a *Pinus sylvestris* seed orchard. In: Muhs, H.-J. (Ed.) *Proceedings of IUFRO joint meeting on biochemical genetics and legislation of forest reproduction material, Grosshansdorf 1985.* BFH-Mitteilungen 154, pp. 21–24.

Yazdani, R., Lindgren, D. and Rudin, D. (1985*a*): Gene dispersion and selfing frequency in a seed tree stand of *Pinus sylvestris* L. In: Gregorius, H.-R. (Ed.) *Population genetics in forestry. Lecture notes in biomathematics.* Springer, Berlin–Heidelberg–New York, 60, pp. 139–154.

Yazdani, R., Muona, O., Rudin, D. and Szmidt, A. E. (1985*b*): Genetic structure of a *Pinus sylvestris* L. seed tree stand and naturally regenerated understory. *For. Sci.* **31, 2**:430–436.

Ziehe, M. (1982): Sexually asymmetric fertility selection and partial self-fertilization. 1. Population genetic impacts on the zygotic genotypic structure. *Silva Fenn.* **16**:94–98.

Ziehe, M. and Gregorius, H.-R. (1981): Deviations of genotypic structures from Hardy–Weinberg proportions under random mating and differential selection between the sexes. *Genetics* **98 (1)**:215–230.

CHAPTER 12

Vegetative propagation

O. Monteuuis and C. Barnéoud

Grafting

Grafting is performed during winter or at the beginning of spring before scion budbreak, taking account of the usual recommendations (Hartmann and Kester, 1975). The more widely used techniques are the side-veneer graft, spliced side graft, and the cleft graft which are easier in practice (Fig. 12.1).

Pinus sylvestris is usually used as the rootstock, although good results have also been obtained with *Pinus nigra* which can increase the vigor of the grafted plant, and at the same time being more adapted to particular edaphic situations such as limestone soils.

Generally, success of grafting in greenhouse conditions is quite satisfactory (over 70%) although some grafts may fail later on, probably as a consequence of incompatibility between the stock and the scion, even if both are of the same species. Taking care that size and vigour of the rootstock correspond to the scion, may reduce much subsequent heterogeneity within the developed grafted plant populations.

The purpose of grafting may also determine the choice of the rootstock. Thus, in order to hasten flowering, rootstocks need not be too young, and the scion be selected as mature as possible within the donor tree. The scion should also be grafted rather far from the root system of the stock.

Grafting should be considered also as a means of inducing an eventual rejuvenation of the mature material with the objective of true-to-type cloning (Franclet et al., 1980; Franclet, 1981). For this purpose, the scions which exhibit the most juvenile characteristics will be preferred, within the selected ortet, to be grafted onto young vigorous seedlings close to their root system. The foliage of the rootstock is to be preserved and it is recommended to minimize as much as possible the size of the scion, which could be reduced to the terminal bud only in a cleft-graft (see Fig. 12.1). The influence of serial (repeated) graftings on the possibilities of rejuvenation could be analyzed as it has been done for other species (Franclet, 1981). Nevertheless, up to now, such rejuvenation techniques have been mainly applied to *Pinus pinaster* (Franclet, 1976, 1977; Chaperon, 1980; Alazard and Kadio, 1984). Moreover, considering *Pinus* species, one has to keep in mind that most of the characteristics of the shoot were formed the previous summer. This could explain why the expected morphological rejuvenation does not generally occur from the first shoot elongation, unless scions have been restricted to apex or meristems (Dumas et al., 1989). Subsequently appropriate treatments and practices such as pruning, growth regulator sprays, and fertilization (Franclet, 1977, 1981; Hillson and Dancik, 1978) can be applied to this grafted material which can be later cultivated as stock plants, and it will increase its ability for true-to-type cloning through cuttings.

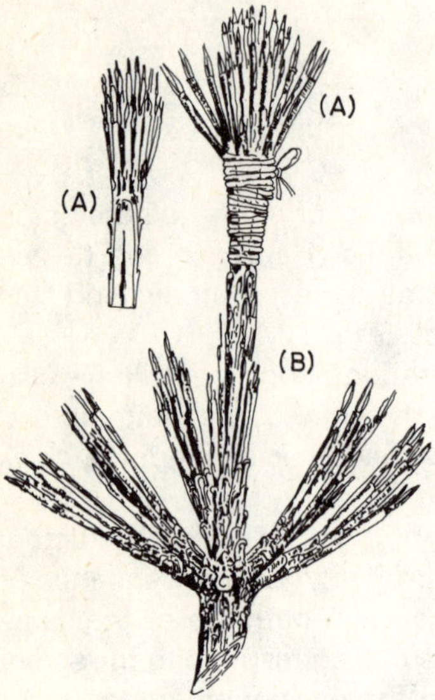

Fig. 12.1. Illustration of the cleft-grafting technique applied to Scots pine, associating the scion (A) with the rootstock (B)

Propagation by cuttings

Maturation state of the ortet

In the same way as for a lot of Gymnosperms, the ability of Scots pine to produce adventitious roots diminishes rapidly as the cutting donor ages, especially when the ortets are growing in severe environmental conditions. In experimental conditions, rooting rates of 3.3% were scored when using cuttings from 8-year-old field-grown plants, as against 43.4% for 3-year-old plants and 70% when removed from 3.5-month-old seedlings (Monteuuis and Pages, 1987). This marked decrease of Scots pine rooting capacity with increasing age of the donor plant justifies the use of juvenile experimental material, some months or even weeks old seedlings, which exhibit much greater ability for organogenesis. On the other hand, this juvenile, soft and fragile material requires more attention in special propagation installations with appropriate humidity, temperature and sanitary conditions.

Genotypic variations

Within seedling populations of the same age, it is noticeable that there are individual differences in rooting capacity leading later to interclonal variation (Berg, 1980; Monteuuis and Pages, 1987). These differences related to the genotypes affect the rooting rate as well as the quality of the new adventitious root system (number and development of roots per cutting).

Seasonal variations

The rooting of Scots pine cuttings can be greatly influenced by the time of removal of ramets from the donor plants. The best results were generally obtained from the end of summer until the beginning of spring (Hansen and Ernstsen, 1982; Monteuuis and Pages, 1987). The observed seasonal variation in rooting can be related mainly to natural light which acts on the physiological state of the donor plant through its three components: photoperiod, irradiance and radiance exposure (Hansen and Ernstsen, 1982). In fact, relevant experiments have shown that the capacity of cuttings for rooting was strongly increased when stock plants had been kept growing under short photoperiod and low irradiance at 10 to 15 $w.\ m^{-2}$ (Strömquist, 1979).

The influence of light on rooting should be interpreted from the standpoint of carbohydrate turnover rates (Hansen et al., 1978; Strömquist, 1979; Ernstsen and Hansen, 1986), or of the endogenous hormone levels (Whitehill and Schwabe, 1975; Wodzicki, 1978; Sandberg and Ericsson, 1987).

Conditioning and treatments of stock plants

From other forest tree species, and especially *Pinus* species (Franclet, 1977, 1981; Chaperon, 1980; Alazard and Kadio, 1984; Fraysse et al., 1985), it seems advantageous to consider propagation of Scots pine through cuttings from stock plants conditioned for this purpose (Hillson and Dancik, 1978). The stock plants are obtained by grafting or cutting ramets taken from ortets selected in situ. Irrigation and fertilization should be provided judiciously with reference to other similar species, in order to stimulate the rootability of the plant material (Franclet, 1977).

Many authors prefer to use short shoots (needle fascicles from brachyblasts) as cuttings rather than long shoots (Boeijink and Van Broekhuizen, 1974; Whitehill and Schwabe, 1975; Hillson and Dancik, 1978). The development of such shoots can be promoted by appropriate hedging operations, which consist mainly in suppressing the terminal bud or the apical portion of the long shoots to hasten the elongation of the previously inhibited brachyblasts, in the same manner as it has been done for *Pinus pinaster* (Franclet, 1976; Chaperon, 1980; Alazard and Kadio, 1984). More generally, hedging reduces the expanding development of shoots related to the root system. The influence of the latter is considered as positive for adventitious root formation. Foliar sprays of cytokinin solution, BAP, have also been satisfactorily used to favour brachyblast proliferation (Whitehill and Schwabe, 1975; Inglis, 1984).

Besides the treatments mentioned it is recommended that stock plants are kept under low light regimes some time before the cutting harvest (Strömquist, 1979; Hansen and Ernstsen, 1982; Ernstsen and Hansen, 1986). It appears from most of the studies that light must be considered a determining factor for propagation of *Pinus sylvestris* by cuttings.

Selection and preparation of the cuttings for rooting

All cuttings used for rooting should bear a terminal apical bud, so-called "head stem cuttings". Short shoots obtained from elongated brachyblasts after suppression of apical dominance by hedging or climatic conditioning are preferred to long shoots for rooting (Strömquist, 1979). Nevertheless, whatever their origin, it is advisable to take the cuttings from the basal part of the ortet or stock plant, close to the root system. In the nursery a rooting rate of 54% was observed for long shoots collected at the base of the ortets, as against 17% for needle fascicles, known to be more reactive to rooting, collected one meter above the root collar (Monteuuis and Pages, 1987).

Subsequent to collection, cuttings can be stored several weeks to a few months at low temperature (0 °C to 5 °C) in darkness, in a cold storage room (Whitehill and Schwabe, 1975). After the cuttings have been sprayed with a suitable fungicide in solution to prevent fungal contaminations, e.g. *Botrytis* spp. (Boeijink and Van Broekhuizen, 1974; Phillion et al., 1983), sealing the ramets into polyethylene bags will avoid damaging desiccation during the storage period. The effect of low temperature storage on further rooting appears to be comparable to that of the winter period in natural conditions (Strömquist, 1979).

Just before lining out on rooting benches, the cuttings are trimmed in order to obtain homogeneous cutting bundles of similar height. Long shoots are usually cut at about 8 to 10 cm below the terminal bud, whereas cuttings from short shoots are about 5 cm long and those obtained from juvenile material bearing primary needles are still shorter, at 3 to 4 cm. A lateral wound at the base of the cuttings (Hartmann and Kester, 1975) seems to be advisable to stimulate the production of adventitious roots (Whitehill and Schwabe, 1975; Berg, 1980; Phillion et al., 1983).

Root-promoting growth regulators

Treating the base of the cuttings with auxin-type growth regulators in powder or liquid form (Hartmann and Kester, 1975) can noticeably promote adventitious root production. Nevertheless, the effects of these exogenous root-promoting substances may vary depending upon their chemical nature and the concentrations of compounds used in relation to the physiological state of the material and the seasonal cycle (Strömquist, 1979; Hansen and Ernstsen, 1982). Thus, when applied at certain periods of the year, the influence of such exogenous rooting hormones can be zero or even negative compared with the untreated controls (Monteuuis and Pages, 1987). Such results suggest that the use of root-promoting growth regulators should be undertaken with caution and discernment and make the choice of a proper chemical substance difficult. Although most *Pinus* species react favourably to both the synthetic auxins, NAA and IBA (Franclet, 1976; Berg, 1980; Chaperon, 1980; Struve and Blazich, 1982; Fraysse et al., 1985), IBA appears to be the most efficient and therefore is widely used for Scots pine (Boeijink and Van Broekhuizen, 1974; Strömquist, 1979; Berg, 1980; Hansen and Ernstsen, 1982; Phillion et al., 1983). The effect of the root-promoting compound may differ depending on whether it has been applied in a powder preparation or in a liquid solution (Hartmann and Kester, 1975). The soaking in diluted solution is more penetrating, but may cause the leaching out of endogenous substances engaged in the process of rhizogenesis (Mounteuuis and Pages, 1987). Nevertheless, the following scheme can be recommended: soaking the wounded base of the cutting for 24 to 48 hours in diluted IBA water solution at a strength of 50 to 150 mg of IBA per litre. Furthermore, Eliasson and Strömquist (1981) pointed out a beneficial influence of catechol (a phenolic compound) on rooting.

Obviously the method of application of root-promoting exogenous substances need to be determined for each case, corresponding to the particular nursery conditions.

Environmental conditions for cuttings during rooting

The porosity and texture of the rooting medium in particular may influence the formation of new adventitious roots. This substrate influence, noticeable on certain occasions (Monteuuis and Pages, 1987), remains often dependent on other environmental parameters of the propagation: relative humidity, temperature, bench size, drainage of the layers (Hartmann and Kester, 1975). Generally substrate combinations which include sand, perlite (Hartmann and Kester, 1975), or other materials chosen for their good drainage properties, are advisable for the rooting of Scots pine cuttings in propagation frames where the relative humidity of the air, surrounding the cuttings, is always maintained at a high level (Boeijink and Van Broekhuizen, 1974; Whitehill and Schwabe, 1975; Berg, 1980). Too much moisture in the rooting substrate, leading rapidly to overwatering and reduced rooting of the cuttings, remains a real risk in the water saturated atmosphere of the propagation installations (Boeijink and Van Broekhuizen, 1974). High air humidity can be maintained by a simple polyethylene cover placed a few centimetres over the top of the cuttings. Proper bottom temperature (20 °C to 24 °C) can be provided by heating ducts below the cuttings (Hartmann and Kester, 1975). Nevertheless, such installations which work very well during winter when the control of proper temperatures remains easy, need to be replaced by spacious propagation cases when outside temperatures increase during spring and summer. In fact, the prevention of high temperatures, which can damage the cuttings in the rooting phase, becomes a crucial problem on hot days. Sophisticated propagation installations such as mist or fog systems (Boeijink and Van Broekhuizen, 1974; Whitehill and Schwabe, 1975) in elaborate greenhouses equipped with a cooling system and with automatic controls and regulations, undoubtedly offer real advantages under such circumstances.

Artificial light has been used by several workers (Whitehill and Schwabe, 1975; Hansen et al., 1978; Hansen and Ernstsen, 1982; Phillion et al., 1983; Ernstsen and Hansen, 1986) during the rooting phase of cuttings. From a physiological point of view it appears from previous experiments that low light regimes are sufficient for the accomplishment of rhizogenesis, in agreement with the results obtained under natural light during the winter. According to Hansen et al. (1978) the influence of light during the rooting phase seems to be less critical than the treatment of the stock plants just before ramet removal, as previously mentioned. Thus, reduced light exposures should be sufficient to ensure the endogenous carbohydrate levels required for good rhizogenesis (Strömquist, 1979). On the other hand, light can be used on certain occasions to control fungal contaminations, especially Botrytis cinerea (Phillion et al., 1983) jointly with fungicide spraying treatments performed weekly (Boeijink and Van Broekhuisen, 1974; Hartmann and Kester, 1975; Berg, 1980).

The duration of the rooting period generally does not exceed three months. Transplanted rooted cuttings are tall enough to be outplanted in the field at the end of one year of cultivation in normal nursery conditions.

Micropropagation

The possibilities of propagating *Pinus sylvestris* through in vitro culture have been investigated for several years.

The age of the ortets remains a crucial factor for success, since the capacity for organogenesis diminishes very soon after germination (David and Thomas, 1979; Tranvan and Thomas, 1981). Thus, although induced adventitious bud-formation can be achieved in the case of very young seedlings of up to 2-week-old (Bornman, 1984), this powerful technique appears to be unsuccessful when applied to older ortets exhibiting shoots with secondary needles. For such developed material, stimulation of axillary buds from long shoots or short shoots grown *in vitro* or even *in vivo* conditions, gives the best propagation rates (Chalupa, 1983). Whatever the option chosen, induction of adventitious buds or accelerated development of axillary buds, in vitro cytokinin induction treatments on adapted culture media are necessary to stimulate explant responsiveness (David and Thomas, 1979). The cytokinin application varies in duration and in concentration depending on tissue maturity (Tranvan and Thomas, 1981). BAP is mostly used as cytokinin (Tranvan, 1979; von Arnold and Eriksson, 1981; Tranvan and Thomas, 1981; Bornman, 1984) sometimes in combination with an auxin (Bornman, 1981; Chalupa, 1983).

Elongation of cytokinin-induced buds occurs on media lacking growth regulators and can be promoted by the presence of activated charcoal, at 0.5% to 1% (Bornman, 1981; Chalupa, 1983). Elongated shoots can be used as the new explant source. To insure further multiplication cycles, shoots are exposed to

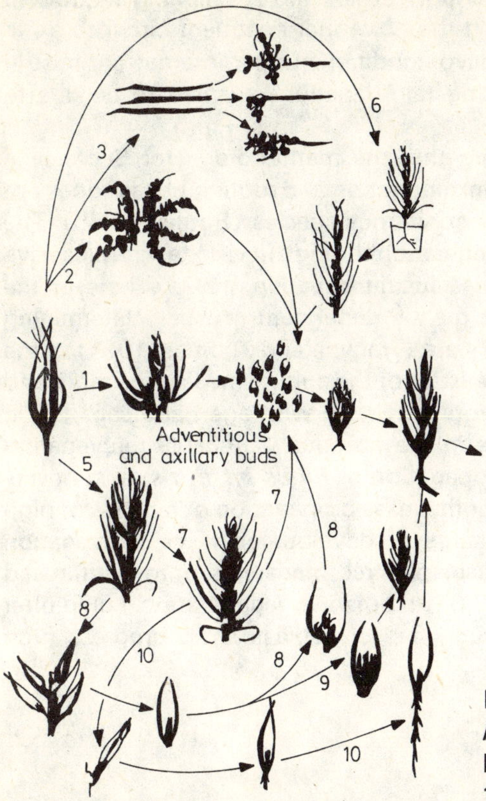

Fig. 12.2. Pathways of regeneration in *Pinus sylvestris.* (The figure and the legend are taken from Bornman, 1984.)

Seedlings can be used in different ways in a micro-multiplication programme. For initiation of adventitious shoot buds ca. 14-day-old seedling explants are cultured with their cotyledons upright (1) or flattened (2) on cytokinin-containing nutrient agar. Alternatively, cotyledons are explanted individually (3) or explants are given a high-concentration, short-duration cytokinin pulse (4) before transferring to hormone-free nutrient agar. Adventitious buds are excised, elongated in vitro and rooted in vivo following a high-concentration, long-duration auxin pulse or in vitro in auxin-containing nutrient agar.

Six- to 7- to 8-week-old intact seedlings can also be given cytokinin sprays or grown in a cytokinin-containing nutrient medium (6), or the explants pulsed (5), all treatments which result in development of axillary buds and also induction of adventitious buds (6, 7).

Seedlings can also be cold-hardened to advance physiological maturity and attain uniformity in physiological response. Removal of buds and spray-treating with cytokinin releases suppression of brachyblast shoot apices (8). Brachyblasts are then cultured in vitro for production of multiple adventitious buds (8) or for elongation of the initial shoot bud (9) which is eventually rooted in vivo. Brachyblasts of non-acclimated seedlings (10) can first be rooted in vitro, but best in vivo, before shoot elongation is induced.

cytokinin treatments, or can be rooted either in vitro or in vivo conditions, followed by application of exogenous auxins in order to promote adventitious root formation (Bornman, 1984; Grönroos, 1987). Chalupa (1983) indicated that in vitro elongated shoots induced for rooting during 20 to 30 days, on an agar adapted mineral medium containing 2.0 mgl^{-1} NAA and 2.0 mgl^{-1} IBA, rooted from 36% to 75% when transferred after the hormonal treatment either to an in vitro auxin-free agar medium, or to in vivo conditions under intermittent mist in a non-sterile solid substrate. The rooting rate variations seemed to be due to genotypic differences.

It should be clear from the foregoing that the methodology for Scots pine micropropagation through multiplication, elongation and rooting phases does not fundamentally differ from that for other coniferous species (Boulay, 1985). The various pathways of *Pinus sylvestris* regeneration through in vitro techniques have been described by Bornman (1984) as illustrated in Fig. 12.2. Whatever the method considered, its effectiveness is greatly dependent on such determining factors as tissue maturity, physiological state (Tranvan and Thomas, 1981) of the propagated material, genotype, characteristics of the culture media. The rapid and early decrease of the potential for micropropagation with age handicaps the in vitro cloning of mature superior ortets in the absence of reliable rejuvenation techniques (Bornman, 1984). Micropropagation of *Pinus sylvestris* must nevertheless be considered a powerful tool for the mass propagation of families of high genetic value selected in their early stage of development. The rejuvenation techniques by repeated micrograftings of old tree meristems onto germinated young seedlings followed by bud induction and plantlet regeneration, could offer a solution. Work on propagation through free protoplasts is in progress (von Kirsten et al., 1986).

Literature

Alazard, P. and Kadio, A. (1984): Croissance juvénile des boutures de pin maritime. *Ann. AFOCEL* **1983:**119–155.

Arnold, S. von and Eriksson, T. (1981): Production of adventitious plants from spruce and pine. In: Eriksson, G., and Lundkvist, K. (Eds) *Proceedings of the symposium of clonal forestry, Uppsala, Sweden, April 8–9, 1981*, pp. 7–31.

Berg, D. J. (1980): The influence of growth regulators, fungicides and media composition on the rooting of *Pinus strobus* L. and *P. sylvestris* L. cuttings. *The Plant Propagator* **26 (4)**:9–18.

Boeijink, D. E. and Van Broekhuisen, J. T. M. (1974): Rooting of cuttings of *Pinus sylvestris* under mist. *N. Z. J. For. Sci.* **4 (2)**:127–132.

Bornman, C. H. (1981): In vitro regeneration potential of the conifer phyllomorph. In: Eriksson, G. and Lundkvist, K. (Eds) *Proceedings of the symposium of clonal forestry, Uppsala, Sweden, April 8–9, 1981*, pp. 43–55.

Bornman, C. H. (1984): Application of in vitro culture technology in clonal forestry. In: Hanover, J., Karnosky, D. and Keathley, D. (Eds) *Proceedings of the international symposium of recent advances in forest biotechnology, Michigan, 10–13 June, 1984*, pp. 178–192.

Boulay, M. (1985): Aspects pratiques de la multiplication in vitro des essences forestières. *Ann. AFOCEL* **1984**:7–43.

Chalupa, V. (1983): Micropropagation of Scots pine (*Pinus silvestris* L.). In: Micropropagation of conifer and broadleaved forest trees. *Comm. Inst. For. Cechoslov.* **13**:12–15.

Chaperon, H. (1980): Nouvelles perspectives d'amélioration génétique induites par le bouturage du pin maritime. *Ann. AFOCEL* **1979**:31–55.

David, A. and Thomas, M. J. (1979): Organogenèse et multiplication végétative in vitro chez les Gymnospermes. *Ann. Biol.* **18 (9–10)**:381–417.

Dumas, É., Franclet, A. and Monteuuis, O. (1989): Microgreffage de méristèmes primaires caulinaires de Pins maritimes (*Pinus pinaster* Ait.) âges sur de jeunes semis cultivés *in vitro. C. R. Acad. Sc. Paris, t. 309, Série* **III**:723–728.

Eliasson, L. and Strömquist, L. H. (1981): Interaction between auxin and catechol in induction of adventitious roots. In: Hanover, J., Karnosky, D. and Keathley, D. (Eds) *Proceedings of symposium on clonal forestry, Uppsala, Sweden, 8–9 April, 1981,* pp. 73–79.

Ernstsen, A. and Hansen, J. (1986): Influence of gibberellic acid and stock plant irradiance on carbohydrate content and rooting in cuttings of Scots pine seedlings (*Pinus sylvestris* L.). *Tree Physiology* **1**:115–125.

Franclet, A. (1976): Recherches de conditions favorables au bouturage du pin maritime. *Ann. AFOCEL* **1975**:99–128.

Franclet, A. (1977): Manipulation des pieds-mères et amélioration de la qualité des boutures. *AFOCEL, Études et Recherches* **8**:20.

Franclet, A. (1981): Rajeunissement et propagation végétative des ligneux. *Ann. AFOCEL* **1980**:11–41.

Franclet, A., David, A., David, H. and Boulay, M. (1980): Première évidence morphologique d'un rajeunissement de méristème primaire caulinaire de Pin maritime âgé (*Pinus pinaster* Sol.). *CR. Acad. Sc. Paris* T. 290 (31 mars 1980) série D: 927–930.

Fraysse, Y., Chaperon, H. and Dupuis, J. M. (1985): Premiers résultats en France de bouturage de pin taeda. *Ann. AFOCEL* **1984**:201–244.

Grönroos, R. (1987): Rooting of *Pinus sylvestris* and *Pinus contorta* hypocotyl cuttings in vitro. *Acta Universitatis Upsaliensis* **88**:26.

Hansen, J. and Ernstsen, A. (1982): Seasonal changes in adventitious root formation in hypocotyl cuttings of *Pinus sylvestris.* Influence of photoperiod during stock plant growth and of indole-butyric acid treatment of cuttings. *Physiol. Plant.* **54**:99–106.

Hansen, J., Strömquist, L. H. and Ericsson, A. (1978): Influence of the irradiance carbohydrate content and rooting of cuttings of pine seedlings (*Pinus sylvestris* L.). *Plant. Physiol.* **61**:975–979.

Hartmann, H. T. and Kester, D. E. (1975): *Plant propagation: principles and practices.* 3rd ed. Prentice-Hall Inc., Englewood Cliffs, N. J., p. 662.

Hillson, R. H. and Dancik, B. P. (1978): Proliferation of short shoots in container-grown Scots pine seedlings: a means toward rapid vegetative multiplication. *Tree Planters' Notes (Spring)* **29 (2)**:9–12.

Inglis, J. E. (1984): Effects of some growth substances on the promotion and rooting of interfascicular shoots in *Pinus caribaea. The Plant Propagator* **30 (3)**:4–6.

Kirsten, U. von, Jacob, H. E. Tesche, M. and Kludge, S. (1986): Isolierung und Elektrofusion von Koniferen Protoplasten. *Silvae Genet.* **35**:5–6, 186–190.

Monteuuis, O. and Pages, C. (1987): Données sur le bouturage du pin sylvestre. *AFOCEL-ARMEF, Informations-Forêt* **321 (3)**:143–153.

Phillion, B. J., de Witt, J. and Bunting, W. R. (1983): Propagation of juvenile Scots pine cuttings under a 24 hour photoperiod. *Tree Planters' Notes (Fall)* **34 (4)**:39–42.

Sandberg, G. and Ericsson, A. (1987): Indole-3-acetic acid concentration in the leading shoot and living stem bark of Scots pine: seasonal variation and effects of pruning. *Tree Physiol.* **3**:173–183.

Strömquist, L. H. (1979): *Some aspects on rooting of Pinus sylvestris (L.) cuttings.* Thesis, University of Umeå, Sweden, p. 28.

Struve, D. K. and Blazich, A. (1982): Comparison of three methods of auxin application on rooting of eastern white pine stem cuttings. *Forest. Sci.* **28 (2)**:337–344.

Tranvan, H. (1979): *In vitro* adventitious bud formation on isolated seedlings of *Pinus silvestris* L. *Biol. Plant.* **21 (3)**: 230–233.

Tranvan, H. and Thomas, M. J. (1981): Aptitudes morphogénétiques des plantules et des fragments de plantules de *Pinus silvestris* L. en culture in vitro. *Bull. Soc. Bot. Fr., 128, Lettres bot.* **3**:151–163.

Whitehill, S. J. and Schwabe, W. N. (1975): Vegetative propagation of *Pinus sylvestris. Physiol. Plant.* **35**:66–71.

Wodzicki, T. J. (1978): Seasonal variation of auxin in stem cambial region of *Pinus silvestris* L. *Acta Soc. Bot. Poloniae* **67 (3)**:225–231.

Throughout this book, there is a very high ratio of references (bibliography) to chapter length!

CHAPTER 13

Regulation of flowering in seed orchards

W. Chałupka

Generative development of Scots pine grafts

Maturity of grafts and intensity of flowering

The vegetative and generative development of Scots pine scions after grafting was investigated by Giertych (1970). The scions were collected from mature trees in the crown zone where female strobili are normally initiated (called "female" scions), and in the zone where only male strobili were distributed ("male" scions). "Female" scions gave more successful unions and had a lower mortality in the first year than the "male" ones.

An inhibition of vegetative development of "female" as well as "male" scions was observed during two consecutive years after grafting. In the first year lateral shoots did not develop at all, and in the second year they appeared in few ramets. In the third growing season the number of lateral shoots on each scion was 5–6 on average (Giertych, 1970). During the next 2–3 years an intensive growth and crown development took place and a distinct enlargement of needle length was observed by Dolgolikov (1970).

After developing lateral shoots the formation of the first strobili is possible. There is a stage in Scots pine graft development when only female strobili are initiated (Bánó and Marjai, 1961; Koski, 1975; Bhumibhamon, 1978). In a small number of "female" grafts the first female strobili were already observed in the second growing season after grafting. "Male" grafts produced first female strobili in the third growing season but already after the fourth growing season the origin of scions from ortet crown part did not influence female strobilus initiation (Giertych, 1970). The number of clones with regular female flowering increases from year to year and all clones initiate female strobili at the age of about ten (Fig. 13.1).

Male strobili appear sporadically in the first three years after grafting (Dolgolikov, 1970; Giertych, 1970; Tuchin and Sirotkin, 1983) regardless of origin. At the age of 4 to 7 years an average of 5 to 10 strobili per graft are observed on lower branches (Dolgolikov, 1970). Until 12 to 15 years grafts produce more female than male strobili (Bhumibhamon, 1978; Efimov, 1980; Nikkanen and Velling, 1987). The production of male strobili distinctly increases around the age of about 15 years (Bánó and Marjai, 1961; Baumeister, 1975), when the flowering stage is attained generally in all grafts (Tuchin and Sirotkin, 1983), and the number of male strobili per graft rises to about 500 (Hadders, 1985). Some years later, at the age of 15 to 20 years, the production of male strobili was about 5 times higher than the number of female strobili (Bhumibhamon, 1978). The production of strobili in Scots pine seed orchards increases with the age of grafts, but nevertheless periodicity of flowering is constantly observed (Wesoły et al., 1984).

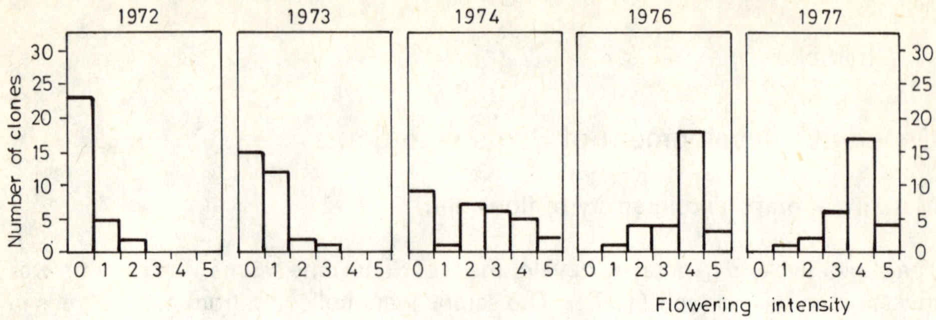

Fig. 13.1. Number of female flowering clones of Scots pine and the number of female strobili per graft in the Grubbenvorst seed orchard (The Netherlands), established in 1968 with 30 clones. *0* = no strobili; *1* = 1 to 5 strobili; *2* = 6 to 10 strobili; *3* = 11 to 25 strobili; *4* = 26 to 50 strobili; *5* = more than 50 strobili per graft (after van't Leven, 1979)

The intensity of flowering is correlated with growth characters such as height and diameter of the grafts (Andersson and Hattemer, 1978; Nikkanen and Velling, 1987). This is especially true of the height between 3 and 6 m (Bhumibhamon, 1978), and DBH above 6 cm (Koski, 1975). The number of female and male strobili correlates positively also with the length, diameter and volume of graft crowns (Bhumibhamon, 1978; Wesoły, 1980).

Distribution of strobili in the crown of grafts

Some differences are observable in the distribution of female and male strobili in the crown of Scots pine grafts. Female strobili are concentrated in the southern part of the crown, while male strobili are distributed more evenly (Baumeister, 1975; Bergman, 1976; Bhumibhamon, 1978). At the level of bud whorls most of the female strobili are initiated in the buds located distally to the trunk of the graft regardless of the cardinal directions (Kosiński and Giertych, 1979).

The vertical distribution of strobili within the crown is related to the age and origin of clones. In younger grafts the percentage of female strobili in the uppermost whorls of the branches is higher than in the older ones (Paal, 1979). In northern Finnish clones both female and male strobili were spread over the whole crown, while in the clones from southern Finland three distinct flowering zones—female, mixed, and male—were distinguishable (Fig. 13.2).

Differences exist also in the distribution of strobili between individuals of the same age in clonal and seedling seed orchards. Efimov (1980) observed two zones in the crown of grafts aged 11 years, a female one (four uppermost whorls with only female strobili), and a mixed one (lower whorls with female as well as male strobili). In contrast with the situation in grafts, only the mixed zone occurred in seedlings (Fig. 13.3).

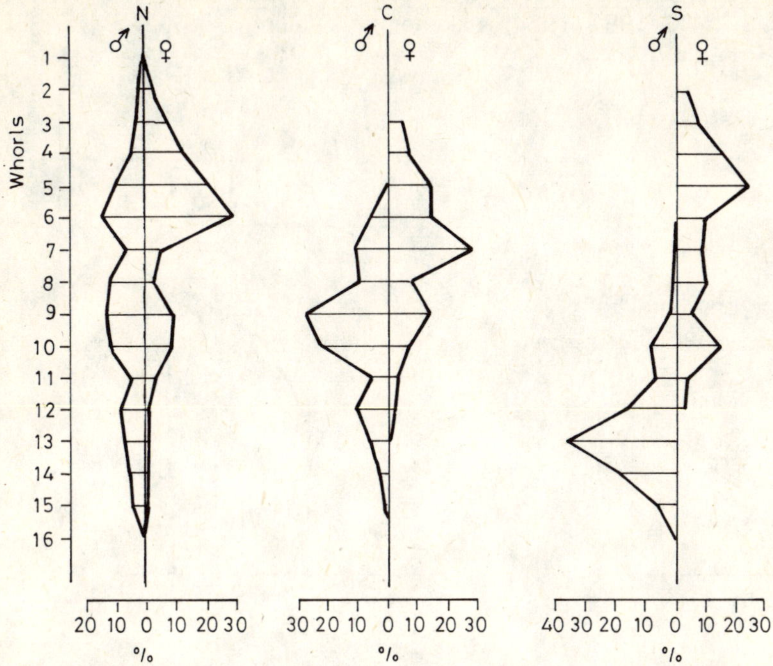

Fig. 13.2. Vertical distribution of female and male strobili in the crowns of 16-year-old grafts of Scots pine in northern (N), central (C) and southern (S) Finland (from data of Bhumibhamon, 1978)

Variation in flowering

Female strobili develop earlier on individual grafts than male do (protogyny), and they are mostly receptive when the pollination begins (Bhumibhamon, 1978; Chung, 1981; Wesoły, 1982). In the southern part of the crown female and male flowering starts several days earlier and the periods of receptivity and pollination are shorter in general than on the northern side (Baumeister, 1975; Jonsson et al., 1976; Wesoły, 1980). The phenology of flowering (i.e., early or late beginning of flowering) is a stable character of most of the Scots pine clones and ranks them in the same order from year to year (Chung, 1981).

Interclonal variation in the time of strobilus initiation is distinct. In Latvian seed orchards the difference between earliest and latest Scots pine clones in the time of male strobili initiation was 5 weeks, but no difference between clones was observed in the time of female strobili initiation (Laura, 1973).

Clones differ also in the further phases of strobili development, namely flower bud burst, beginning of receptivity and time of pollination (Hagman, 1972; Jonsson et al., 1976; Chung, 1981), as well as in the abundance of flowering (Chałupka, 1978, 1980, 1981, 1984; Nikkanen and Velling, 1987).

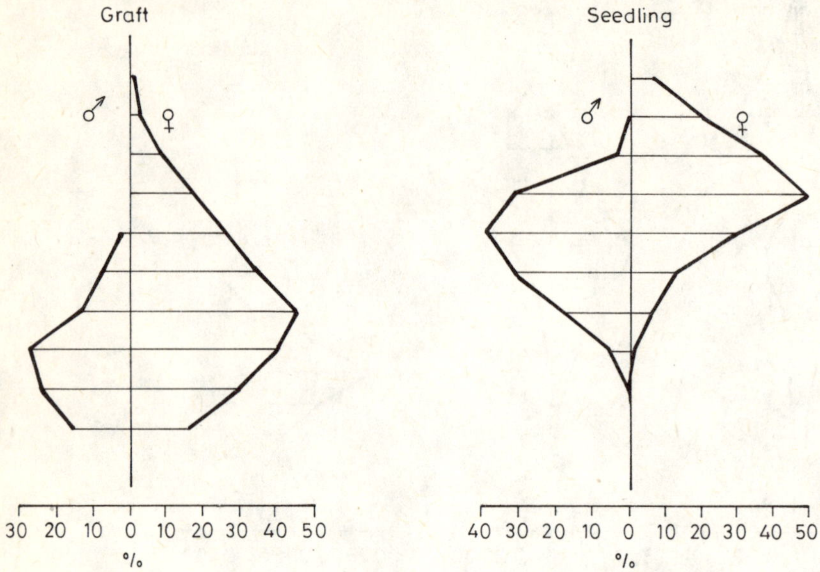

Fig. 13.3. Vertical distribution of female and male strobili in the crown of 11-year-old grafts and seedlings in clonal and seedling seed orchards (from data of Efimov, 1980)

In seed orchards most strobili are produced by a rather small number of clones (Fig.13.4 *a* and *b*). More than half of the female as well as of the male strobili is produced by no more than 25% of the best flowering clones, and half of the clones produces more than 80% of strobili of both sexes (Jonsson et al., 1976; Bhumibhamon, 1978; Wesoły et al., 1984; Danusevicius, 1987).

The ability of clones to flower is a stable character and usually a good positive correlation exists between the subsequent years in the intensity of female, as well as male flowering (Andersson and Hattemer, 1975; Jonsson et al., 1976; Wesoły et al., 1984; Dreimanis, 1985). At an age of 12 to 15 years the number of female strobili per graft was positively correlated with the number of male strobili in the same year (Nikkanen and Velling, 1987). This correlation was not observed earlier at age of 7 to 10 years (Andersson and Hattemer, 1975).

Interclonal variation in flowering intensity did not appear every year (Hattemer et al., 1977), and it was noticed more frequently for male strobilus production than for the female (Chałupka, 1985). There are some suggestions that the interclonal variation in flowering will decrease with age of the seed orchard, and this expectation concerns primarily female flowering (Bhumibhamon, 1978; van't Leven, 1979; Wesoły et al., 1984).

Interclonal differences were also observed for the phenomenon of strobilus and conelet abortion, which may range between 1.7 and 72.2% (Brown, 1971; Chung, 1981).

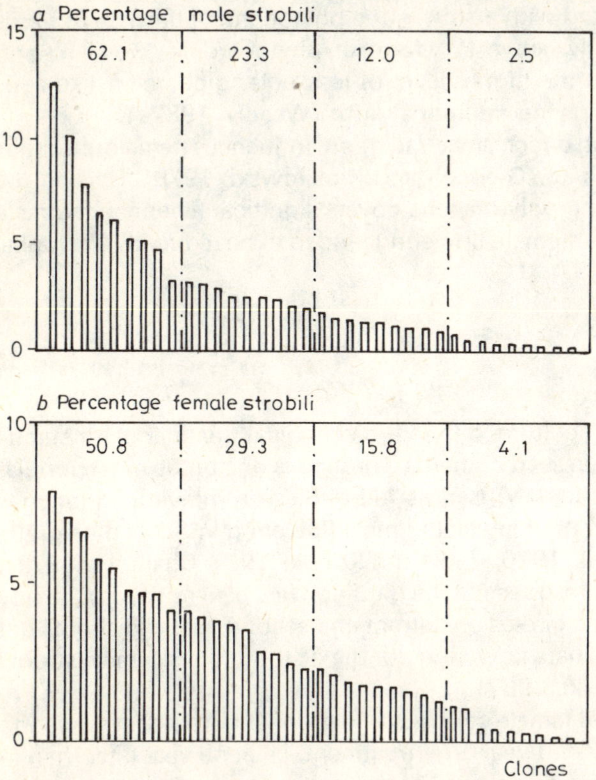

Fig. 13.4. Mean percentage of male *(a)* and female *(b)* strobili of 36 clones growing in a seed orchard (average of the years from 1973 to 1975) (after Jonsson et al., 1976)

Stimulation of flowering in Scots pine grafts

Regulation of light and heat conditions

Direct solar radiation seems to be one of the most important external factors responsible for the induction of generative structures in apical meristems of Scots pine. Shading buds with tinfoil screen from the south (i.e., buds illuminated with indirect sunlight only) significantly reduced female flowering on Scots pine grafts (Giertych and Królikowski, 1978). Studies on light conditions inside buds indicated a strong light filtering role of bud scales (Pukacki and Giertych, 1982). As the authors suggested, the bud scales prevent flowering by limiting transmission of sunlight to the apical meristem. This conclusion was tested by transporting additional amounts of sunlight through bud scales using optical fibres. This treatment increased significantly the number of female strobili initiated on Scots pine (Kosiński and Giertych, 1982; Pukacki et al., 1985). It is possible that this flowering effect was connected with a shift in the phytochrome balance towards an increase in the biologically active P_{730} form of phytochrome (Kosiński and Giertych, 1982), and/or with the higher level of less polar gibberellin-like substances in the buds after supplying additional light (Wesoły, 1987*a*).

There is a suggestion that direct solar radiation influences female strobilus initiation through the heat effect (Giertych and Królikowski, 1978). Raising the temperature around buds using polyethylene covers significantly enhanced male flowering on lateral shoots, but female flowering tended to be reduced (Chałupka, 1981).

Mechanical methods

Following experiences in horticulture, different mechanical treatments for stimulation of flowering were used in seed orchards. Girdling is one of the most popular methods. This treatment was mostly used as full girdles on individual branches and it increased significantly both female and male flowering in Scots pine grafts (Faulkner, 1966; Faulkner et al., 1970; Ronis and Kodola, 1977; Chałupka, 1978). Some interclonal variations in response to the girdling were also noticed (Chałupka, 1978; Wesoły, 1985). Semi-girdles on current and second year-old shoots did not affect the abundance of female flowering nor the distribution of female strobili in the whorls of buds (Kosiński, 1987).

Strangulation also promoted female strobilus initiation. The strongest flowering effect was observed in the second year after treatment, i.e., one year later than in the case of girdling (Ronis and Kodola, 1977).

Trials were also made with crown pruning. Restricted pruning enhanced both female and male flowering in Scots pine grafts. Such treatment was especially effectively promoting male flowering when applied in February (Melchior and Heitmüller, 1961; Melchior 1962). However, intensive pruning of crowns significantly decreased female strobili initiation (Retkes, 1969). (For further details see Chapter 10.)

In an interesting experiment the removal of female strobili and previous year's conelets was tried to improve the nutritional status of shoot apices, but this treatment did not affect flowering next year (Faulkner, 1966).

Hormonal stimulation of flowering

Trials with growth regulators to stimulate flowering in Scots pine grafts were very rare until the seventies. Melchior (1962) was probably the first to use the synthetic auxin NAA, but without any observable results. Later studies, applying mostly gibberellins, were successful.

The most effective reagent was found to be a mixture of two gibberellins, GA_4 and GA_7 ($GA_{4/7}$). These less polar gibberellins strongly promoted male as well as female flowering when applied by spraying both in water and ethanolic solutions (Chałupka, 1978, 1980; Luukkanen, 1980; Luukkanen and Johansson, 1980*a* and *b;* Wesoły, 1987*b*). The injection of $GA_{4/7}$ to individual branches on grafts was also a very effective treatment (Chałupka, 1987).

Good correlation was observed between the time of $GA_{4/7}$ application and the sex of strobili initiated. $GA_{4/7}$ significantly promoted male flowering when applied at the end of May and beginning of June, while application of $GA_{4/7}$ in July and August resulted in a significant increase of female flowering (Chałupka, 1980, 1984, 1987; Luukkanen, 1980; Luukkanen and Hagqvist, unpublished data). The results are probably dependent on whether $GA_{4/7}$ is applied during the initiation of short shoot primordia or first long shoots (Hejnowicz, 1982).

Great variability was revealed in the reaction of individual clones to treatments. It was observed that poorly flowering clones tended to respond more effectively to $GA_{4/7}$ application (Luukkanen, 1980, 1981; Luukkanen and Johansson, 1980*b;* Chałupka, 1981, 1985; Wesoły, 1987*b*). Similarly the $GA_{4/7}$ was more effective in the lower branches than in the upper ones (Chałupka, 1980). The same was observed by Kurm (1987) who applied GA_3 to different whorls in Scots pine grafts.

Supplying the more polar GA_3 from May to mid-August caused significant reduction in the proportion of shoots with male strobili (Chałupka, 1980). Results of Finnish experiments suggest also that GA_3 may partially replace $GA_{4/7}$ in promoting flowering in Scots pine grafts (Luukkanen, 1980; Luukkanen and Johansson, 1980*a* and *b*).

In several experiments NAA was used to stimulate flowering but without any distinct effect (Chałupka, 1981). Also the combination of $GA_{4/7}$ with NAA was not better than $GA_{4/7}$ alone (Luukkanen and Johansson, 1980*a* and *b*).

Research on the stimulation of flowering of Scots pine grafts was also successful in the (former) German Democratic Republic, but the authors did not publish any information on the growth regulators used (Kubitz et al., 1977).

Combined methods

Different cultural treatments were frequently used together with growth regulators for stimulating flowering in many coniferous species (Bonnet-Masimbert, 1987). However, there were only few such experiments on Scots pine grafts. A slight synergism was observed in the joint action of $GA_{4/7}$ and branch girdling, but this interaction was statistically not significant (Chałupka, 1978). In other experiments no interaction was found between the growth regulators used ($GA_{4/7}$ and NAA) and the treatment with polythene covers (Chałupka, 1981).

Conclusions

More basic research is needed including anatomical, physiological and biochemical studies on the formation of shoot apices and the differentiation of generative structures. The clonal differences in phenology of flowering must be considered when setting up a seed orchard.

The proper selection of the site conditions, as well as the application of suitable stimulative measures (fertilization, soil cultivation, crown shaping, etc.) strongly influence the flowering and cone yield of grafts. Technological details are given in Chapter 10 on Seed orchards.

Most of the mechanical methods of flower stimulation diminish somewhat the vigour of grafts and, therefore, are not recommended for long-term use in the Scots pine seed orchards.

The results of experiments with hormonal stimulation of flowering indicate that some possibilities of flower initiation and regulating the sex ratio of strobili do exist.

Literature

Andersson, E. and Hattemer, H. H. (1975): Growth and flowering of "primary" and "secondary" grafts of Scots pine. *Silvae Genet.* **24 (2–3)**:49–54.

Andersson, E. and Hattemer, H. H. (1978): Variation among clones and ortet-ramet relationship in Scots pine (*Pinus silvestris* L.). *Stud. For. Suec.* **148**:1–31.

Bánó, I. and Marjai, Z. (1961): Erdeifenyő oltványok növekedése fejlődése és magtermelése (Growth, development and seed production of Scots pine grafts). *Erdészeti Kutatások* **57**:135–164.

Baumeister, G. (1975): Möglichkeiten der Früherkennung quantitativer Saatgutertragsleistungen bei Klonen von *Pinus silvestris* L. in Samenplantagen. *Silvae Genet.* **24 (5–6)**:177–187.

Bergman, F. (1976): Kott- och froegenshager i skilda krondelar hos tall (*Pinus silvestris* L.) i norra Sverige (Characteristics of cone and seed from different parts of the crown in Scots pine (*Pinus silvestris* L.) in Northern Sweden). Institutionen för Skogsföryngring, Rapporter och Uppsatser, Stockholm, Nr. 68: 1–162.

Bhumibhamon, S. (1978): Studies on Scots pine seed orchards in Finland with special emphasis on the genetic composition of the seed. *Comm. Inst. For. Fenn.* **94 (4)**:1–118.

Bonnet-Masimbert, M. (1987): Flower induction in conifers: A review of available techniques. *For. Ecol. Manag.* ***19**:135–146.*

Brown, I. R. (1971): Flowering and seed production in grafted clones of Scots pine. *Silvae Genet.* **20 (4)**:121–132.

Chałupka, W. (1978): Effect of growth regulators on the flowering in Scots pine (*Pinus sylvestris* L.) grafts. *Silvae Genet.* **27 (2)**:62–65.

Chałupka, W. (1980): Regulation of flowering in Scots pine (*Pinus sylvestris* L.) grafts by gibberellins. *Silvae Genet.* **29 (3–4)**:118–121.

Chałupka, W. (1981): Influence of growth regulators and polythene covers on flowering of Scots pine and Norway spruce grafts. *Silvae Genet.* **30 (4–5)**:142–146.

Chałupka, W. (1984): Time of $GA_{4/7}$ application may affect the sex of Scots pine flowers initiated. *Silvae Genet.* **33 (4–5)**:173–174.

Chałupka, W. (1985): Regulacja kwitnienia na plantacjach nasiennych sosny zwyczajnej (*Pinus sylvesteris L.*) i swierka pospolitego (*Picea abies* (L.) Karst.) (Regulation of flowering in seed orchards of Scots pine (*Pinus sylvestris* L.) and Norway spruce (*Picea abies* (L.) Karst.)). Polska Akademia Nauk, Instytut Dendrologii, Kórnik, p. 146.

Chałupka, W. (1987): Stimulation of flowering in Scots pine *(Pinus sylvestris)* grafts by gibberellin injection. *For. Ecol. Manag.* **19**:177–181.

Chung, M. S. (1981): Flowering characteristics of *Pinus sylvestris* with special emphasis on the reproductive adaptation to local temperature factor. *Acta For. Fenn.* **169**:1–69.

Danusevicius, J. (1987): Flowering and seed production of clones and their stimulation in seed orchards. *For. Ecol. Manag.* **19**:233–240.

Dolgolikov, V. I. (1970): O reproduktivnoj sposobnosti molodykh privivok sosny i eli (On reproductive capacity of young Scots pine and Norway spruce grafts). In: Ermakov, V. I. et al. (Eds) *Lesnaya genetika, selektsiya i semenovodstvo.* Kareliya, Petrozavodsk, pp. 483–487.

Dreimanis, A. A. (1985): Faktory vliyayushchie na polnotsennoe razvitie shishek i semyan na lesosemennykh plantatsiyakh (Factors influencing beneficial development of cones and seeds in forest seed orchards). In: Pirags, D. M. et al. (Eds) *Semennye plantatsii v lesnom semenovodstve.* Zinatne, Riga, pp. 71–79.

Efimov, J. P. (1980): Some results on the regularities of seed crops in Scots pine seed orchards. *Silvae Fenn.* **14 (1)**:63–70.

Faulkner, R. (1966): A review of flower induction experiments and trials 1948–1963. *For. Comm. Rep. on For. Res.* for the year ended March 1965, pp. 207–219.

Faulkner. R., Fletcher, A. M., Longman, K. A. and Johnstone, R. C. B. (1970): Forest Genetics — Flowering. *For. Comm. Rep. on Forest Research* for the year ended March 1970 pp. 108–110.

Giertych, M. (1970): Generative development of Scots pine (*Pinus silvestris* L.) grafts. *Arbor. Kórnickie,* **XV**:71–92.

Giertych, M. and Królikowski, Z. (1978): Importance of bud insolation on female flower induction in pine (*Pinus silvestris* L.). *Arbor. Kórnickie* **XXIII**:161–169.

Hadders, G. (1985): Massovoe dopolnitel'noe opylenie na semennykh plantatsiyakh sosny obyknovennoj (Supplemental mass pollination in seed orchards of Scots pine). In: Pirags, D. M. et al. (Eds) *Semennye plantatsii v lesnom semenovodstve.* Zinatne, Riga, pp. 80–88.

Hagman, M. (1972) On some factors influencing the yield from seed orchards of *Pinus sylvestris* L. and their interclonal and intraclonal variation. *For. Tree Improv.* **4**:67–83.

Hattemer, H. H., Andersson, E. and Tamm, C. O. (1977): Effects of spacing and fertilization on four grafted clones of Scots pine. *Stud. For. Suec.* **141**:1–31.

Hejnowicz, A. (1982): Budowa i rozwój wegetatywnych pąków sosny zwyczajnej *Pinus sylvestris* L. (Structure and development of vegetative Scots pine, *Pinus sylvestris* L., buds). Polska Akademia Nauk, Instytut Dendrologii, Kórnik, p. 105.

Jonsson, A., Ekberg, I. and Eriksson, G. (1976): Flowering in a seed orchard of *Pinus sylvestris* L. *Stud. For. Suec.* **135**:1–38.

Kosiński, G. (1987): Effect of semi-girdles and gravimorphic treatments on female flower distribution in Scots pine *(Pinus sylvestris)* crowns. *For. Ecol. Manag.* **19**:57–63.

Kosiński, G. and Giertych, M. (1979): Female flowering in Scots pine (*Pinus sylvestris* L.) crowns in relation to the trunk and cardinal directions. *Silvae Genet.* **28 (2–3)**:111–114.

Kosiński, G. and Giertych, M. (1982): Light conditions inside developing buds after floral induction. *Planta* **155**:93–94.

Koski, V. (1975): Natural pollination in seed orchards with special reference to pines. In: Faulkner, R. F. (Ed.) *Seed orchards. Forestry Comm. Bull.,* **54**:83–91.

Kubitz. H., Matschke, J. and Eifler, I. (1977): Verfahren zur Induktion von Blüten bei Kiefer unter besonderer Berüksichtigung des Einsatzes von Bioregulatoren. *Soz. Forstwirtschaft* **27 (2)**:89–90.

Kurm, M. (1987): Stimulation of Scots pine seed crop with bioactive substances in Estonia. *For. Ecol. Manag.* **19**:173–176.

Laura, M. (1973): Razvitie generativnykh organov sosny obyknovennoj na semennykh plantatsiyakh Latvii (The development of Scots pine reproductive organs in the seed orchards of Latvia). In: Nekrasova, T. M. et al. (Eds) *Polovaya reproduktsiya khvoinykh.* Vol. 1. Nauka, Novosiborsk, pp. 117–121.

Leven, E. N. van't (1979): De zaadgaarden van Staatsbosbeheer (The seed orchards of the State Forest Service). Rijksinstituut voor onderzoek in de bos- en landschapsbouw "De Dorschkamp", Wageningen, Mededeling, Nr. 178: 1–32.

Luukkanen, O. (1980): Promotion of flowering in Scots pine grafts by means of gibberellins. In: Białobok, S. (Ed.) *Proceedings of IUFRO symposium on Scots pine forestry of the future, 29 September–4 October 1980, Kórnik, Poland,* pp. 249–257.

Luukkanen, O. (1981): Effects of gibberellins GA_4 and GA_7 on flowering in Scots pine grafts. *Silva Fenn.* **15 (4)**:359–365.

Luukkanen, O. and Johansson, S. (1980*a*): Flower induction by exogenous plant hormones in Scots pine and Norway spruce grafts. *Silva Fenn.* **14 (1)**:95–105.

Luukkanen, O. and Johansson, S. (1980*b*): Effect of exogenous gibberellins on flowering in *Pinus sylvestris* L. grafts. *Physiol. Plant.* **50**:365–370.

Melchior, G. H. (1962): Weitere Untersuchungen zur Förderung der Blütenbildung an Kiefer durch Rück-schnitt. *Silvae Genet.* **11 (1)**:11–15.

Melchior, G. H. and Heitmüller, H.-H. (1961): Erhöhung der Zahl der männlichen Blüten an *Pinus silvestris* — Pfropflingen durch Rückschnitt. *Silvae Genet.* **10 (6)**:180–187.

Nikkanen, T. and Velling, P. (1987): Correlations between flowering and some vegetative characteristics of grafts of *Pinus sylvestris. For. Ecol. Manag.* **19**:35–40.

Paal, C. (1979): Raspolozhenie shishek v kronakh privitykh sosen (Distribution of cones in crowns of Scots pine grafts). *Metsanduslikud uurimused, Metsaselektsioon (Tallin)* **XV**:46–59.

Pukacki, P. and Giertych, M. (1982): Seasonal changes in light transmission by bud scales of spruce and pine. *Planta* **154**:381–383.

Pukacki, P., Kosiński, G. and Giertych, M. (1985): Light conditions inside Norway spruce and Scots pine buds influence strobili formation. *Acta Physiol. Plant.* **7 (1)**:3–12.

Retkes, J. (1969): Pruning experiments in Scots pine seed orchards. *Erdészeti Kutatások* **65 (1)**:129–132.

Ronis, Ye. Ya. and Kodola B. Zh. (1977): Stimulyatsiya tsveteniya sosny (Stimulation of flowering in Scots pine). *Lesnoe Khozyaistvo* **4**:58–59.

Tuchin, P. V. and Sirotkin, Ya. D. (1983): Dinamika muzhskogo tsveteniya klonov sosny na privivochnykh plantatsiyakh (Male flowering of Scots pine clones in seed orchards). In: Ermakov, V. I. et al. (Eds) *Vsesoyuznye soveshchanie po lesnoj genetike, selektsii i semenovodstvu,* 1–4 XI 1983, Petrozavodsk, tezisy dokladov, Vol. 2, pp. 107–108.

Wesoły, W. (1980): Studia nad intensywnością kwitnienia szczepów sosny zwyczajnej (*Pinus silvestris* L.) (Studies on the intensity of flowering and seed bearing in Scots pine, *Pinus silvestris* L., grafts). Instytut Przyrodniczych Podstaw Leśnictwa, Akademia Rolnicza, Poznań, Ph. D. thesis, p. 113.

Wesoły, W. (1982): Analiza fenologii kwitnienia szczepów sosny zwyczajnej (*Pinus silvestris* L.) na plantacji nasiennej w Siemianicach (Analysis of flowering phenology of Scotch pine (*Pinus silvestris* L.) grafts in a seed orchard of Siemianice). *Roczniki Akademii Rolniczej (Poznań)* **140**:149–159.

Wesoły, W. (1985): Effect of girdling on flowering and levels of endogenous growth regulators in embryonic shoots of Scots pine (*Pinus sylvestris* L.). *Acta Physiol. Plant.* **7 (4)**: 171–179.

Wesoły, W. (1987*a*): Wpływ warunków oświetlenia wewnątrz pączków na zawartość endogennych substancji giberelinopodobnych u szczepów sosny zwyczajnej (The influence of light conditions inside buds on the content of gibberellin-like substances in grafts of Scots pine). *Sylwan* **1**:69–75.

Wesoły, W. (1987*b*): Effect of spraying with gibberellin $A_{4/7}$ on flowering and endogenous gibberellin content in buds of Scots pine *(Pinus sylvestris)* grafts. *For. Ecol. Manag.* **19**:121–127.

Wesoły, W., Urbański, K. and Barzdajn, W. (1984): Kwitnienie i obradzanie sosny zwyczajnej (*Pinus silvestris* L.) na plantacjach nasiennych (Flowering and seed bearing in Scots pine (*Pinus silvestris* L.) seed orchads). *Sylwan* **128**:33–42.

CHAPTER 14

Hybridization

G. Kosiński

Poland

Natural hybrids

The widespread range of Scots pine (*Pinus sylvestris* L.) overlaps several other species of the subgenus *Diploxylon* (*P. pinea* L., *P. nigra* Arnold, *P. heldreichii* Christ., *P. mugo* Turra., *P. pinaster* Ait., *P. halepensis* Miller, and *P. brutia* Tenore; Critchfield and Little, 1966). However, very few data have been found on natural hybrids between the mentioned species.

The best-known spontaneous hybrids with *P. nigra* (Wettstein, 1951; Vidaković, 1958; Staszkiewicz, 1970; Boydak, 1977) have been located in the natural range of both species. Some intermediate characters of the foliage and chromosomes, fertile pollen, normal meiosis and embryo development have been observed. Introgression between *P. nigra* and *P. sylvestris* has also been described (Pravdin, 1964). However, needle resin ducts alone cannot be considered as sufficient proof of hybrid origin.

Another important natural hybrid has been described between *P. sylvestris* and *P. mugo* (Dobrinov, 1965; Marcet, 1967; Montacchini and Caramiello, 1968; Staszkiewicz and Tyszkiewicz, 1969; Prus–Głowacki and Szweykowski, 1983). In most cases the opinion of taxonomists has been based on foliage, resin ducts and morphology of cones.

According to Bobrov (1980; 1983), during the late Pleistocene and Holocene, *P. sylvestris* introgression in North-East Asia resulted in the hybridization with *P. densiflora.*

Artificial hybrids

The results of controlled crossing experiments of *P. sylvestris* with other pine species are presented in Table 14.1. The hybrid with *P. nigra* was the one most often produced. However, many crosses were unsuccessful because of incompatibility, yielding empty seeds. Vidaković (1986) assumes a relation between the likelihood of interspecific fertilization and heterozygosity of female parents. Compatibility seems to be restricted to certain combinations (Vidaković and Borzan, 1973; Vidaković, 1986). The success of hybridization may vary each year and depend on races and genotypes (Moulalis et al., 1976). To overcome incompatibility between *P. sylvestris* and *P. nigra,* the use of mentor pollen irradiated with

Table 14.1. Attempted controlled crossings of *Pinus sylvestris* with other pine species (after Giertych 1976 with more recent data added)

P. sylvestris as mother	P. sylvestris as father	References
	Reported as successful	
P. nigra, P. mugo	P. mugo	Johnson, 1939
P. mugo	P. mugo	Dengler, 1942
P. nigra	P. nigra	Johnson and Heimburger, 1946
P. contorta, P. murrayana		Kotelova, 1956
P. nigra, P. patula, P. radiata P. palustris, P. pinaster		Schmidt, 1956
	P. densiflora	Wright, 1956
P. nigra, P. pinaster		Jamblinne, 1957
P. nigra, P. mugo P. densiflora × sylvestris	P. densiflora	Wright and Gabriel, 1958
P. nigra		Schütt and Hattemer, 1959
P. densiflora × nigra	P. nigra	Vidaković, 1963
	P. nigra, P. contorta	Benea et al., 1963
P. nigra, P. mugo P. heldreichii	P. nigra, P. heldreichii	Moulalis et al., 1976
P. nigra	P. nigra*	Vidaković, 1977*a,b,c*
P. massoniana, P. nigra P. densiflora × sylvestris P. densiflora × thunbergii	P. densiflora	Garrett, 1979
P. mugo		Kormutak, 1984
	P. nigra	Vidaković, 1986
P. mugo	P. mugo, P. nigra	Kormutak and Lanakova, 1988
	Reported as unsuccessful	
P. mugo	P. mugo	Liese, 1927
P. banksiana		Dengler, 1942
P. mugo		Johnson and Heimburger, 1946
P. attenuata, P. resinosa P. thunbergii, P. banksiana P. contorta, P. muricata	P. nigra, P. contorta	Duffield, 1952

Table 14.1. Cont.

P. sylvestris as mother	P. sylvestris as father	References
	Reported as successful	
P. resinosa, P. thunbergii P. densiflora, P. massoniana P. tabulaeformis P. yunnanensis, P. taivanensis P. thunbergii × densiflora	P. resinosa, P. thunbergii P. tabulaeformis, P. nigra	Wright and Gabriel, 1958
	P. thunbergii	Wright, 1962
P. nigra, P. mugo P. brutia, P. halepensis P. nigra × densiflora	P. nigra	Vidaković, 1963
P. contorta	P. contorta	Hagman, 1964
	P. nigra	Vidaković and Jurković-Bevilaqua, 1970
P. nigra		Vidaković and Borzan, 1973
	P. pinea, P. pinaster	Moulalis et al., 1976
P. banksiana	P. banksiana	Belostotskaya, 1979
	P. nigra P. nigra × sylvestris	Vidaković, 1983
P. banksiana	P. contorta, P. nigra	Kormutak, 1984
P. thunbergii, P. virginiana	P. massoniana	Nakai, 1986
	P. thunbergii	Nakai, 1987
P. nigra, P. thunbergii P. rigida	P. rigida	Kormutak and Lanakova, 1988

* Paternal characters identified.

gamma rays, irradiating pollen of the incompatible partner or treatment by the chemomutagenic ethylnitrosourea were to some degree successful (Vidaković, 1977; Kormutak, 1984).

Besides those with *P. nigra* Arnold, fully documented hybrids were presented only with *P. mugo* Turra and *P. densiflora* Sieb. et Zucc. In spite of the difficulties, many other crosses were carried out, although only very few progenies have been raised. Some hybrids involving *P. patula* Schlechtendal et Chamisso, *P. palustris* Mill., *P. radiata* D. Don and *P. pinaster* Ait. have not been confirmed because of lack of paternal characters in the seedlings, and no available information on their performance after the first year (Schmidt, 1956; Jamblinne, 1957). The latter hybrid (*P. sylvestris* x *P. pinaster*) reported as successful has not been confirmed in reciprocal crosses either (Moulalis et al., 1976).

In contrast, in crosses with *P. densiflora,* Scots pine was fertile only as a father (Wright, 1956; Wright and Gabriel, 1958; Garrett, 1979). In a series of experiments, Wright and Gabriel (1958) found the crosses of *P. sylvestris* with *P. massoniana* as unsuccessful, whereas Garrett (1979) described the hybrid as very promising, growing faster in height and diameter than Scots pine. The hybrid *P. contorta* x *P. sylvestris* has been identified by a single biochemical character only (Benea et al., 1963).

It is worth mentioning, that even though difficult to achieve, some hybrids are fertile and the production of back-crosses and F_2 generations is possible. In a series of experiments in the USA the following multiple crosses have been successful: (*P. densiflora* x *P. sylvestris*) x (*P. densiflora* x *P. sylvestris*), *P. densiflora* x (*P. densiflora* x *P. sylvestris*), (*P. densiflora* x *P. sylvestris*) x *P. densiflora* (Wright, 1956), *P. sylvestris* x (*P. densiflora* x *P. sylvestris*) (Wright and Gabriel, 1958; Garrett, 1979). In the back-cross (*P. nigra* x *P. sylvestris*) x *P. nigra* differences in combining ability of a series of male parents have been observed (Vidaković, 1983). The reciprocal crosses have been more successful, with much less evidence of incompatibility (Vidaković, 1983; 1986). The back-crosses *P. nigra* x (*P. sylvestris* x *P. nigra*) and the open pollination of the hybrids (*P. nigra* x *P. sylvestris*) and (*P. densiflora* x *P. sylvestris*) resulted in viable offspring (Vidaković et al., 1986*a, b*). Some of the mentioned hybrids had a promising growth rate and stem form. Triple hybrids of *P. sylvestris* x (*P. densiflora* x *P. nigra*) have also been obtained by Vidaković (1963) and *P. sylvestris* x (*P. densiflora* x *P. thunbergii*) by Garrett (1979).

In studying hybridization failure, both incompatibility and sterility have been observed. Inhibition of pollen tube growth has been presented as the main reason (McWilliam, 1959; Hagman, 1964; Kormutak, 1984). Out of the crosses of *P. sylvestris* with 22 other pine species, in 19 cases very strong incompatibility was found, as well as in 4 reciprocal crosses (Hagman, 1975).

Self-pollination

There is no evidence for self-incompatibility in Scots pine preventing self-fertilization (Hagman, 1967; 1975). However, self-pollination usually yields a much higher proportion of empty seeds (Plym-Forshell, 1953; 1974; Nilsson, 1964; Koski, 1971; 1973), and in some cases all seeds have been empty (Wright and Gabriel, 1958; Koski, 1971). The amount of empty and filled seeds together is the same after self- and cross-pollination (Plym-Forshell, 1953; 1974), and their size is also the same (Sarvas, 1962).

Without pollination, ovules do not develop into seeds. After self-pollination empty seeds can be distinguished only if the megasporangium attains full size about the time of fertilization. X-ray photography and microscopic observations allow the identification of a much greater number of underdeveloped embryos after selfing than after cross- and open-pollination (Plym-Forshell, 1953; 1974; Ehrenberg and Simak, 1957). Also the percentage of germination of filled seeds is much lower after self-pollination than after cross- or open-pollination (Plym-Forshell, 1953; Ehrenberg, 1963; Nilsson, 1964). The selfed progeny shows higher mortality and slower growth, and includes dwarfs as well as various other abnormalities (Scamoni, 1950; Sarvas, 1962; Ehrenberg, 1963; Nilsson, 1964).

Literature

Belostotskaya S. Kh. (1979): Osobennosti razvitiya muzhskogo i zhenskogo gametofitov sosny obyknovennoj pri vnutri- i mezhvidovoj gibridizatsii (Peculiarities of the development of male and female gametophytes of Scots pine in intra- and interspecific crosses). *Lesovedenie* **5**:61–72.

Benea V., Leandru L. and Nitu C. (1963): Etudes physiologiques et biochimiques dans travaux d'amelioration du pin. *FAO/FORGEN, 63* **2**:5-11.

Bobrov E. G. (1980): Nekotorye cherty noveishej istorii flory i rastitel'nosti yuzhnoj chasti dal'nego vostoka (Some aspects of the recent history of the flora and vegetation of the southern part of the Far East). *Bot. Zh.* **65 (2)**:1972–1984.

Bobrov E. G. (1983): Introgressivnaya gibridizaciya i geoistoricheskiye smeny formacii taezhnoj zony SSSR (Introgressive hybridization and paleobotanical changes in the vegetation of the taiga zone in the USSR). *Bot. Zh.* **68 (1)**:3–9.

Boydak M. (1977): Possibilities of natural and artificial hybridization between *Pinus silvestris* L. and *Pinus nigra* Arn. var. *caramanica* Schn., and between *Pinus nigra* Arn. var. *caramanica* and *Pinus brutia* Ten. in Turkey. Orman Fakultesi Dergisi, Review of the Faculty of Forestry, Univ. of Istanbul, A., **27 (2)**;340–357.

Critchfield W. B. and Little E. L. (1966): Geographic distribution of the Pines of the World. U. S. Dep. of Agric. Miscl. Publ. 991.

Dengler, A. (1942): Herkunfts- und Kreuzungsversuche im Versuchsgarten des Waldbauminstituts Eberswalde. *Mitt. Dendr. Ges.* **55**:152–169.

Dobrinov I., (1965). Study on the natural hybrids between *Pinus sylvestris* and *P. mugo* var. *mughus* in Bulgaria. *Nauch. Trud. Lesoteh. Inst. Sofia* **13**:39–48. *For. Abs.* 27, No 5388.

Duffield J. W. (1952): Relationships and species hybridization in the genus *Pinus. Z. Forstgenetik* **1**:93–97.

Ehrenberg, C. E. (1963): Genetic variation in progeny test of Scots pine (*Pinus sylvestris* L.). *Studia For. Suec.* **10.**

Ehrenberg C. E. and Simak M. (1957): Flowering and pollination in Scots pine (*Pinus sylvestris* L.). *Medd. Statens Skogsforskningsinst.* **46 (12)**:1–27.

Garrett P. W. (1979): Species hybridization in the genus *Pinus.* Forest Service, U. S. Dep. of Agriculture, NE For. Exp. Sta., Res. Paper NE-436.

Giertych M. (1976): Hybridization. In: Przybylski T., Giertych M., Białobok S.; Genetics of Scots pine (*Pinus sylvestris* L.). *Ann. Forest.* **7**/**3**:59–105.

Hagman M. (1964): Incompatibility in *Pinus. Hereditas* **52 (2)**:245.

Hagman M. (1967): Genetic mechanisms affecting inbreeding and outbreeding in forest trees; their significance for microevolution of forest tree species. *14th IUFRO Congress, Munich 1967,* **Sect. 22**:347–365.

Hagman M. (1975): Incompatibility in forest trees. *Proc. R. Soc. Land. B.* **188**:313–326.

Jamblinne A. T. de (1957): Percentage de survie differentiel chez les hybrides du genre *Pinus. Züchter. Spec.* **4**:71–73.

Johnson L. P. V. (1939): A descriptive list of natural and interspecific hybrids in North American forest-tree genera. *Can. J. Res. C.* **17 (2)**:411–444.

Johnson L. P. V. and Heimburger C. (1946): Preliminary report on interspecific hybridization in forest trees. *Can. J. Rec. C.* **24**:308–312.

Kormutak A. (1984): Some cytological and biochemical aspects of interspecific incompatibility in pines *(Pinus sp.). Acta Dendrobiologica* **7**:1–92.

Kormutak A. and Lanakova M. (1988): Biochemistry of reproductive organs and hybridological relationships of selected pine species *(Pinus sp.). Acta Dendrobiologica* 1–119.

Koski V. (1971): Embryonic lethals of *Picea abies* and *Pinus sylvestris. Comm. Inst. For. Fenn.* **75 (3)**:1–30.

Koski V. (1973): On self-pollination, genetic load, and subsequent inbreeding in some conifers. *Comm. Inst. For. Fenn.* **78 (10)**:1–42.

Kotelova, N. V. (1956): K voprosu o biologii oplodotvoreniya sosny obyknovennoj (Questions of reproductive biology of Scots pine). *Nauchnotekhn. informatisiya* **23:** 13–27.

Liese J. (1927): Zur Frage der Bastardbildung zwischen *Pinus silvestris* und *montana. Forstarchiv* **12**:1–4.

Marcet E. (1967): Über den Nachweis spontaner Hybriden von *Pinus mugo* Tur. und *Pinus sylvestris* L. auf Grund von Nadelmerkmalen. *Bull. Soc. Bot. Suisse* **77**:314–361.

McWilliam J. R. (1959): Interspecific incompatibility in *Pinus. Am. J. Bot.* **46**:425–433.

Montacchini F. and Caramiello R. (1968): *Pinus mugo* and *P. uncinata* in Piemont. Notes on taxonomy and distribution. *G. bot. ital.* **102 (6)**:529–539. In: *For. Abs.* **30,** No 479.

Moulalis D., Bassiotis C. and Mitsopoulos D. (1976): Controlled pollinations among pine species in Greece. *Silvae Genet.* **25 (3–4)**:95–107.

Nakai I. (1986): Crossability in interspecific pollinations among species of hard pine in the genus *Pinus. J. Jpn. For. Soc.* **68**:406–416.

Nakai I. (1987): The affinity of Japanese red and black pines with other pine species. *J. Jpn. For. Soc.* **69**:251–257.

Nilsson B. (1964): Studies on inbreeding. *Arsb. Foren. Skogstradsforadl.:* 31–33. In: *For. Abs.* **26,** No 1782.

Plym-Forshell C. (1953): Kottens och froets utbildning efter sjalv- och korsbefrunktning hos tall (*Pinus silvestris* L.) *Medd. Skogsforskn. Inst. Stockholm* **43 (10)**:1–42.

Plym-Forshell C. (1974): Seed development after self-pollination and cross-pollination of Scots pine, *Pinus sylvestris* L. *Stud. For. Sue.* **118**:1–37.

Pravdin L. F. (1964): *Sosna obyknovennaya (Scots pine).* Nauka, Moscow.

Prus-Głowacki W. and Szweykowski J. (1983): Studies on isoenzyme variability in populations of *Pinus sylvestris* L., *Pinus mugo* Turra, *Pinus uliginosa* Neumann and individuals from a hybrid swarm population. *Bull. Soc. Amis Sci. Lett. Poznań, Ser. D.* **22**:107–122.

Sarvas R. (1962): Investigations on the flowering and seed crop of *Pinus silvestris. Comm. Inst. For. Fenn.* **53 (4)**:1–198.

Scamoni A. (1950): Über die weitere Entwicklung künstlicher Kiefernkreuzungen in Eberswald. *Züchter* **20 (1/2)**:39–42.

Schmidt W. (1956): Kreuzungsversuche zur Erzielung von Kieferhybriden. *Forsch. u. Berat. Nordrhein Westfalen* nr 2: 134–137. In: *For. Abs.* **18,** No 2500.

Schütt P. and Hattemer H. H. (1959): Die Eignung von Merkmalen des Nadelquerschnitts für die Kiefern-Bastarddiagnose. *Silvae Genet.* **8**:93–99.

Staszkiewicz J. (1970): Systematyka i zmienność (Systematics and variability). In: Białobok, S. (Ed.) *Sosna zwyczajna Pinus silvestris* L. Panstw, Wyd. Nauk, Poznań, pp. 55–77.

Staszkiewicz J. and Tyszkiewicz S. (1969): Les hybrides naturels de *Pinus silvestris* L. dans le bassin de Nowy Targ. *Bul. Acad. Sci. Ser. Sci. biol.* **17 (9)**:579–584.

Vidaković M. (1958): Investigation on the intermediate type between the Austrian and the Scots pine. *Silvae Genet.* **7**:12–19.

Vidaković M. (1963): Interspecific hybridization of several pine species from the sub-genus *Diploxylon Koehne. FAO-FORGEN* **63, 1**:2b/5.

Vidaković M. (1977): Savladivanje inkompatibilnosti pri krizanju nekih borova (Incompatibility in crossing some pine species). *Genetika* **9/1**:51–63.

Vidaković M. (1983): Viability of seeds, obtained from back crosses, as an indicator of the degree of incompatibility between *Pinus nigra* and *P. sylvestris. Ann. Forest.* **10/4**:81–88.

Vidaković M. (1986): Relation of self-fertilization and interspecific incompatibility by crossing *Pinus nigra* x *P. sylvestris. Ann. Forest.* **12/1**:1–14.

Vidaković M. and Jurković-Bevilacqua B. (1970): Observations on the ovule development following cross pollination between Austrian and Scots pines using irradiated and non-irradiated pollen. *Proceedings of IUFRO working group on sexual reproduction of forest trees. Sect. 22, Varparanta, Finland.* Finnish Forest Research Institute, pp. 16–21.

Vidaković M. and Borzan Z. (1973): Contribution to the investigations of incompatibility by crossing Scots pine and European black pine. In: Bialobok, S. (Ed.) *Proceedings of the IUFRO symposium on genetics of Scots pine, Warsaw, Kórnik,* pp. 1–19.

Vidaković M., Borzan Z., Komlenović N. and Krstinić A. (1986a): Testiranje nekih familija ćistih vrsta i pozratnih križanaca dvoiglićavih borova na krsu (Testing some families of two needle pines and their back crossed hybrids in karst). *Ann. Forest.* **12/3**:57–69.

Vidaković, M., Krstinić, A., Halambek, M. and Borzan, Z. (1986*b*): Uspijevanje nekih vrsta i hibrida dvoiglićavih borova na Durdevaćkim pijescima (Growth of some two needle pines and their hybrids on sandy soil in Durdevac). *Ann. Forest.* **12/4**:71–87.

Wettstein W. von (1951): Über Bastarde *Pinus nigra* var. *austriaca* x *silvestris. Z. Pflanzenzüchtung* **30**:473–477.

Wright J. W. (1956): Cone characteristics and natural crossing in a population of F_2 pine hybrids. *Z. Forstgenetik* **5**:45–48.

Wright J. W. (1962): Genetics of forest tree improvement. *FAO Symposium, 1962, Rome,* pp. 399.

Wright J. W. and Gabriel W. J. (1958): Species hybridization in the hard pines, series Silvestris. *Silvae Genet.* **7**:109–115.

CHAPTER 15

Progeny testing

D. Lindgren

Introduction: purpose of progeny testing

There is no generally accepted delineation between progeny testing and the wider concept of breeding. Sometimes progeny testing is given a narrow definition constraining the term to GCA-ranking of parents, but this narrow definition will not be used in this paper. (GCA stands for General Combining Ability.)

Progeny testing is concerned with the real backbone of breeding: creation of progenies; establishment of experiments with a family structure; statistical evaluation; genetic interpretation of results and breeding decisions.

An important reason for arranging tests with progenies is to get fresh genotypes for the next cycle of breeding population. The phrase progeny testing has often been interpreted in such a way that this aspect was overlooked or postponed to an unfortunate degree. Because of these undesirable consequences "parental test" may be a more adequate term for the narrow-sense interpretation of progeny testing.

Mating and mating designs

Parents

The mating design is affected by the availability of genotypes to mate. Scots pine does not seem to be a species suitable for field crosses, or for the use of mature trees and young trees in tests. In forest conditions, female flowers usually develop at a late age. They are widely dispersed and difficult to reach in the crown. Pollen collection may often be possible, although timing is considerably difficult as the time available for pollen collection is short. Collection of wind-pollinated cones is usually simpler.

Scots pine grafts function reproductively rather well. Female flowers develop relatively early on grafts, and it is often possible to complete crossing programmes with a limited amount of time-consuming climbing. The risk of ontogenetically caused differences between families ought to be lower in grafts than in forest conditions because of the more uniform environment.

It is also often worthwhile to apply treatment to promote early flowering on grafts or seedlings.

Wind pollination

Wind pollination has many advantages. The method is simple from a practical point of view. Seed in large quantities can be obtained cheaply and fast from many trees.

But there are some disadvantages and uncertainties connected with wind-pollinated seeds. The pollen parents are not selected, thus wind-pollinated progenies have a limited value as a source for second generation selections. As half of the genes in the progeny originates from an unknown father, it means that half of the potential information is not used, but transformed into statistical noise. A reason for using the term "wind pollination" or "open pollination" rather than "half sibs" is just this uncertainty about the pollen parents. Neither the amount nor the effect of selfing or inbreeding will be constant between genotypes. What is interpreted as GCA effects may actually be caused by selfing or by mating with relatives. The fertilizing pollen mix will be different for each seed lot according to the location of tree, strobilus, ovule, phenology, year, etc. Sampling effects may be interpreted as differences between mothers. Sometimes one may have to wait for seeds.

Wind pollination has its major advantage in rapid parental ranking. Seeds can be collected and the testing started soon after the identification of a good phenotype. At the same time some grafts can be made and stored in an archive or used in a seed orchard. When the progeny test has been evaluated, superior grafted clones could be used in seed orchards or as parents for controlled crosses. Wind-pollinated seeds can be collected from young grafts before male flowering has started. This has been done on a large scale in the Finnish breeding programme. Theoretically the pollen parents are expected to be similar for all clones at one location, and no disturbing selfing will occur.

Selection from wind-pollinated families is often applied. Wind-pollinated families carry genes from more pollen parents than the same number of full-sib families, and thus this is a way of broadening the breeding population. For the superior seed parents it is likely that the top ranking phenotype in a wind-pollinated progeny is inferior to that of the original seed parent, thus this method of second generation selections is usually not efficient.

Scots pine is probably a species for which meaningful information for ranking of the breeding values of the seed parents can be obtained based on the performance of wind-pollinated progenies. Isozyme studies indicate that the amount of selfing is not high, and thus not likely to cause variation of the same magnitude as that caused by genetic variation among the parents. One third, but not much more, of a progeny may be full sibs (Shen et al., 1981). Probably a higher ratio would be necessary to seriously disturb the ranking. Much of the pollen comes from some distance and can be expected to be similar for all seed parents. Neighbours in a stand are not likely to be close relatives (Yazdani et al., 1989).

Controlled crossings

The pollen parent is kept under control by isolation and application of known pollen. Scots pine breeding has probably to rely on controlled crosses in regions where it is a common tree, as this is the only realistic way to avoid unimproved and unknown pollen.

The difficulties connected with the production of controlled crosses are large for Scots pine. The costs and the demand for a well-administered organization are high. The exact date when an operation like pollination is needed cannot be accurately predicted. The time of the season is extremely busy. The field work is often carried out in remote places. Suitable equipment must be arranged. Delays and complications occur because of reasons like lack of updated information, inaccessibility of certain genotypes, insufficient flowering, or the pollen becomes available too late or turns out to be non-viable. Mistakes leading to erroneous pedigrees are easily made. Weather and animals reduce cone and seed yield in an unpredictable way, causing imbalance in the experiments. The long generative cycle makes it time-consuming to make complementary crossings.

Selfing

There are theoretical advantages with selfing, and it is possible to get selfed progenies from Scots pine (Lindgren, 1975). As selfing conserves the effective population size, and it does not increase coancestry and does not "dilute" valuable genes, it may be used for recombining the genes of the best general combiners (Lindgren, 1986). Theoretically selfing has one big advantage for parental ranking. The progeny gets all its genes from the parent tested. There are considerable practical difficulties in producing sufficient amount of seeds and plants from selfing of Scots pine, as well as difficulties generated by the low vigour of many S_1 plants. There are few examples of successful applications of selfing in conifer breeding, and it is not justified to recommend this method for more than limited use. There are large differences between genotypes in yield of sound seeds following selfing. Sometimes selfing has been used for selection against self-fertile clones in seed orchards.

Polycross

The term refers to the intentional mixing of several sources of pollen into a pollen-cocktail. This design is efficient for ranking parents. As the pollen parents of individual members in the progeny are unknown, it is not attractive for selection forwards. It is, however, possible to envisage breeding systems where the coancestry is managed by keeping sublines separate, and where complete pedigrees are not kept within sublines. In this situation the lack of knowledge of the exact father may not prevent selection forwards. Different pollen parents may contribute differently to the progeny. Polycross shares with wind pollination the disadvantage that the influence of the pollen parent on the progeny is reduced from valuable information to statistical noise.

Several crosses per parent

Mating designs with several mates per parent are required to get information about genetic parameters. If there are many crosses per parent the price will be high. For creating the next breeding generation it is desirable to choose a design which enables many unrelated selections. Common ancestry (connectedness) is necessary for some types of estimates and some procedures for ranking and evaluation. However, a breeding population carrying genes from many founders is highly desirable, and may be worth the price of constraints in parameter estimation. As it is more expensive and complicated to handle many pollen parents than many seed parents, it might be a good idea to use the same pollen parents in more crosses than the seed parents.

For progeny testing Scots pine in Sweden in the period before 1980, the most used mating designs were (Hadders, 1977):
Common testers where all mothers in a seed orchard are mated to the same set of four pollen parents. This mating design was used for nine seed orchards. Only four unrelated selections can be made from such a progeny trial. If only progeny of the better half of pollen parents is desired, only two unrelated selections can be made.
Partial dialleles according to Kempthorne and Curnow (1961) were used for 22 seed orchards. This is a connected design, and it is possible to make a fair amount of unrelated selections from the progeny.

Single pair mating

In this design each parent is mated to a single partner only. The design is used in Sweden for mating parents with known GCA for the purpose of creating a new generation of breeding material. There has been much hesitation to use it in an early stage of breeding programmes because it is not very reliable for GCA-ranking.

Field design for progeny test

Conditions during seeding, germination and plant raising are not likely to be uniform, thus it is recommended to introduce replications already at the nursery stage.

The experimental unit may be single tree plots (STP), where each tree is surrounded by trees from other genetic entries. A variation is "non-contiguous plots", where different trees belonging to the same entry are regarded as a plot in spite of their being planted separately from each other. Alternatively the plot configuration may be multi-tree row or block plots.

For progeny-testing Scots pine in Finland breeders have generally used family plots of 16 or 25 trees and 5 to 7 replications per trial. The use of single tree plot designs are likely to increase in the future (Mikola, 1985). In Sweden multi-tree plots were used initially, but later single tree plots have become standard (Werner et al., 1981). There are experiments where comparisons of single tree plots and multi-tree plots are possible (Johnsson, 1973), but it is not easy to evaluate the best design from experimental observations.

STP's are statistically efficient. The statistical sample of the environment will be large and representative. Areas not suitable for planting may be left out anywhere based on decisions on the spot. The STP design is suitable for adjusting to the environment by comparing to a response surface or a surrounding formed for a particular tree. The competing environment is similar to a forest, as a forest is a mixture of descendants from different parents rather than a single full-sib family. Wind pollination from an STP progeny test will constitute the best pollen cloud for GCA-ranking, as well as for seed harvest if the experiment is converted to a seed orchard. The design is robust to accidents affecting parts of the experiment.

The STP design will give priority to genotypes starting fast. This is a reason for choosing it for pioneer species like Scots pine. It is more suitable for assessing early individual characters than late developing characters influenced by interactions between trees. Most selection decisions will probably be made before competition between trees has got sufficient time to change rankings, so this disadvantage has limited practical importance. There are ways to follow trees for a long time and to apply thinning also in STP experiments.

Genotype – environment interactions and transfer boundaries

The same genetic material is not well adapted everywhere. Thus, breeding efforts must be targeted to geographic areas. One of the purposes of a progeny test may be to decide where to use a particular genetic material. For this purpose it is desirable that replications of the progeny test be spread over areas large enough to support such decisions. This type of problems may be approached by GEI-techniques (Genotype–Environment Interaction). Matheson and Raymond (1985) have summarised this field. There are two kinds of interactions: one which is predictably related to environment, and one more erratic. The authors suggest that, in general provenance–environment interactions are often significant, but rarely have a great effect on gain. An index was constructed expressing the difference in predicted gain as compared to assuming the interaction component of variance as zero. Reviewing non-tropical pines, the values ranged between 0.15% and 4.4%. Frequently one or a few provenances perform well in many environments and would make a reasonable selection for growing almost anywhere.

The authors suggest to reduce GEI by discarding unstable families, but identifying unstable families requires large experiments. Adding stability to the list of selection criteria will also reduce the gain for all other characters, and therefore may be uneconomical. If the breeding population is organized in sublines, there may be a possibility of considering GEI on the subline level. GEI may also differ for characters, e.g. Gullberg and Vegerfors (1987) suggest that GEI for Scots pine in southern and central Sweden should be considered for height and diameter, but not for survival and spike knots.

A model to consider the GEI which is related to the environment was constructed by Raymond and Lindgren (1990). The production of a seed source compared with a hypothetical best seed source will vary with the environment according to the following function:

$$Y = A/(1-((X-B)/C)^2),$$

where Y is the expected production of the same seed source at an environment characterized by the value X; A is the production of the seed source at the location where it performs best; X is the environment for which the production is predicted; B is the environment where the seed source is best; C is the change in environment from the optimum which will cause a 50% reduction in production.

The formula can also be used for predicting performance of different seed sources at a fixed environment using other definitions. Raymond and Lindgren (1990) applied the model to Swedish Scots pine provenance data. The environment (X) was described by latitude adjusted to a constant altitude. It was found that C = 4 latitudes is in good agreement with experimental data. For more southern localities it is likely that latitude is relatively less important, and thus larger C values should be applied. This function will help in deciding the range across which it is meaningful to test the same progeny. Based on the same Scots pine data but a different model Roberds and Namkoong (1989) made a study about the partition of the breeding population. Models for the more erratic type of GEI have been developed by Lindgren (1984, 1985) and are discussed later.

Maternal, paternal and ontogenetic effects

Cytoplasmatic DNA is not inherited in the Mendelian way as nuclear DNA. It seems that, in contrast to other plants, in conifers the chloroplast DNA is paternally inherited, while the mitochondrial DNA is mainly maternally inherited. One may question the implications these findings will have for breeding. Perhaps we should not just select for general and specific combining ability, but also for maternal and paternal effects. On the other hand, polymorphism in cytoplasmatic DNA may be of limited importance on the phenotypic variance for important characters. There is a possibility that factors like seed weight and physiological status of the seeds are important for the performance of a seed lot for Scots pine. Giertych (1974) presented results indicating that differences related to seed weight may remain for decades. Based on a study of behaviour of the same Scots pine full sib-crosses produced on three different localities situated in southern, middle and northern Sweden, Lindgren and Wang (1986) concluded that the difference between full-sib families by genetic causes are larger than the possible differences associated with the locality where the seeds were produced. There is, however, an indication that maternal effects may be important enough for contributing some uncertainty in breeding value estimates. Andersson (1989) showed that systematic effects caused by differences in soil treatment where the seed parents grow were small compared to differences between full-sib families.

Age at evaluation

There are possibilities of increasing gain per generation by making decisions early. There are also "early tests" by which growth rhythm can be evaluated in one or a few seasons. These subjects are treated in another chapter in this book (Nilsson, 1990). There are also drawbacks and risks with early selection. The relation between V_{GCA} and other variance components may change with age of the test, which affects the efficiency of selection procedures. For decisions based on field trials of Scots pine, Lindgren (1984) suggests that they be taken when the trees are around 4 m. Genetic progress may be unnecessarily slow if the breeder waits much longer than that. For some characters, such as survival and growth rhythm, decisions may be made earlier. Seed orchard rogueing may be based on earlier decisions than plus-tree selection.

Standard lots

A progeny test includes a large number of test families. There are good reasons to add further entries to make the test families comparable to other regeneration materials.

Commercial standards. For the purposes of demonstration and public relations it is of value to compare the improved materials with non-improved seed sources. The relevance of the selected commercial check lots will be uncertain.

Gain estimate standards. To get a valid estimate of the selection gain it would be a good practice to create controls. When plus trees are selected, grafted and progeny tested, the same procedures ought to be made with a few random comparison trees.

Genetic standards. The same reproducible full-sib families could be included in all tests. Then comparisons can be made of seed lots tested in different experiments at different times. For Sweden such a set of standard test families is described by Hadders (1982).

Provenance array. There may by reasons to have a standard array of checklots (provenances) covering a range of origins (Mikola, 1986).

Data handling and estimation procedures

Data are collected in the field. This may be done manually on sheets, which have been prepared for the task. The trend is towards replacing pencil by portable electronic devices. The data should be checked for different types of errors.

Backup storage, identification and documentation are important procedures. The aims and procedures of data handling are discussed in papers compiled by Lindgren (1977*b*) including examples with Scots pine. Prediction of breeding values of both the measured trees and their parents (or other relatives) are usually of primary interest. The efficiency of selection depends much on the methods, models and techniques for calculation, e.g. the following statement has often not been well understood: "Breeding values should be considered as random variables to be predicted rather than as fixed effects to be estimated." Animal breeders have been more successful in applying relevant methods for prediction of breeding values than forest geneticists.

Modern methods of variance component estimation and prediction of breeding values are described by Henderson (1984) and Kennedy (1981), both written for animal breeders. Henderson (1984) is rather complete, but not easy to read and demands good knowledge of matrix algebra, while Kennedy (1981) is more of a review article. Applications for forest tree improvement are discussed by White et al. (1989).

One particular difficulty is the unavoidable lack of balance always occurring in experiments with forest trees. With modern computers and methods these difficulties are less of a constraint.

Variance components

Variance components are based on statistical models. Sometimes it may be difficult to separate causes for variance, particularly if the design is unbalanced. Thus, maternal or reciprocal effects may be confounded into variance of general or specific combining ability.

Earlier it was regarded as important to get estimates of variance components for choosing breeding strategy. However, for this purpose estimates need not be exact for reasons given below:

— The interpretation of variance components for the estimation of genetic parameters is based on assumptions which never are fulfilled. Such assumptions are the Hardy–Weinberg equilibrium, linkage equilibrium, random sampling, etc.

— The genetic variance and covariance components depend much on the sites, populations, management regimes, traits, ages, etc. Thus, estimates cannot be reliably generalized.

— Fortunately, it seems sufficient to have an idea of the magnitude of the genetic variance components for proper strategic decisions, e.g. Lindgren (1985) found the optimal number of test sites for parental ranking to be much more dependent on the cost components than on the variance components. Matheson and Lindgren (1985) found that the degree of dominance variance is not critical for the choice between clonal forestry and seed orchards.

There is a reason why accurate estimates of variance components are important. Good estimates of variance components for a particular test, especially genetic covariances, are required for reliable predictions of breeding values. A particular need exists in situations when there is much imbalance in the data or when pooling of different sources of data becomes important. Such situations occur when information from several localities or information from several types of relatives is utilized, and also in multitrait selections or when effects of selection are quantitatively predicted. In balanced simple situations, such as ranking of parents tested by common testers, the ranking need not be strongly dependent on estimates of the variance components. However, such situations are rare and cannot be reliably predicted. Thus, it seems wise to assume that variance component estimation will remain an important part of the evaluation of many progeny tests.

If variance component estimation is the reason for progeny tests, the following points could be considered:

The required number of plants per family is not large, e.g. El-Kassaby et al. (1987) found an optimal family size of three or four trees replicated three times optimal for precise information on the genetic parameters based on a Douglas-fir trial. It is desirable to have a large number of parents, and for most purposes it is necessary that each parent be mated to more than one partner. However, in general, many parents are preferable to many crosses per parent.

To facilitate strategy decisions and experimental planning it is desirable that variance components be published. It is actually surprising how few values are reported, and how incomplete the reports are, considering the huge sums invested in establishing and measuring experiments suitable for such evaluation. Variance components in progeny tests with Scots pine were published among others by Johnsson (1973; 1977), Krusche et al. (1980), Wilhelmsson and Lindgren (1984), Andersson (1986), Park and Gerhold (1986), Eriksson et al. (1987), Velling (1988), Jiang (1988), Gullberg and Vegerfors (1987).

The variance component for SCA is usually not bigger than that for GCA for Scots pine (Johnsson, 1973; Wilhelmsson and Lindgren, 1984; Park and Gerhold, 1986; Velling, 1988).

Parental ranking

Parental ranking is required for the following types of decisions for parental genotypes:

— rogueing of existing seed orchards;
— selective harvest for cones;
— selecting for new seed orchards;
— use as parents for the next generation of the breeding population.

Rank values have been used for procedures referred to as 'culling', 'rogueing' or 'selection of genotypes'. Usually these procedures have been applied in a non-optimal way. For reasons of preserving diversity and avoiding selfing a certain number of genotypes is regarded as desirable. However, the same level of diversity could often be obtained through a larger number, but allowing the higher ranking genotypes a higher representation. The latter approach will result in a higher gain at same level of diversity or selfing. An algorithm for calculating the optimal proportions of genotypes with known breeding values in clonal seed orchards, with an application to a Scots pine seed orchard were described by Lindgren and Matheson (1986). More detailed calculations of the numerical consequences were presented by Lindgren et al. (1989).

Lindgren (1977*a*, 1984, 1985) presented models for optimization of progeny testing for parental ranking. Applied to Scots pine breeding programmes, the suggestions are:

— If resources are scarce, wind pollination with around 10 plants per progeny is the only affordable alternative. If the resources are somewhat better, polycross with around 20 plants is the method yielding the best ranking.

— Suitable family size may be 15–50 trees. It is better to use resources on creating more full-sib families per tested parent than expanding the size of individual families.

— It is better to test more parents and create possibilities to get gain by higher selection intensities than to invest on testing each parent with more than 30–100 offspring.

— Single pair mating is not the best method for parental ranking, but in many situations still gives a useful ranking of the parents at a low price (see below):

— Lindgren (1977*a*) exemplified comparison of different mating designs using variance components from a ten-year-old field trial with Scots pine (Johnsson, 1973). The variance components for height were: $V_{SCA} = 0.398\ V_{GCA}$ and V_W = variance within a full sib family = $27.8\ V_{GCA}$. The relative gain is expressed on a scale where 1,000 is the gain obtained if the true GCA were known instead of predicted. The alternatives have the same costs resulting in different progeny size.

Mating design	Progeny per parent	Relative gain
Single pair mating	143	0.621
Double crosses	43	0.703
Polycross	71	0.831

— There is an optimal number of plants per site and progeny (for a typical Scots pine progeny test this number may be in the range from 5 to 15). If funds permit more plants, it is better to increase the number of sites than the number of plants per site.

— Understanding of the cost components of the testing system is as important for an efficient design as the knowledge of the genetic variance components.

Creating a new generation of the breeding population

In the long term this is the most important purpose of generating and testing progenies. There are certain requirements the breeding population has to meet:

— Coancestry must be controlled to ensure the possiblity of non-related selections for future seed orchards.

— The effective populations size should be kept large. Genes from many initial founders should remain in the breeding population.

An additional advantage with known pedigrees is that information from relatives can be utilized for selection. It contributes to safety and decreases the risk of overemphasis of juvenile behaviour if the performance of old relatives is considered. However, keeping the pedigree by controlled crosses is expensive, administratively demanding and time-consuming. Recently it was suggested that keeping control of coancestry by multilining or sublining was better than relying on pedigree of individuals (Barnes, 1984; Cotterill, 1986). Wind pollination should probably not be trusted for recombination within a breeding population of Scots pine within its native range. Wind pollination may be acceptable where Scots pine is a rare or exotic tree of limited economic importance, provided places well-isolated from unimproved pollen can be found.

Selection is an act of balance between choosing the genotypes with the highest predicted breeding values, and of maintaining diversity. Selection between families may theoretically give more gain than selection within families. This cannot be utilized to the full extent in long-term breeding as family selection reduces effective population size considerably (Lindgren, 1986). Thus, selection within families is more important than between-family selection.

Single pair mating is an attractive choice for long-term breeding, as it is the most efficient for keeping a high effective population size. As strong between-family selection will probably not be used it is unavoidable that sometimes a good parent will be discarded because of mating with an inferior partner or because of a high negative SCA.

Literature

Andersson, B. (1986): Möjligheter till urval med avseende på virkeskvalitet i unga avkommeförsök av tall (Possibilities with selection for quality in young progeny trials of Scots pine). Institutet för Skogsförbättring. Årsbok 1985; pp. 58–80.

Andersson, B. (1989): Effect of maternal soil treatment on first year growth and autumn frost hardiness of *Pinus sylvestris* L. full-sib families. *Scand. J. For. Res.* **4 (2)**: 163–174.

Barnes, R. D. (1985): A multiple population breeding strategy for Zimbabwe. In: Barnes R. D. and Gibson, G. L. (Eds) *Provenances and genetic improvement strategies in tropical forest trees.* Department of Forestry, Commonwealth Forestry Institute, University of Oxford, pp. 619–632.

Cotterill, P. (1986): Breeding strategy: Don't underestimate simplicity! In: Weir, R. J. (Ed.) *Proceedings of a joint meeting of IUFRO Workshop on breeding theory, progeny testing and seed orchards. Virginia, October 13–17, 1986.* pp. 8–23.

El-Kassaby, Y. A., Fashler, A. M. K. and Sziklai, O. (1987): Effect of family size and number on the accuracy and precision of the estimates of genetic parameters in the IUFRO Douglas-fir provenance — progeny trial. *For. Ecol. Manag.* **18**:35–48.

Eriksson, G., Ilstedt, B., Nilsson, C. and Ryttman, H. (1987): Within- and between-population variation of growth and stem quality in a 30-year-old *Pinus sylvestris* trial. *Scand. J. For. Res.* **2**:301–314.

Giertych, M. (1974): Inadequacy of early tests for growth characters as evidenced by a 59-year-old experiment. In: Andersson, E. (Ed.) *Proceedings of IUFRO workshop on population and ecological genetics, breeding theory and progeny testing. Stockholm 1974.* pp. 237–242.

Gullberg, U. and Vegerfors, B., (1987): Genotype-environment interaction in Swedish material of *Pinus sylvestris. Scand. J. For. Res.* **2 (4):** 417–432.

Hadders, G. (1977): Avkommeprövning av i fröplantager ingående kloner. Lägesrapport hösten 1976. In: Lindgren, D. (Ed). *Evaluation of mating designs.* Department of Forest Genetics. Research Notes 23 pp. 5–18.

Hadders, G. (1982): Referensmaterial för fältförsök med tall och gran (Reference material for field experiments of pine and spruce). Institutet för skogsförbättring. Information nr 5, 1982/83, p. 4.

Henderson, C. R. (1984): *Application of linear models in animal breeding.* University of Guelph, Guelph, Canada, p. 462.

Jiang, I., B.-J. (1988): Analysis of multidata sets with examples from genetic tests of provenances of *Pinus contorta* (lodgepole pine) and full-sib progenies of *Pinus sylvestris* (Scots pine). Swedish University of Agricultural Sciences. Department of Forest Genetics. Research Notes 40. Thesis.

Johnsson, H. (1973): Avkommeprövning av en tallfröplantage baserat på tioåriga avkommor (Progeny test of a pine seed orchard, based on ten year old test families). Institutet för Skogsförbättring Årsbok 1972, pp. 140–159.

Johnsson, H. (1977): Varians- och heritabilitetsberäkningar ur försöksmaterial av olika sammansättning. In: Lindgren, D. (Ed.) Utvärdering av korsningssystem (Evaluation of mating designs). Skogshögskolan, Institutionen för skogsgenetik, Rapporter och Uppsatser 23, pp. 61–109.

Kempthorne, O. and Curnow, R. N. (1961): The partial diallel cross. *Biometrics* **17**:229–250.

Kennedy, B. W. (1981): Variance component estimation and prediction of breeding values. *Can. J. Genet. Cytol.* **23**:565–578.

Krusche, D., Das, B. L. and Stephan, B. R. (1980): Results of a progeny test with *Pinus sylvestris* and estimation of genetic gain for different selection methods. *Silvae Genet.* **29**:122–129.

Lindgren, D. (1975): Use of selfed material in forest tree improvement. Department of Forest Genetics (Stockholm). Research Notes 15, pp. 1–76.

Lindgren, D. (1977*a*): Genetic gain by progeny testing as a function of mating design and cost. *Third world consultation on forest tree breeding (Canberra),* pp. 1223–1225.

Lindgren, D. (Ed.) (1977*b*): Utvärdering av korsningssystem (Evaluation of mating designs). Skogshögskolan, Institutionen för skogsgenetik, Rapporter och Uppsatser, p. 23.

Lindgren, D. (1984): Prediction and optimization of genetic gain with regard to genotype × environment interactions. *Studia Forest. Suec.* **166**:15–24.

Lindgren, D. (1985): Cost-efficient number of test sites for field trials. *Biometrics* **41**:887–893.

Lindgren, D. (1986): How should breeders respond to breeding values? In: Weir, R. J. (Ed.) *Proceedings of a joint meeting of IUFRO workshop on breeding theory, progeny testing and seed orchards. Virginia, October 13–17. 1986.* pp. 361–372.

Lindgren, D. and Matheson, A. C. (1986): Increasing the genetic quality of seed from seed orchards by using the better clones in higher proportions. *Silvae Genet.* **35**:173–177.

Lindgren, D. and Wang, Q. (1986): Are genetic results influenced by the environment during seed maturation. In: Weir, R. J. (Ed.) *Proceedings of a joint meeting of IUFRO workshop on breeding theory, progeny testing and seed orchards. Virginia, October 13–17. 1986.* pp. 192–199.

Lindgren, D., Libby, W. L. and Bondesson, L. (1989): Deployment to plantations of numbers and proportions of clones with special emphasis on maximizing gain at a constant diversity. *Theor. Appl. Genet.* **77**:825–831.

Matheson, A. C. and Lindgren, D. (1985): Gains from the clonal and seed orchard options compared for tree breeding programs. *Theor. Appl. Genet.* **71**:242–249.

Matheson, A. C. and Raymond, C. A. (1985): Provenance x environment interaction; its detection, practical importance and use with particular reference to tropical forestry. In: Barnes R. D. and Gibson G. L. (Eds) *Provenances and genetic improvement strategies in tropical forest trees.* Department of Forestry, Commonwealth Institute, University of Oxford, pp. 81–107.

Mikola, J. (1985): Methods used for the genetic evaluation of tree breeding material in Finland. In: Tigerstedt, P. M. A., Puttonen, P. and Koski, V. (Eds) *Crop physiology of forest trees.* Helsinki University Press, Helsinki, pp. 225–231.

Mikola, J. (1986): Reference material as an aid in interpreting progeny test results. In: *Proceedings of a joint meeting of IUFRO workshop on breeding theory, progeny testing and seed orchards. Virginia, October 13–17. 1986.* pp. 98–106.

Park, Y. S. and Gerhold, H. D. (1986): Population hybridization in Scots pine. I. Genetic variance components and heterosis. *Silvae Genet.* **35**:159–165.

Raymond, A. C. and Lindgren, D. (1990): Genetic flexibility. A model for determining the range of suitable environments for a seed source. *Silvae Genet.* **39**:112–120.

Roberds, J. H. and Namkoong, G. (1989): Population selection to maximize value in an environmental gradient. *Theor. Appl. Genet.* **77**:128–134.

Shen, H. H., Rudin, D. and Lindgren, D. (1981): Study of the pollination pattern in a Scots pine seed orchard by means of isozyme analysis. *Silvae Genet.* **30**:7–15.

Velling, P. (1988): The relationships between yield components in the breeding of Scots pine. Thesis. University of Helsinki, Helsinki.

Werner, M., Hadders, G. and Rosvall O. (1981): Förädlingsprogram för tall, gran och contortatall (Breeding program for Scots pine, Norway spruce and lodgepole pine). Institutet för Skogsförbättring Årsbok 1980. pp. 59–111.

White, T. L. and Hodge, G. R. (Eds) (1989): *Predicting breeding values with applications in forest tree improvement.* Kluwer Academic Publishers, Dorndrecht–Boston–London, p. 367.

Wilhelmsson, L. and Lindgren, D. (1984): Variance components in some Swedish trials with *Pinus sylvestris* (L.). Summary of lecture In: Cotterill, D. (Ed.) *Proceedings of RWG1 (forest genetics) meeting, Perth, Australia 1983.* p. 70.

Yazdani, R., Lindgren, D. and Stewart, S. (1989): Gene dispersion within a population of *Pinus sylvestris* (L.). *Scand. J. For. Res.* **4**:295–306.

CHAPTER 16

Inheritance of resistance to biotic factors

B. R. Stephan

Introduction

Scots pine (*Pinus sylvestris* L.) is host for many different biotic agents. Under natural conditions of virgin forests a dynamic equilibrium can be assumed between populations of host and parasites (Segal et al., 1982). This balance is mostly disturbed and shifted in favour of the parasite by manifold interventions, particularly by establishment of artificial, even-aged and genetically more or less uniform plantations, often outside the area of natural distribution of the host plant. Populations are used which are not adapted to the respective plantation site or are confronted in a new habitat with new biotic agents, often resulting in serious epidemics with high losses. Therefore, the choice of well-adapted, healthy and resistant Scots pine populations must have high priority. That requires sufficient information about the biology and life cycle of the biotic agent, and a good knowledge about the resistance or susceptibility of the host plant or population against the respective agent. Resistance for the host plant is the positive result of an interaction with the biotic factor, influenced by the environment.

Resistance is considered here as the genetically controlled ability of the host to resist disease development and insect, and mammal attacks. In the literature there is much information on the behaviour of Scots pine populations towards pests and diseases under various environmental conditions, but it is sometimes difficult to distinguish between environmentally influenced absence of injury and genetically controlled resistance. Although the genetics of Scots pine has been under study for decades, information about the inheritance of resistance to biotic factors is very meagre, but this is not exceptional for forest tree species.

Resistance to fungal pathogens, insects and other animals

Provenance trials offer a good means for comparing various populations under the same environmental conditions of a given site for reactions against fungal pathogens, insects and mammals. The trials allow judgement about the adaptation of a provenance to the new habitat and possibly to the local population of pathogens (Mátyás, 1987). There are numerous references with regard to differences in the behaviour of Scots pine provenances against parasites at many different locations, as shown, e.g. by Schütt (1958/59), Dietrichson (1968), Lanier (1969), Lanier et al. (1973), Stephan (1975), Stephan and Scholz (1975), Mańka (1976), Martinsson (1979), Bridgen and Hanover (1982) and others. In general, differences in resistance exist against nearly all important biotic agents, whether they are fungal diseases or pests, although resistance to fungal pathogens has been found more often than that to insects or mammals.

Resistance to fungal pathogens,
especially to *Lophodermium* needle cast

Ample information is available on the reaction of Scots pine provenances to needle cast diseases, particularly those caused by *Lophodermium seditiosum* Minter, Staley et Millar and *Phacidium infestans* Karst. (According to Minter et al. (1978). *Lophodermium seditiosum* is the new name of the parasite which causes needle cast and all earlier information on *L. pinastri* which seems to be only a saprophyte, refer to *L. seditiosum.*) It appears that a transfer of populations from south to north increases susceptibility. Thus, use of northern and northeastern populations, e.g. from Scandinavia or Poland and northwestern USSR, further south seems to be advisable as many examples of this transfer confirm for the *Lophodermium* needle cast (Dengler, 1955; Schütt, 1964*b*; Rohmeder and Eicke, 1969; Przybylski, 1972) and for the *Phacidium* needle cast (Björkman, 1963; Ehrenberg, 1973). However, there are also contrary examples where northern provenances proved less resistant against needle cast (Przybylski and Sztuka, 1968; Fedorov et al., 1978).

These interactions could be confirmed by further examples and this complicates generalized recommendations. The reasons for such obvious contradictions are mostly unknown. Some information is provided by the general survival data from old provenance experiments (Giertych, 1979; Giertych and Oleksyn, 1981; Oleksyn and Giertych, 1984). Populations which have in general the lowest mortality are those from the Baltic region and from central west USSR, from Bialorussia to Volynia in the Ukraine. This is an indication of a general resistance to a variety of biotic and abiotic factors.

Progenies from crossings, selected mostly for other characters than for resistance, and the offspring of single trees likewise show differences in behaviour towards pathogens. For instance, a strong genetic effect responsible for 58% of the total variation between single tree progenies in susceptibility to *Lophodermium* needle cast disease was found in the Netherlands (Squillace et al., 1975). Dutch progenies, particularly those from northwestern Netherlands, were significantly more resistant than German ones. A natural and artificial selection during past generations can be assumed, for Scots pine is not native in the Netherlands and was presumably imported in the past from Germany and elsewhere. The result emphasizes the significance of adaptability. In various cases the local populations were more adapted to a given site than provenances from other regions. Full-sib progenies of Scots pine trees selected for their resistance to *Lophodermium* needle cast at higher elevations of the Black Forests (southwestern Germany) were infested very seriously when transferred to lower elevations in the Rhine valley, whereas the adjacent local Scots pines remained healthy (Stephan, 1984).

It is a common observation that in stands of attacked Scots pine a few individual trees are not infested or are attacked to a lesser extent compared with the others. However, such trees can be healthy because, by chance, they have not been infected, or because they possess genetic resistance. This has to be checked with clonal material of those single trees under controlled conditions which are similar to nature. Korczyk (1984) observed that differences in resistance are evident only to a certain threshold of infection intensity.

In relation to *Lophodermium* resistance Schütt (1957*a,b*) found more than 100 healthy trees with an estimated frequency of about one tree per 50 ha of young Scots pine stands. Clonal material of these single trees was inoculated artificially and showed a similar behaviour as the original trees in the stands. Successful selection of obviously resistant trees is reported also for the fungal parasites *Endocronartium pini* (Pers.) Hiratsuka (=*Peridermium pini* (Pers.) Lev.) (Klingström, 1967), *Heterobasidion annosum* (Fr.) Brefeld (=*Fomes annosus* (Fr.) Coke) (Chudnyi, 1976; Ladejtschikova and Pasternak, 1982), *Lophodermella sulcigena* (Rostr.) Höhnel (Jalkanen, 1981; 1982), *Melampsora pinitorqua* Rostrup (Jonsson et al., 1978), and in *Cyclaneusma minus* (Butin) DiCosmo, Peredo et Minter (=*Naemacyclus minor* Butin) (Merrill and Slover, 1983).

Resistance to insects

There are only a few examples of resistance of Scots pine to insects. In American field trials, especially northern- and Central-European provenances were highly susceptible to the root collar weevil (*Hylobius radicis* Buchanan) (Wright and Wilson, 1972) and the shoot and twig feeding Zimmerman pine moth (*Dioryctria zimmermani* Grote) (Wright et al., 1975), whereas the attack of terminal shoots by the white pine weevil (*Pissodes strobi* Peck) (Wright et al., 1976) and by the pine shoot borer (*Eucosma gloriola* Heinrich) (Steiner, 1974) was the lowest. Such examples indicate that resistance is specific for each pest. Those populations which were least attacked by one pathogen or pest might be the most heavily attacked ones by another. But selection of individual trees resistant to insect species can be successfully made (e.g. Bridgen and Hanover, 1982).

Resistance to other animals

Some information show also a relation between resistance and attack by mammalian herbivores. In northern Finland, Rousi and Häggman (1984) observed in a clonal collection and a natural stand that certain individual trees were browsed by the arctic hare (*Lepus timidus* L.), while others were not. There is also information from eastern Finland about browsing preferences between individual trees or clones by moose (*Alces alces* L.) (Löyttyniemi and Hiltunen, 1978; Haukioja et al., 1983).

Conclusions

Although, as mentioned above, generally valid recommendations from the field trials with provenances, progenies or single trees cannot be given because these results were obtained at different test locations, in different years, from trees of various ages and often with not comparable populations, the following conclusions can be made:

1. Between populations there exists a great and inherent variation in resistance against many important diseases and pests which attack Scots pine. On the same site certain populations show significantly less attack than others.

2. Within populations, even within susceptible ones, there are individual trees with a higher resistance. In some cases the variation is even greater between individual trees within a population than between populations (e.g. Merrill and Slover, 1983).

3. Populations or individual trees with an absolute resistance or immunity have not been observed so far. Even resistant populations or single trees become severely infested when a certain threshold of infection intensity is exceeded.

4. Resistance against one biotic factor does not necessarily imply resistance against another pathogen or pest.

5. Results of field test studies are generally applicable only for the site in question and its surroundings and can strongly be influenced by genotype–environment interactions (e.g. Hattemer, 1965; Siwecki, 1967; Przybylski, 1972).

Present knowledge about resistance mechanisms

For a long time special interest was directed at the mechanisms of resistance against diseases and pests. Emphasis has been laid on studies of preformed or induced factors. Some of these morphological, anatomical, physiological or biochemical factors, which were shown to have a relation with resistance, are listed in Table 16.1.

In several cases a correlation exists between a specific character of the host plant and resistance against a biotic factor. In other cases, and with other material, the relation between the same characters could, or could not be confirmed. Therefore, one can state that the present knowledge about resistance mechanisms is not well grounded enough for generalized recommendations. This is also true of our knowledge about the suitability of resistance factors for early diagnosis.

Resistance mechanisms against fungal pathogens

Some growth characters appear to be related with resistance. But many growth characters mentioned as related to resistance can be by-products of climatic adaptation, e.g. characters connected to the annual growth rhythm, the degree of

Table 16.1. Traits with relation to resistance of Scots pine against biotic agents (? = relation not absolutely clear)

Traits related with resistance	Efficient against	References
General characters		
Slow growth	*Endocronartium pini*	Martinsson and Nilsson, 1987
Dwarf growth	*Phacidium infestans*	Roll-Hansen, 1975
Fast growth	*Lophodermium seditiosum*	Squillace et al., 1975
Winter foliage colour, date of bud burst	*Eucosma gloriola*	Steiner, 1974
Anatomy, physiology	*Heterobasidion annosum*	Ladejtschikova and Pasternak, 1982
Low vigour of terminal growth	*Pissodes strobi*	Wright et al., 1976
Ultrastructure	*Melampsora pinitorqua*	Jonsson et al., 1978
Three different traits	*Lophodermium seditiosum*	Siwecki et al., 1975
High needle weight	*Lophodermium seditiosum*	Janson, 1975
High dry matter content of	*Phacidium infestans*	Björkman, 1963
needles	*Alces alces*	Haukioja et al., 1983
High seed weight	*Melampsora pinitorqua*	Gavris, 1939; Martinsson, 1980
Physiological characters		
Less nutritious foliage	*Neodiprion sertifer*	Wright et al., 1967
Water-retaining capability of needles	*Lophodermium seditiosum*	Kotov and Kotova, 1981
Press sap of needles	*Lophodermium seditiosum*	Schütt, 1964*a*, 1966
Buffering of pH in needles	*Lophodermium seditiosum*	Review by Scholz and Stephan, 1982; Schachler and Bolland, 1986
Biochemical characters		
Erucic acid in black seed shells	damping-off fungi	Grzywacz and Rosochacka, 1977, Rosochacka and Grzywacz, 1980
Toxic substances of needles	*Neodiprion sertifer*	Wright et al., 1976
High phenol content	*Lepidus timidus*	Rousi and Häggman, 1984
Cuticular waxes of needles (?)	*Lophodermium seditiosum*	Schütt, 1971
C_9-alcohol, C_{31}-paraffin content of waxes (?)	*Lophodermium seditiosum*	Schuck, 1972
Acid ether extract (?)	*Melampsora pinitorqua*	Klingström, 1969
Resin acid concentration	various insects	Bridgen and Hanover, 1982
High resin content	*Aradus cinnamomeus*	Baumanis et al., 1982
Terpente content	*Heterobasidion annosum*	Chudny, 1976
	Rhyacionia buoliana, Aradus cinnamomeus, Blastophagus piniperda	Smelyanets, 1977
	Alces alces	Haukioja et al., 1983
High α-pinene content	*Neodiprion sertifer, Diprion pini*	Chudny et al., 1974
Low Δ3-carene and terpinolene content	*Dioryctria zimmermani*	Wright et al., 1975

winter hardiness, the time of shoot flushing, or, in general, the geographic origin. A differentiation is mostly difficult, as the following examples show.

Slow-growing Scots pines were found to be more resistant to the blister rust fungus *Endocronartium pini* (Martinsson and Nilsson, 1987), and dwarf plants to the needle cast fungus *Phacidium infestans* (Roll-Hansen, 1975). In the latter case a connection may exist with the higher dry matter content of needles (Björkman, 1963; Janson, 1975). Fast-growing trees were more resistant to *Lophodermium* needle cast in Dutch studies (Squillace et al., 1975), and a similar result was obtained with progenies of controlled crossings (Stephan and Krusche, 1986). The faster growing progenies were obviously better adapted to the test locations than the slow-growing ones, and that was manifested simultaneously in a higher resistance to *Lophodermium.* Also such characters like high seed weight, or various anatomical and ultrastructural traits are sometimes correlated with resistance, but we do not know the causal connections (see Table 16.1).

Physiological and biochemical processes are of greatest importance for the interaction between host and parasite. Following the infection host and parasite are in close contact. The interaction between metabolic processes of virulence and resistance determines pathogenesis and decides the health or infestation. But for Scots pine, there is not much information available about the physiological reactions in relation to resistance (see Table 16.1).

A close relation has been shown between a higher water retention capacity of needles and *Lophodermium* resistance in half-sib families (Kotov and Kotova, 1981). On the level of provenances Hattemer (1964) could not find any relation between the pH or the osmotic pressure of the cytoplasm and *Lophodermium* resistance. The investigation of selected clones under controlled or semi-controlled conditions revealed more detailed relations. Needle extract of resistant clonal material inhibited the growth of *Lophodermium in vitro* (Schütt, 1964*a*). In other investigations a relation between *Lophodermium* resistance, buffering capacity of homogenized needles and the pH-regulating ability of *Lophodermium seditiosum* could be found (Scholz and Stephan, 1974). These results were confirmed with other Scots pine material (see review by Scholz and Stephan, 1982; Schachler and Bolland, 1986), but not in all cases (e.g. Martinsson, 1979; Stephan and Scholz, 1981), demonstrating that buffering capacity is not the only trait controlling resistance.

The specific impedance of needles and the resistance of Finnish provenances against the shoot disease *Gremmeniella abietina* (Lagerb.) Morelet (= *Scleroderris lagerbergii* Gremmen) has been investigated by Repo et al. (1984), but according to their studies it was not possible to identify susceptible seedlings and provenances.

The fact that fungal spores of needle parasites germinate and grow on the needle surface inspired investigations on the chemistry of the cuticle. But the composition of cuticular waxes showed a weak relation to the *Lophodermium* resistance on the level of provenances, although particularly the C_9-alcohol and C_{31}-paraffin content seemed to have some influence (Schuck, 1972). Some components of the cuticular waxes influence ascospore germination of *Lophodermium* (Schütt, 1971).

Especially in the case of resin acids a great qualitative and quantitative variation between seed sources has been demonstrated. They are under strong genetical control, are hardly influenced by the environment, and are therefore useful markers of indirect selection of resistance (see review by Bridgen and Hanover, 1982). The high monoterpene content of trees inhibits the development of *Heterobasidion annosum* (Chudnyi, 1976). The effect of various terpene components is different and cannot be generalized (see Table 16.1). In some instances no relation with resistance could be found (Gref, 1982). Other biochemical traits such as the content of erucic acid in black seed shells, as well as some unknown toxic substances in needles and extracts of pine shoots can play a significant role in the resistance to damping-off pathogens (Grzywacz and Rosochacka, 1977; Rosochacka and Grzywacz, 1980).

Resistance mechanisms to animals

As mentioned above, the annual growth rhythm, the nutrient content, and many other pecularities of a tree can result in resistance. Steiner (1974) found highly significant differences between Scots pine varieties in resistance to a pine shoot borer *(Eucosma gloriola).* Resistance was correlated with latitude of origin and, consequently, with winter foliage colour and date of bud burst. Yellow varieties from northern regions with earliest bud burst were the most resistant. Green varieties from southern regions burst bud latest and were least resistant. The preference of the insect *Pissodes strobi* to lay its eggs on vigorous leader shoots, can be considered as a kind of resistance in the case of slow-growing Scots pine races (Wright et al., 1976). A lower nutrient content and the obvious toxicity of needles result in reduced attacks of the Scots pine variety *uralensis* by the previous year's foliage feeding European pine sawfly (*Neodiprion sertifer* Geoff) (Wright et al., 1967).

Regarding the biochemical characters, the significance of concentration and composition of the resin acid content, and particularly the terpenes, in connection with the resistance or susceptibility to insects must be mentioned (Bridgen and Hanover, 1982). Terpenes can have an attractive or a repellent effect on various insects (Smith, 1972). In many cases a high content of resin or of individual terpene compounds have a positive effect on resistance (Smelyanets, 1977; Chudnyi et al., 1974; Baumanis et al., 1982; Bridgen and Hanover, 1982). But resistance to the Zimmerman pine moth *(Dioryctria zimmermani)* in varieties from southern Europe and Asia Minor may be due to low levels of Δ 3-carene and terpinolene in the cortical oleoresin.

Apparently, there is also resistance against mammalian herbivores. A high total phenol content was found in Scots pines not browsed by the arctic hare *(Lepus timidus)* (Rousi and Häggman, 1984). Scots pine clones not browsed by moose *(Alces alces)* had a higher dry matter percentage, higher concentrations of Ca and Mg, and higher terpene contents of abienol, ß-guaiene, ß-elemene and α-copaene (Haukioja et al., 1983). On the other hand, Löyttyniemi and Hiltunen (1978) found that terpene content apparently plays no important role in the browsing preference by moose.

Conclusions

With respect to the knowledge of resistance mechanisms one should add that some relations were found between a specific trait and resistance, detailed information about the precise genetic interactions even for serious pathosystems with Scots pine are still lacking. In most cases it is not yet clear whether resistance is caused by an incompatibility between host and parasite or by a hypersensitive reaction, of the host. One can state that a large spectrum of both qualitative and quantitative reactions of the host can be observed.

The following conclusions can be drawn:

1. Resistance to biotic factors is related to different morphological, anatomical, physiological or biochemical characters. In some cases a close correlation has been shown, but this correlation need not be valid for all Scots pine populations and not for all sites.
2. In general, resistance and/or susceptibility show a wide variation with gradual transitions between the extremes.
3. Resistance to a specific pathogen or pest can be governed by several different traits.
4. The present knowledge about resistance mechanisms in Scots pine is not sufficient for early diagnosis.

Inheritance and breeding

Summarizing all the information on Scots pine it seems that the variation in resistance observed between populations is often genetically controlled. However, at present the genetic background of resistance to biotic factors is not yet known well enough to breed resistant progenies in Scots pine. Although resistance mechanisms and their expressions have been investigated for some time, no specific genes have been identified so far. Nevertheless, a few data exist with respect to the inheritance of resistance in Scots pine which allow some conclusions. Most information is available on the pathosystem of Scots pine/*Lophodermium* needle cast disease. Dengler (1942) was one of the first to cross the needle cast susceptible Scots pine with the closely related and fairly resistant *Pinus mugo.* Unfortunately the progeny generally showed a combination of slow growth and poor form (from the parent *P. mugo*) and high susceptibility to *Lophodermium* (from the parent *P. sylvestris*).

Analyses of tests with provenances and crossings show that the inheritance of resistance to biotic factors is probably polygenic (Langner, 1952; Bolland, 1957, 1971; Hattemer, 1965; Šindelář, 1981).

In a few crossing experiments estimation of heritability was possible. Bolland (1968; 1971) calculated in the case of resistance to resin-top disease *(Endocronartium pini)* heritability values between 0.38 and 0.79 and showed a polygenic inheritance. In a progeny trial with 36 crossbred families, Johnsson (1976) found in the resistance to *Lophodermium* needle cast an unusually high heritability value of 0.66 and assumed cumulative effects of several recessive gene loci. Another progeny trial with a total of 96 families at two locations, for *Lophodermium* needle cast resistance, yielded heritability values from 0.32 to 0.57 for four different intraprovenance and interprovenance-mating groups (Stephan and Krusche, 1986). In a progeny trial with progenies of plus and minus trees Ehrenberg (1973) found a high amount of heterosis in the *Phacidium infestans* resistance of hybrids. They exceeded by far the midparent population in survival and resistance.

In all these studies continuous variation of the resistance trait to fungal pathogens has been found even within plots, and a low or lacking dominance variance, indicating polygenic inheritance and additive effects. Therefore, one can assume that resistance of Scots pine to biotic agents, at least against needle cast and resin-top pathogens, is horizontal in the sense of Robinson (1976). This type of resistance provides a much higher security if variation within the parasite population is taken into consideration.

Breeding programmes for resistance are very rare in Scots pine. It has been a secondary objective, if any, within programmes directed to other, economically more important characters like performance, wood quality, form, etc. A strong additive genetic variance together with a relatively high heritability, as shown in the above-mentioned examples, indicate that selection of resistant trees and their random mating in a seed orchard would give improved resistance in the offspring. Resistant parents can be expected to yield a high frequency of resistant offspring (Bolland, 1968, 1971; Ehrenberg, 1973; Johnsson, 1976; Martinsson, 1979; Stephan and Krusche, 1986). Therefore, there is a need for further breeding to incorporate the very important trait resistance into improvement programmes with Scots pine.

One has to ask whether the current conservative search for resistance on the level of provenances, open-pollinated progenies or single trees will be sufficient for future resistance breeding work. In vitro culture techniques for early tests (e.g. Werner, 1987) and the induction and utilization of somaclonal variation will offer new possibilities. The main focus of resistance identification, early testing, infection methods and incorporated breeding work should be on the standardization of both materials and methods. Not only the host plant material for testing has to be standardized, but also the pathogenic material. Studies in this field must be intensified. When it will be clarified what causes the resistance of the host and the virulence of the parasite and which genes are involved, the wide diversity of resistance in Scots pine will be effectively utilized for the maintenance of healthy forests. "Despite all difficulties: Breeding for resistance to pests in forestry is absolutely mandatory" (Zobel, 1982).

Literature

Baumanis, I., Pirāgs, D. and Spalvinš, Z. (1982): Resistance trials of Scotch pine clones in the Latvian SSR. In: Heybroek, H. M., Stephan, B. R. and von Weissenberg, K. (Eds) *Resistance to diseases and pests in forest trees.* PUDOC, Wageningen, pp. 448–449.

Björkman, E. (1963): Resistance to snow blight (*Phacidium infestans* Karst.) in different provenances of *Pinus silvestris* L. *Stud. For. Suec.* **5**:1–16.

Bolland, G. (1957): Resistenzuntersuchungen vor allem über Kienzopf und Schütte an der Kiefer. *Der Züchter* **27**:38–47.

Bolland, G. (1968): Untersuchungen über Möglichkeiten zur Verringerung der Schütteanfälligkeit der Kiefer durch züchterische Maßnahmen. F/E-Bericht des Instituts f. Forstwissensch., Eberswalde. 22 pp. (cit. in Schachler and Bolland, 1987).

Bolland, G. (1971): Über Bedeutung, erbliche Abhängigkeit und Bekämpfungsmöglichkeiten des Kienzopfes *Peridermium pini* (Willd.) Kleb. *Beitr. Forstwirtsch.* **5 (3)**:157–158.

Bridgen, M. R. and Hanover, J. W. (1982): Indirect selection for pest resistance using terpenoid compounds. In: Heybroek, H. M., Stephan, B. R. and von Weissenberg, K. (Eds) *Resistance to diseases and pests in forest trees.* PUDOC, Wageningen, pp. 161–168.

Chudny, A. V. (1976): O razmeshchenii derev'ev v populatsiyakh sosny obyknovennoj (The distribution of trees in Scots pine stands). *Lesovedenie* **5**:63–68.

Chudnyi, A. V., Maslov, A. D. and Saradshisvili, K. G. (1974): Zavisimost' ustojchivosti sosny k pilil'shchikam ot sostava terpentinnykh masel (The dependence of resistance of Scots pine to Diprion pini on the composition of terpene oils). *Lesnoe Khozyaistvo* **7**:81–83.

Dengler, A. (1942): Herkunfts- und Kreuzungversuche im Versuchsgarten des Waldbauinstituts Eberswalde. *Mitt. Deutsch. Dendrol. Ges.* **55**:157–169.

Dengler, A. (1955): Schütteversuch mit finnischen und märkischen Kiefern. *Archiv Forstwes.* **4**:4–8.

Dietrichson, J. (1968): Provenance and resistance to *Scleroderris lagerbergii* Gremmen *(Crumenula abietina* Lagerb.). The international Scots pine provenance experiment of 1938 at Matrand. *Medd. Norske Skogforsøksves.* **25**:395–410.

Ehrenberg, C. (1973): The effect of increased heterozygocity in pine plants (*Pinus sylvestris* L.). *Proceedings of the international symposium on genetics of Scots pine, IUFRO* S2. 03. 05. *8–18 October, 1973, Warszawa—Kórnik,* p. 15.

Fedorov, N. I., Mantsevich, E. D. and Yakimov, N. I. (1978): Porazhennost' hvoi shyutte obyknovennym v geograficheskih kul'turakh sosny (Needle cast caused by *Lophodermium pinastri* in provenance plantations of Scots pine). *Lesnoj Zh.* **21 (2)**:10–12.

Gavris, V. P. (1939): Selekciya na vysokuyu ustojchivost' form sosny obyknovennoj (On the selection of rust-resistant forms of Scots pine). *Proc. Lenin Acad. Agric. Sci.* **10**:29–30.

Giertych, M. (1979): Summary of results on Scots pine (*Pinus sylvestris* L.) height growth in IUFRO provenance experiments. *Silvae Genet.* **28 (4)**:136–152.

Giertych, M. and Oleksyn, J. (1981): Summary of results on Scots pine (*Pinus sylvestris* L.) volume production in Ogievskij's pre-revolutionary Russian provenance experiments. *Silvae Genet.* **30 (2–3)**:56–74.

Gref, R. (1982): Diterpenoids and some water-soluble compounds of *Pinus sylvestris* and their possible relationship to resistance to fungal parasites. Thesis, Sveriges Lantbruksuniversitet, Umeå, p. 29.

Grzywacz, A. and Rosochacka, J. (1977): Attempt at elucidation of the role of fatty acids in the resistance of *Pinus silvestris* L. seeds to infection by damping-off fungi in dependence on the colour of their seed shells. *Acta Soc. Bot. Pol.* **46**:569–575.

Hattemer, H. H. (1964): Die Reaktion und der osmotische Wert des Nadelzellsafts von Kiefern (*Pinus silvestris* L.) verschiedener geographischer Herkunft im Zusammenhang mit deren Anfälligkeit gegen die Schütte (*Lophodermium pinastri* (Schrad.) Chev.). *Mitt. Bundesforschungsanst. Forst-u. Holzwirtsch.* **56**:104.

Hattemer, H. H. (1965): Das Verhalten einiger Kiefern-Herkünfte gegenüber der Schütte. *Allg. Forst. Jagdztg.* **136**:242–248.

Heybroek, H. M., Stephan, B. R. and von Weissenberg, K. (Eds) (1982): *Resistance to diseases and pests in forest trees.* PUDOC, Wageningen. p. 503.

Haukioja, E., Huopalahti, R., Kotiaho, J. and Nygrén, K. (1983): Millaisia männyntaimia hirvi suosii? (The kinds of pine preferred by moose?). *Suomen Riista* **30**:22–27.

Jalkanen, R. (1981): *Lophodermella sulcigena* on pines. A literature review. *Folia Forest.* **476**:15 pp.

Jalkanen, R. (1982): *Lophodermella sulcigena* in clones and progenies of Scots pine in Finland. In: Heybroek, H. M., Stephan, B. R. and von Weissenberg, K. (Eds) *Resistance to diseases and pests in forest trees.* PUDOC, Wageningen, pp. 441–447.

Janson, L. (1975): Czynniki warunkujace stopien porazenia mlodnikow sosnowych przez osutke (*Lophodermium pinastri* (Schrad.) Chev.) (Factors conditioning the extent of the infestation of pine thickets by *Lophodermium pinastri*). *Sylwan* **119 (3)**:52–66.

Johnsson, H. (1976): Attack by *Lophodermium pinastri* in a 6 × 6 diallel of *Pinus sylvestris.* Årsbok, Föreningen Skogsträdsförädling Institutet för Skogsförbättring, Årsbok 1975, pp. 74–78.

Jonsson, A., Holmvall, M. and Walles, B. (1978): Ultrastructural studies of resistance mechanisms in *Pinus sylvestris* L. against *Melampsora pinitorqua* (Braun) Rostr. (pine twisting rust). *Studia For. Suec.* **145**:28.

Klingström, A. (1967): Current research on *Peridermium pini. XIV. IUFRO Congress, München.* Vol. V, Section 24, pp. 375–381.

Klingström, A. (1969): *Melampsora pinitorqua* on progenies of *Pinus sylvestris* and in relation to growth regulating substances. *Studia For. Suec.* **69**:74.

Korczyk, A. (1984): Badania nad podatnoscia potomstw drzew doborowych *Pinus sylvestris* L. 8 proweniencji z Polski na porazenie przez *Lophodermium pinastri* (Schrad.) Chev. (Studies on the susceptibility of the progeny of plus trees of 8 Polish Scots pine provenances to attack by *Lophodermium pinastri*). *Sylwan* **128 (1)**:43–57.

Kotov, M. I. and Kotova, L. I. (1981): Ob otbore sejancev sosny na ustojcivost' protiv sjutte (Selecting Scots pine seedlings for resistance to needlecast). *Lesnoe Khozyaistvo* **11**:55–58.

Ladejtschikova, E. I. and Pasternak, G. M. (1982): Biochemical aspects of the resistance of *Pinus sylvestris* to *Fomes annosus.* In: Heybroek, H. M., Stephan, B. R. and von Weissenberg, K. (Eds) *Resistance to diseases and pests in forest trees.* PUDOC, Wageningen, pp. 198–205.

Langner, W. (1952): Züchtung auf Resistenz gegen *Lophodermium pinastri* bei *Pinus silvestris. Zeitschr. Forstgenetik* **1**:59.

Lanier, L. (1969): Contribution à l'étude du rouge cryptogamique dû pin sylvestre dû au *Lophodermium pinastri* (Schrad.) Chev. Thesis, Nancy. p. 177.

Lanier, L., Lacaze, J. F. and Millier, C. (1973): Lutte par voie génétique contre le rouge cryptogamique des pins. *Eur. J. For. Path.* **3**:97–105.

Löyttyniemi, K. and Hiltunen, R. (1978): Monoterpenes in Scots pine in relation to browsing preference by moose (*Alces alces* L.). *Silva Fenn.* **12**:85–87.

Mańka, K. (1976): Resistance to pathogenic agents. In: Białobok, S. and Zelawski, W. (Eds) *Outline of physiology of Scots pine.* PWN, Warszawa, pp. 331–342.

Martinsson, O. (1979): Testing Scots pine for resistance to *Lophodermium* needle cast. *Studia For. Suec.* **150**:63.

Martinsson, O. (1980): Testing Scots pine for resistance to pine twist rust. *Folia Forest.* **422**:25–31.

Martinsson, O. and Nilsson, B. (1987): The impact of *Cronartium flaccidum* on the growth of *Pinus sylvestris. Scand. J. For. Res.* **2**:349–357.

Mátyás, C. (1987): Adaptation processes in forest tree populations. Thesis, Sopron, p. 164.

Merrill, W. and Slover, S. (1983): *Naemacyclus* needlecast resistance of twelve Scots pine seed sources. *Am. Christ. Tree J.,* 1983: 33–34.

Minter, D. W., Staley, J. M. and Millar, C. S. (1978): Four species of *Lophodermium* on *Pinus sylvestris. Trans. Br. Mycol. Soc.* **71 (2)**:295–301.

Oleksyn, J. and Giertych, M. (1984): Results of a 70-year old Scots pine (*Pinus sylvestris* L.) provenance experiment in Pulawy, Poland. *Silvae Genet.* **33 (1)**:22–27.

Przybylski, T. (1972): Variability of Scots pine (*Pinus silvestris* L.) of Polish provenances. *Arboretum Kórnickie* **17**:121–167.

Przybylski, T. and Sztuka, J. (1968): Doświadczenie proweniencyjne z sosną zwyczajną (*Pinus silvestris* L.) w Nadleśnictwie Lubień (Provenance experiments with Scots pine (*Pinus sylvestris* L.) in the forest district Lubień). *Arboretum Kórnickie* **13**:261–274.

Repo, T., Mela, M. and Valtanen, J. (1984): Männynversosyövälle alttiiden ja vastustuskykyisten taimialkuperien erottaminen neulasten ominaisimpedanssin mittauksella (Separation of susceptible and resistant provenances of Scots pine to *Gremmeniella abietina* by specific needle impedance). *Fol. For.* **610**:11.

Robinson, R. A. (1976): *Plant pathosystems.* Springer, Berlin–Heidelberg–New York, p. 184.

Rohmeder, E. and Eicke, G. (1969): Ergebnisse eines über sechzigjährigen Kiefern-Provenienzversuches im Forstamt Hagenbach/Pfalz. *Forstwiss. Centralbl.* **88**:338–350.

Roll-Hansen, F. (1975): *Phacidium infestans* on dwarf plants of *Pinus sylvestris. Eur. J. For. Path.* **5**:1–7.

Rosochacka, J. and Grzywacz, A. P. (1980): The colour of *Pinus silvestris* seeds and their susceptibility to damping-off. 2. Colour of seed coats and their chemical composition. *Eur. J. For. Path.* **10**:193–201.

Rousi, M. and Häggman, J. (1984): Short note: relationship between the total phenol content of Scots pines and browsing by the arctic hare. *Silvae Genet.* **33 (2–3)**:95–97.

Schachler, G. and Bolland, G. (1986): Zu den Beziehungen zwischen der Widerstandsfähigkeit der Kiefer gegenüber *Lophodermium pinastri* (Schrad.) Chev. und der Pufferkapazität von Nadelhomogenaten. *Beitr. Forstwirtsch.* **20**:81–87.

Scholz, F. and Stephan, B. R. (1974): Physiologische Untersuchungen über die unterschiedliche Resistenz von *Pinus sylvestris* gegen *Lophodermium pinastri.* I. Die Pufferkapazität in Nadeln. *Eur. J. For. Path.* **4**:118–126.

Scholz, F. and Stephan, B. R. (1982): Buffering of pH in plant organs and resistance against fungi. In: Heybroek, H. M., Stephan, B. R. and von Weissenberg, K. (Eds) *Resistance to diseases and pests in forest trees.* PUDOC, Wageningen, pp. 176–186.

Schuck, H. J. (1972): Die Zusammensetzung der Nadelwachse von *Pinus silvestris* in Abhängigkeit von Herkunft und Nadelalter sowie ihre Bedeutung für die Anfälligkeit gegenüber *Lophodermium pinastri* (Schrad.) *Chev. Flora* **161**:604–622.

Schütt, P. (1957*a*.): Über Aussichten und erste Maßnahmen einer züchterischen Bekämpfung der Kiefernschütte. *Allg. Forstzeitschr.* **12**:13–15.

Schütt, P. (1957*b*): Untersuchungen über Individualunterschiede im Schüttebefall bei *Pinus silvestris. Silvae Genet.* **6**:109–112.

Schütt, P. (1958/59): Züchtung mit Kiefern. Teil I und II. Mitt. *Bundesforschungsanst. Forst-u. Holzwirtsch.* **40**:65; **42**:40.

Schütt, P. (1964*a*): Eine mykologische Methode der Resistenzprüfung entwickelt für den *Lophodermium*-Befall der Kiefer. *Phytopath. Zeitschr.* **51**:62–72.

Schütt, P. (1964*b*): Der Schüttebefall der Kiefer in Abhängigkeit von Herkunft und Anbauort. *Forstwiss. Centralbl.* **83**:140–163.

Schütt, P. (1966): A new approach to the development of a mycological method of resistance testing, In: Gerhold, H. D., Schreiner, E. J., McDermott, R. E. and Winieski, J. A. (Eds) *Breeding pest-resistant trees.* Pergamon Press, London, pp. 437–440.

Schütt, P. (1971): Untersuchungen über den Einfluß von Cuticularwachsen auf die Infektionsfähigkeit pathogener Pilze. *Eur. J. For. Path.* **1**:32–50.

Segal, A., Manisterski, J., Browning, J. A., Fischbeck, G. and Wahl, I. (1982): Balance in indigenous plant populations. In: Heybroek, H. M., Stephan, B. R. and von Weissenberg, K. (Eds) *Resistance to diseases and pests in forest trees.* PUDOC, Wageningen, pp. 361–370.

Šindelář, J. (1981): Genetically fixed variability of Scotch pine *(Pinus sylvestris)* population in the territory of the CSSR. *Lesnictvi (Prague)* **27**:385–408.

Siwecki, R. (1967): Badania nad porażeniem sosny przez *Lophodermium pinastri* (Schrad.) Chev., w zależnosci od proweniencji nasion żywiciela (Studies on the infection of pine by *Lophodermium pinastri* (Schrad.) Chev. in relation to geographic origin). *Sylwan* **111**:53–60.

Siwecki, R., Chwalinski, K. and Kaczmarek, Z. (1975): An attempt to determine the characters that differentiate provenances of Scots pine in resistance to *Lophodermium pinastri* (Schrad.) Chev. *Mitt. Bundesforschungsanst. Forst- u. Holzwirtsch.* **108**:121–129.

Smelyanets, V. P. (1977): Mechanisms of plant resistance in pine trees *Pinus sylvestris.* Part 1–5. *Zeitschr. Angew. Entomol.* **83**:225–233, 337–344; **84**:113–123, 232–241, 344–353.

Smith, R. H. (1972): Xylem resin in the resistance of the *Pinaceae* to bark beetles. USDA Forest Service, General Technical Report PSW-1: 7 pp.

Squillace, A. E., LaBastide, J. G. A. and Vredenburch, C. L. H. van (1975): Genetic variation and breeding of Scots pine in the Netherlands. *Forest Sci.* **21**:341–352.

Steiner, K. (1974): Genetic differences in resistance of Scotch pine to eastern pineshoot borer. *The Great Lakes Entomologist* **7**:103–107.

Stephan, B. R. (1975): Resistance in pine species to *Lophodermium pinastri. Mitt. Bundesforschungsanst. Forst- u. Holzwirtsch.,* **108**:105–111.

Stephan, B. R. (1984): Sicherung der Stabilität von Waldbeständen durch Resistenzzüchtung. *Mitt. Bundesforschungsanst. Forst- u. Holzwirtsch.* **146**:37–48.

Stephan, B. R. and Krusche, D. (1986): Genetic variation of resistance to *Lophodermium needle* cast in Scots pine progenies of intraprovenance and interprovenance crossings. USDA, Forest Service, General Technical Report WO-50: 28–34.

Stephan, B. R. and Scholz, F. (1975): Current status of research concerning resistance of Scots pine to needle casts. Special paper for Second FAO/IUFRO World Technical Consultation on Forest Diseases and Insects, New Delhi, April 1975. 7 pp. FAO 29627.

Stephan, B. R. and Scholz, F. (1981): Preliminary results on *Lophodermium* needle cast resistance and buffering of pH in Scots pine crosses. *Proceedings of meeting IUFRO working party* S2.06.04, Sarajevo, 1980, pp. 81–84.

Werner, A. (1987): Responses of in vitro grown pine seedlings to infection by four strains of *Heterobasidion annosum. Eur. J. For. Path.* **17**:93–101.

Wright, J. W., Lemmien, W. A., Bright, J. N., Day, M. W. and Sajdak, R. L. (1976): Scotch pine varieties for Christmas tree and forest planting in Michigan. Michigan Agr. Exp. Sta. Research Report, 293, p. 16.

Wright, J. W. and Wilson, L. F. (1972): Genetic differences in Scotch pine resistance to pine root collar weevil. Michigan Agr. Exp. Sta. Research Report, 159, p. 6.

Wright, J. W., Wilson, L. F. and Bright, J. N. (1975): Genetic variation in resistance of Scotch pine to Zimmerman pine moth. *The Great Lakes Entomologist* **8**:231–236.

Wright, J. W., Wilson, L. F. and Randall, W. K. (1967): Differences among Scotch pine varieties in susceptibility to European pine sawfly. *Forest Sci.* **13**:175–181.

Zobel, B. J. (1982): The world's need for pest-resistant forest trees. In: Heybroek, H. M., Stephan, B. R. and von Weissenberg, K. (Eds) *Resistance to diseases and pests in forest trees.* PUDOC, Wageningen, pp. 1–8.

CHAPTER 17

Inheritance of resistance to abiotic factors

J. Oleksyn

Low temperature stress

Frost damages are frequently observed when populations are moved northwards, or from a mild climate to a more continental one. This type of observations has been made in various provenance experiments in Scandinavia and the USSR (Eiche, 1966; Cherepnin, 1980). The considerable interpopulational variability in sensitivity of Scots pine to the action of low temperatures led Dietrichson (1961) to postulate that the basic aim of tree improvement, higher volume productivity and better timber quality cannot be achieved in northern countries without populations or individuals sufficiently resistant to early, late and winter frosts.

Langlet (1936) has also shown that the southern races of Scots pine are characterized by greater water content in needles during the autumn compared to northern races which appears to be an adaptation to a warmer climate. According to Dietrichson (1961) this relationship demonstrated by Langlet could be used as a criterion in determining autumn resistance of pine to frost injuries.

Significant differences in sensitivity to the action of low temperatures of various clones and provenances of *Pinus sylvestris* have been observed by van Haverbeke (1979) on 3- to 6-year-old trees growing in a seed orchard in eastern Nebraska (USA). He found that about 35% of variation in frost injury is associated with differences between provenances. The mean degree of injury of individual clones in this experiment varied from 0% to 50%. In spite of considerable, and statistically significant ($P \leq 0.01$) variation in injury between clones, this variance component amounted to only 10%.

In studies conducted on provenances from a smaller part of the range (Sweden, 60–65 °N) Norell et al. (1986) have found a large and statistically significant, intrapopulation variation in sensitivity of Scots pine seedlings to the action of low temperatures. According to them "the large within-population variation offers good opportunities for autumn hardiness breeding". The authors have found no significant interaction between the sensitivity of *P. sylvestris* to autumn frosts and the distance in latitude separating the location of origin and the test site.

Field and laboratory experiments in which several Swedish clones and provenances of *Pinus sylvestris* were subjected to the action of a series of low temperatures have shown that the northern populations were less injured than southern sources (Ericsson, 1986; Lindström and Nyström, 1987). It was also found (Andersson, 1986) that the progeny from seed orchards is more sensitive to frost than the progeny from natural stands. The authors concluded that pollen contamination and to some extent after-effects (probably caused by higher seed weight) were the main reasons for late hardening of one-year-old seed orchard progenies.

In northern Sweden, where the main limiting factor to pine cultivation is low temperature, detailed studies on the mortality of transferred Scots pine populations have been conducted by Eriksson et al. (1980). They have found that at constant altitude each degree of latitudinal transfer southwards increases survival by 10.8% (Fig. 17.1). At the same latitude, altitudinal transfer of pine increases survival by 3% for every 100 m downwards.

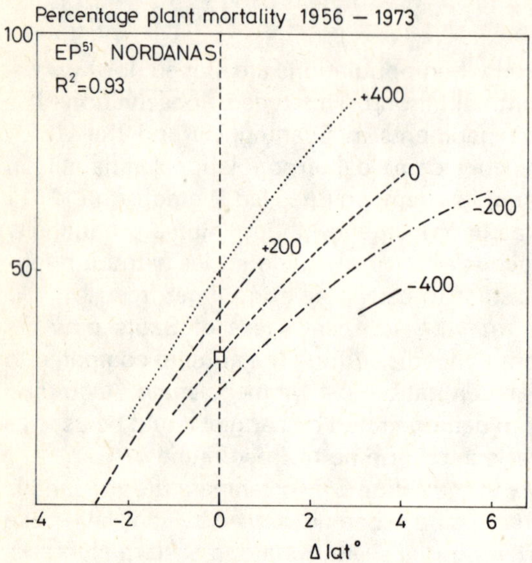

Fig. 17.1. The effect of latitudinal (in degrees) and altitudinal (in metres) transfer on mortality at Nordanas (Sweden). Plus values indicate a transfer upwards or in a northward direction. The quadratic correlation coefficient is indicated (from Eriksson et al., 1980)

From the studies conducted by Nilsson and Aman (1986) it appears that it is possible to select Scots pine for frost resistance on the basis of an analysis of phenological rhythms and the growth of parent clones. According to these authors clones unsuitable for frost resistance breeding have needles that are loosely attached to the leader shoot in late July/early August (possibly relative to measurements later in the autumn), have a large annual height growth and finally have a low dry matter content of the needles in the summer or autumn.

A significant correlation between laboratory frost resistance test results of one-year-old full-sib families and their survival in the conditions of northern

Sweden indicates the possibility of relying on laboratory testing as a method for selecting plus trees for these regions (Nilsson and Andersson, 1987).

Results of laboratory tests for frost resistance of Scots pine seedlings obtained from intrapopulation and outbred progenies indicate that this trait is expressed through polygenic inheritance (Norell et al., 1986).

Drought stress

Studies conducted by Biryukova and Harlamova (1981) in northern Khazakstan (USSR) on a provenance experiment have shown that ecotypes and individuals of Scots pine react differently to summer droughts. From the results presented by the authors it appears that the content of free proline in buds and needles is a good indicator when testing various populations of Scots pine for resistance to drought, but it needs to be stressed that the most drought-resistant populations had the lowest growth increments in the trial.

Ionizing radiation stress

Seeds of Scots pine and other conifers are more sensitive to the action of ionizing radiation than seeds of mono- and dicotyledonous plants (Lehtiniemi, 1976; Kudinov, 1986). There is considerable variation in sensitivity between seed lots from individual trees or between populations (Simak et al., 1961).

A detailed analysis of the effect of various doses of X-rays on the germination and development of seedlings of 10 Swedish provenances from various latitudes (57°–66°N) and altitude (45–490 m) has been made by Ohba and Simak (1961). The seeds were exposed to radiation doses of 0, 600, 2 400, 3 600, and 4 800 R. Studied traits were seed germination, duration of germination, seedling mortality, length of hypocotyl and length of roots. Doses up to 1 200 R did not cause significant damage. At higher doses negative effects were observed, and the individual provenances differed in sensitivity to the treatment. A dose of 3 600 R resulted in greater mortality in populations from regions with a longer growing season (Fig. 17.2). The same populations had a lower germination per cent, shorter hypocotyls and shorter roots relative to the non-irradiated lots (Fig. 17.2). This could indicate that the northern provenances, coming from regions with a shorter growing season but with a longer summer insolation are less sensitive to X-rays.

Environmental pollution stress

Since the early sixties a number of reports have appeared on the variability in sensitivity of Scots pine to various kinds of toxic gases (Table 17.1). The majority of the reports indicate the presence of variation in sensitivity to sulphur dioxide, ozone, hydrogen fluoride and some mixtures of gases. Differences were observed at individual, progeny and population levels.

The sensitivity of Scots pine, before or after being subjected to toxic gases was correlated with net photosynthesis, photorespiration, dark respiration and the CO_2 compensation point (Oleksyn, 1981; Lorenc-Plucińska and Oleksyn, 1982; Oleksyn and Białobok, 1986), height (Oleksyn, 1988), anatomic structure of needles (Huttunen, 1978), peroxidase activity in needles (Kieliszewska-Rokicka, 1979), level of free proline in the needles (Karolewski, 1989), and allele and/or genotype frequencies for such gene loci as glutamate dehydrogenase (GDH) and glutamate oxylacetate transaminase (GOT) (Geburek et al., 1987). A review of information on metabolic reactions of plants differing in sensitivity to toxic gases can be found in several papers (Reinert et al., 1982; Heck et al., 1986; Tingey and Olszyk, 1985).

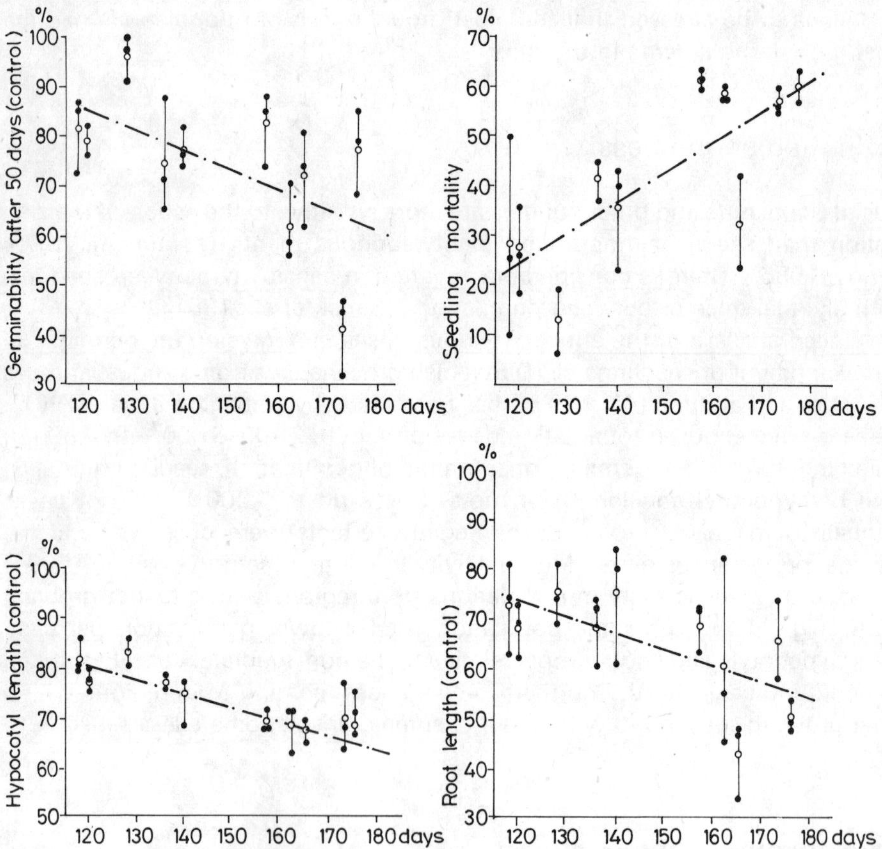

Fig. 17.2. Relationship between the length of the vegetative season (in days) and the sensitivity of seeds of Swedish provenances of Scots pine subjected to X-rays of 3600 R (from Ohba and Simak, 1961). (o — indicates replicate means)

Table 17.1. Observed resistance differences of *Pinus sylvestris* L. to air pollutants

Differences among	Pollutant	References
Clones	HF, SO_2	Rohmeder et al., 1962
Clones	SO_2	Vogl and Börtitz, 1965
Populations	SO_2	Vogl, 1969
Populations	SO_2	Schütt et al., 1970
Clones, families	SO_2	Vogl, 1970
Populations	O_3	Demeritt, 1977
Clones, seedlings*	SO_2, O_3, SO_2+O_3	Białobok and Karolewski, 1978
Populations	O_3, SO_2	Genys and Heggestad, 1978
Populations	mixture (SO_2, NO_X)	Huttunen, 1978
Individuals	SO_2	Lorenc-Plucińska, 1978*a,b*
Clones	SO_2	Mejnartowicz et al., 1978
Populations	SO_2	Smith and Davis, 1978
Populations, families	mixture (SO_2, heavy metals)	Niemtur, 1979
Clones	SO_2	Kieliszewska-Rokicka, 1979
Individuals	O_3	Lorenc-Plucińska, 1979
Individuals	mixture (SO_2, heavy metals)	Mamaev and Shkarlet, 1979
Populations	O_3, SO_2, O_3+SO_2	Weidensaul, 1979
Populations	O_3, SO_2, O_3+SO_2	Weidensaul and Darling, 1979
Clones, seedlings*	SO_2, NO_2, HF, O_3, SO_2+O_3	Białobok et al., 1980
Seedlings*	HF	Lorenc-Plucińska, 1980
Individuals	SO_2	Makarov, 1981
Individuals	SO_2	Oleksyn, 1981
Populations	mixture (SO_2, NO_X)	Huttunen and Törmälehto, 1982
Seedlings*	SO_2	Lorenc-Plucińska, 1982
Seedlings*	HF	Lorenc-Plucińska and Oleksyn, 1982
Individuals**	SO_2	Tzschacksch, 1982
Populations	mixture (SO_2, NO_X)	Oleksyn, 1983
Seedlings*	SO_2	Lorenc-Plucińska, 1984
Individuals	mixture (SO_2, fluorides)	Mejnartowicz and Łukasiak, 1985
Individuals	mixture (SO_2, fluorides)	Przymusiński and Mejnartowicz, 1985
Seedlings*	SO_2	Lorenc-Plucińska, 1986
Populations	SO_2, mixture (SO_2, fluorides)	Oleksyn and Białobok, 1986

Table 17.1. Cont.

Differences among	Pollutant	References
Seedlings*	mixture (mainly SO_2)	Geburek et al., 1987
Populations	SO_2	Siwecki and Karolewski, 1987
Seedlings*	NO_2	Lorenc-Plucińska, 1988
Populations	SO_2, NO_2, SO_2+NO_2, HF, mixtures (SO_2, heavy metals) and (SO_2, fluorides)	Oleksyn et al., 1988
Populations	SO_2, NO_2, HF	Karolewski, 1989
Populations	mixture (SO_2, NO_x)	Oleksyn, 1989
Populations	mixture (SO_2, fluorides)	Oleksyn, 1988

* Open pollinated progenies of Scots pine clones.
** 10-year-old progeny from controlled crosses of parents of known resistance to the development of SO_2-induced needle necroses.

Heritability of *P. sylvestris* needle sensitivity to sulphur dioxide has been studied by Tzschacksch (1982). In the laboratory he exposed detached shoots from 10-year-old progenies of controlled crosses between parents of known needle sensitivity to SO_2. A repeatability of 0.60 was obtained. According to the author the result justifies the undertaking of a selection programme for resistance to SO_2 and the establishment of a seed orchard with tolerant individuals selected in regions under strong industrial pollution.

The studies conducted by Białobok et al. (1980) have also shown that there exists a significant correlation ($r=0.64$, $P\leq0.01$) between the SO_2-sensitivity of clonal parents and their open-pollinated seedling progenies. A similar relationship was observed for the gases NO_2 ($r=0.72$, $P\leq0.01$) and O_3 ($r=0.48$, $P\leq0.1$).

Results of studies conducted so far (Demeritt, 1977; Białobok et al., 1980; Tzschacksch, 1982) indicate that the sensitivity of *P. sylvestris* to SO_2 is under stronger genetic control than sensitivity to O_3.

The laboratory sensitivity to HF and SO_2 of seedlings from 20 Scots pine populations from 11 European countries compared with the results obtained in the field under strong fluoride and SO_2 pollution or SO_2 and heavy metal pollution has shown that the responses are significantly correlated (Oleksyn and Białobok, 1986; Oleksyn et al., 1988). The repeatability of results was also very good in these trials established near various pollution sources. It needs to be pointed out that differing results were obtained by Tzschacksch (1987) who studied the resistance of Scots pine under laboratory conditions and in a polluted environment. In his study the correlation between field and laboratory tests was negative.

Elimination of sensitive genotypes of forest trees from populations subjected to the action of industrial pollution may lead to a reduction of their genetic variability. Changes in the genetic structure under the influence of toxic gases have been shown experimentally with the help of isozyme studies on *Pinus sylvestris* (Szmidt, 1978; Mejnartowicz, 1983; 1984; 1986; Geburek et al., 1987).

This phenomenon may be quite serious for Scots pine since there are data indicating that pollution pressure acts against the most productive populations (Huttunen and Törmälehto, 1982; Oleksyn, 1983; Oleksyn, 1987).

Genetic variation is limited further by the disturbances caused in the reproductive processes. As the studies of Scholz et al. (1985) have shown, under the influence of air pollution the germinative capacity of the pollen of Scots pine trees is affected to various extents.

Studies on the influence of SO_2 and fluorides on the germination of *P. sylvestris* pollen from trees differing in sensitivity have been also conducted by Mejnartowicz and Lewandowski (1985). They have found that both sensitive and tolerant trees had a 50% lower pollen germinability after fumigation. Pollen from more sensitive trees had longer pollen tubes and underwent ageing earlier. Thus it appears that the sensitivity of vegetative organs to pollution need not always be paralleled by the sensitivity of generative organs.

Compared with other tree species Keller and Beda (1984) found that pollen of *P. sylvestris* is more sensitive to fumigation with SO_2 than pollen of *P. mugo* and *Abies alba.*

Information is lacking about the rate of changes in the genetic structure of Scots pine populations under the influence of non-lethal doses of pollution. From computer simulation studies conducted by Bergmann and Scholz (1985) on *Picea abies* it is known, however, that a loss of some genetic information can occur already after few generations.

Countries with heavy environmental pollution should urgently start with the active protection of gene resources of *Pinus sylvestris,* as proposed for FRG by Melchior et al. (1986).

The sensitivity differences between clones of Scots pine could be used for the bioindicative monitoring of industrial pollution. The role of bioindication is to supplement and, in some instances, to substitute for technical monitoring. This purpose could well be served by appropriately selected, sensitive and tolerant clones of Scots pine.

According to many authors the genetic variation in sensitivity to pollution is not a sufficient argument for resistance breeding. There are several reasons for this, such as the forseeable exhaustion of the known resources of brown and black coal (the main sources of pollution in Europe), lack of stability of the chemical composition of immissions, ecological variability and first of all ethical considerations (Scholz, 1986; van Buijtenen and Giertych, 1987).

Thus the problem should be approached with caution. Resistance breeding cannot be justified as an alternative to emission control, but as a component of a comprehensive pollution control strategy that can be recommended only under certain circumstances (Gerhold, 1977).

Literature

Andersson, B. (1986): Freezing tests of Scots pine (*Pinus sylvestris* L.) seed orchard crops. In: Lindgren, D. (Ed.) Provenances and forest tree breeding for high latitudes. *Proceedings of the Frans Kempe symposium at Umeå, June 10–11, 1986.* Sveriges Lantbruksuniversitet, Rapport 6, 99–111.

Bergmann, F. and Scholz, F. (1985): Effects of selection pressure by SO_2 pollution on genetic structures of Norway spruce *(Picea abies).* In: Gregorius, H.-R. (Ed.) *Population genetics in forestry. Lecture notes in biomathematics.* Springer, Berlin, pp. 267–275.

Bergmann, F. and Scholz, F. (1987): The impact of air pollution on the genetic structure of Norway spruce. *Silvae Genet.* **36 (2)**: 80–83.

Białobok, S. and Karolewski, P. (1978): Ocena stopnia odporności drzew matecznych sosny zwyczajnej i ich potomstwa na działanie SO_2 i O_3 oraz mieszaniny tych gazów (Evaluation of the degree of resistance of maternal Scots pine trees and their progeny to the action of SO_2, O_3 and a mixture of these gases). *Arbor. Kórn.* **23**:299–310.

Białobok, S., Karolewski, P. and Oleksyn, J. (1980): Sensitivity of Scots pine needles from mother trees and their progenies to the action of SO_2, O_3 a mixture of these gases, NO_2 and HF. *Arbor. Kórn.* **25**:289–303.

Biryukova, Z. P. and Harlamova, N. V. (1981): Geograficheskaya izmenchivost' zasukhoustojchivosti i zimostojkosti sosny obyknovennoj (Geographic variability of drought resistance and winter-hardiness of Scots pine). *Ekologija* **4**:42–27.

Cherepnin, V. L. (1980): Izmenchivost' semyan sosny obyknovennoj (Variation in seed of Scots pine). Nauka, Novosibirsk, p. 182.

Demeritt, M. E. (1977): Genetic evaluation of two-year height and ozone tolerance in Scotch pine (*Pinus sylvestris* L.). Ph. D. Thesis. The Pennsylvania State University, p. 69.

Dietrichson J. (1961): Breeding for frost resistance. *Silvae Genet.* **10**:172–184.

Eiche, V. (1966): Cold damage and plant mortality in experimental provenance plantations with Scots pine in northern Sweden. *Studia Forest. Suec.* **36**:1–219.

Ericsson, T. (1986): Pine seed orchard tests in northern Sweden. Preliminary results from an assessment in summer 1984 of a series of field trials planted 1973–75. In: Lindgren, D. (Ed.) *Provenances and forest tree breeding for high latitudes. Proceedings of the Frans Kempe symposium at Umeå, June 10–11, 1986.* Sveriges Lantbruksuniversitet, Rapport 6, 113–126.

Eriksson, G., Andersson, S., Eiche, V., Ifver, J. and Persson, A. (1980): Severity index and transfer effects on survival and volume production of *Pinus sylvestris* in Northern Sweden. *Studia Forest. Suec.* **156**:1–32.

Geburek, Th., Scholz, F., Knabe, W. and Vornweg, A. (1987): Genetic studies by isozyme gene loci on tolerance and sensitivity in an air polluted *Pinus sylvestris* field trial. *Silvae Genet.* **36 (2)**:49–53.

Genys, J. B. and Heggestad, E. (1978): Susceptibility of different species, clones and strains of pines to acute injury caused by ozone and sulfur dioxide. *Plant Diesease Rep.* **62 (8)**:687–691.

Gerhold, H. D. (1977): Effect of air pollution on *Pinus strobus* L. and genetic resistance. A literature review. Corvallis Environ. Res. Lab., EPA 600/3–77–002, p. 44.

Heck, W. W., Heagle, A. S. and Shriner, D. S. (1986): Effects on Vegetation: Native, Crops, Forests. In: Stern A. S. (Ed.), *Air Pollution.* Vol. 6, Academic Press, New York, pp. 247–350.

Huttunen, S. (1978): The effects of air pollution on provenances of Scots pine and Norway spruce in Northern Finland. *Silva Fenn.* **12 (1)**:1–16.

Huttunen, S. and Törmälehto, H. (1982): Air pollution resistance of some Finnish *Pinus sylvestris* L. provenances. *Aquilo Ser. Bot.* **18**:1–9.

Karolewski, P. (1989): Free proline content of 18 European provenances of Scots pine (*Pinus sylvestris* L.) and their susceptibility to the action of SO_2, NO_2 and HF. *Fol. Dendrol.* **16**:365–382.

Keller, T. and Beda, H. (1984): Effects of SO_2 on the germination of conifer pollen. *Environ. Pollut. Ser. A.* **33**:237–243.

Kieliszewska-Rokicka, B. (1979): Peroxidase activity in varieties of *Weigela* and *Pinus silvestris* resistant and susceptible to SO_2. *Arbor. Kórn.* **24**:313–320.

Kudinov, M. A. (1986): Vneshnaya sreda i formirovanie ustojchivosti u drevesnykh rastenij (External environment and development of resistance in woody plants). Nauka i Tehnika Publishers, Minsk, p. 167.

Langlet, O. (1936): Studier över tallens fysiologiska variabilitet och dess samband med klimatet. Ett bidrag till kännedomen om tallens ekotyper (Studien über die physiologische Variabilität der Kiefer und deren Zusammenhang mit dem Klima. Beiträge zur Kenntnis der Okotypen von *Pinus silvestris* L.). *Meddel fran Statens Skogsförsöksanstalt* **29**:219–470.

Lehtiniemi, T. (1976): Ionisoivan säteilyn vaikutus varastokuivien ja liotettujen metsäpuiden siementen idäntään ja taimien alkukehitykseen (Effect of ionizing radiation on the germination of storage-dry and soaked forest tree seeds and on the initial development of seedlings). *Silva Fenn.* **10 (1)**:20–31.

Lindström, A. and Nyström, C. (1987): Seasonal variation in root hardiness of container-grown Scots pine, Norway spruce, and lodgepole pine seedlings. *Can. J. For. Res.* **17 (8)**:787–793.

Lorenc-Plucińska, G. (1978*a*). The effect of SO_2 on the photosynthesis and dark respiration of larch and pine differing in resistance to this gas. *Arbor. Kórn.* **23**:121–132.

Lorenc-Plucińska, G. (1978*b*): Effect of sulphur dioxide on photosynthesis, photorespiration and dark respiration of Scots pines differing in resistance to this gas. *Arbor. Kórn.* **23**:133–144.

Lorenc-Plucińska, G. (1979): The effect of ozone on photosynthesis and respiration in Scots pines differing in resistance to this gas. *Arbor. Kórn.* **24**:329–338.

Lorenc-Plucińska, G. (1980): Influence of hydrogen fluoride on the rate of CO_2 exchange in Scots pines of different susceptibility to this gas. *Arbor. Kórn.* **25**:269–276.

Lorenc-Plucińska, G. (1982): Effect of sulphur dioxide on CO_2 exchange in SO_2-tolerant and SO_2-susceptible Scots pine seedlings. *Photosynthetica* **16 (1)**:140–144.

Lorenc-Plucińska, G. (1984): The uptake and translocation of ^{14}C in Scots pine seedlings of differing susceptibility to sulphur dioxide. *Eur. J. For. Path.* **14 (4–5)**:310–318.

Lorenc-Plucińska, G. (1986): Effect of sulphur dioxide on the partitioning of assimilates in Scots pine (*Pinus sylvestris* L.) seedlings of different susceptibility to this gas. *Eur. J. For. Path.* **16 (5–6)**: 266–273.

Lorenc-Plucińska, G. (1988): Effect of nitrogen dioxide on CO_2 exchange in Scots pine seedlings. *Photosynthetica* **22 (1)**:85–86.

Lorenc—Plucińska, G. and Oleksyn, J. (1982): Effect of HF on photosynthesis, photorespiration and dark respiration in Scotch pine. *Fluoride* **15 (3)**:149–166.

Makarov, N. M. (1981): Endogennaya izmenchivost' sosny obyknovennoj po gazoustojchivosti (Endogenic differentation in fume resistance of Scots pine). In: Mamaev, S. A. and Saburov, V. I. (Eds) *Issledovanie form vnutrividovoj izmenchivosti rastenij.* UNC Publishers, Sverdlovsk, pp. 76–83.

Mamaev, S. A. and Shkarlet, O. D. (1979): Individualnaya izmenchivost' sosny obyknovennoj po gazoustojchivosti v svjazi s sokhraneniem khvojnykh nasazhdenij v uslovijah tekhnogennykh landshavtov (Individual differentiation of Scots pine in fume resistance in comparison with protection of coniferous forests in conditions of industrial regions). In: Mamaev, S. A. and Mahnev, A. K. (Eds) *Problemy sozdaniya zashchitnykh nasazhdenij v uslovyah tekhnogennykh landshavtov.* Trudy instituta ekologii rastenij i zhivotnykh, Sverdlovsk, **129**:58–65.

Mejnartowicz, L. E. (1983): Changes in genetic structure of Scots pine (*Pinus silvestris* L.) population affected by industrial emission of fluoride and sulphur dioxide. *Genetica Polonica* **24 (1)**:41–50.

Mejnartowicz, L. E. (1984): Enzymatic investigations on tolerance in forest trees. In: Kozioł, M. J. and Whatley, F. R. (Eds) *Gaseous air pollutants and plant metabolism.* Butterworths, London–New York–Boston, pp. 381–398.

Mejnartowicz, L. E. (1986): Genotypes of Scots pine trees differing in resistance to the action of fluoride and sulphur dioxide described on the basis of glutamic-oxalacetic-transaminase isoenzymes. *Acta Univ. Lodziensis, Folia Physiol. Genet.* **1**:139–146.

Mejnartowicz, L. and Lewandowski, A. (1985): Effects of fluorides and sulphur dioxide on pollen germination and growth of the pollen tube. *Acta. Soc. Bot. Pol.* **52 (2)**:125–129.

Mejnartowicz, L. and Łukasiak, H. (1985): Level of sugars in Scots pine trees of different sensitivity to fluoride and sulphur dioxide. *Eur. J. For. Path.* **15 (4)**:193–198.

Mejnartowicz, L., Białobok, S. and Karolewski, P. (1978): Genetic characteristics of Scots pine specimens resistant and susceptible to SO_2 action. *Arbor. Kórn.* **23**:233–238.

Melchior, G. H., Muhs, H.-J. and Stephan, B. R. (1986): Tactics for the conservation of forest gene resources in the Federal Republic of Germany. *For. Ecol.* Manag. **17**:73–81.

Niemtur, St. (1979): Influence of zinc smelter emissions on peroxidase activity in Scots pine needles of various families. *Eur. J. For. Path.* **9**:142–147.

Nilsson, J.-E. and Aman, M.-J. (1986): The phenological rhythm of *Pinus sylvestris* clones and its relationship with damage of one-year-old offsprings after freeze testing. In: Lindgren, D. (Ed.) *Provenances and forest tree breeding for high latitudes. Proceedings of the Frans Kempe symposium at Umeå, June 10–11, 1986.* Sveriges Lantbruksuniversitet, Rapport 6, 127–147.

Nilsson, J.-E. and Andersson, B. (1987): Performance in freezing tests and field experiments of full-sib families of *Pinus sylvestris* (L.) *Can. J. For. Res.* **17 (11)**:1340–1347.

Norell, L., Eriksson, G., Ekberg, I. and Dormling I. (1986): Inheritance of autumn frost hardiness in *Pinus sylvestris* L. seedlings. *Theor. Appl. Genet.* **72**: 440—448.

Ohba, K. and Simak, M. (1961): Effects of X-rays on seed of Scots pine from different provenances (*Pinus silvestris* L.). *Silvae Genet.* **10 (3)**:84–90.

Oleksyn, J. (1981): Effect of sulphur dioxide on net photosynthesis and dark respiration of Scots pine individuals differing in susceptibility to this gas. *Archiwum Ochrony Środowiska* **(2–4)**:49–58.

Oleksyn, J. (1983): Effect of industrial air pollution from a fertilizer factory on the growth of 70-years-old Scots pine in a provenance experiment. *Aquilo Ser. Bot.* **19**:332–341.

Oleksyn, J. (1987): Air pollution effects on 15 European and Siberian scots pine (*Pinus sylvestris* L.) provenances growing in a 75-year-old experiment. *Arbor. Kórn.* **32**:151–162.

Oleksyn, J. (1988): Height growth of different European Scots pine *Pinus sylvestris* L. provenances in a heavily polluted and a control environment. *Environ. Pollut.* **55 (4)**:289–299.

Oleksyn, J. and Białobok, S. (1986): Net photosynthesis, dark respiration and susceptibility to air pollution of 20 European provenances of Scots pine *Pinus sylvestris* L. *Environ. Pollut. Ser. A.* **40 (4)**:287–302.

Oleksyn, J., Karolewski, P. and Rachwał, L., (1988): Susceptibility of European *Pinus sylvestris* L. populations to SO_2, NO_2, SO_2+NO_2 and HF under laboratory and field conditions. *Acta Soc. Bot. Pol.* **57 (1)**:107–115.

Przymusiński, R. and Mejnartowicz, L. (1985): Localization of peroxidase in needles of Scots pine trees differing in tolerance to fluoride and sulphur dioxide pollution. *Arbor. Kórn.* **30**:257–262.

Reinert, R. A., Heggestad, H. E. and Heck, W. W. (1982): Response and genetic modification of plants for tolerance to air pollutants. In: Christiansen, M. L. and Lewis, C. F. (Eds) *Breeding plants for less favorable environments.* John Wiley, New York, pp. 259–292.

Rohmeder, E., Merz, W. and Schönborn, A. (1962): Züchtung von gegen Industrieabgase relativ resistenten Fichten- und Kiefernsorten *Forstwiss. Cbl.* **81**:321–332.

Scholz, F. (1986): Luftverunreinigungen — ein Umweltfaktor, der Forstgenetik und Forstpflanzenzüchtung vor neue Aufgaben stellt. *Allg. Forstz.* **41 (51/52)**:1288–1294.

Scholz, F. and Bergmann, F. (1984): Selection pressure by air pollution as studied by isozyme-gene-systems in Norway spruce exposed to sulphur dioxide. *Silvae Genet.* **33 (6)**:238–241.

Scholz, F., Vornweg, A. and Stephan, B. R. (1985): Wirkungen von Luftverunreiningungen auf die Pollenkeimung von Waldbäumen. *Forstarchiv* **56 (3)**:121–124.

Schütt, P., Lang, K.-J. and Margait, D. (1970): Ein Schnelltest zur Ermittlung der individuellen SO_2-Empfindlichkeit bei Kiefern. *Forstwiss. Cbl.* **89**:153–161.

Simak, M., Ohba, K. and Suszka, B. (1961): Effect of X-irradiation on seeds of different weight from individual trees of Scots pine (*Pinus silvestris* L.). *Botaniska Notiser* **114**:300–312.

Siwecki, R. and Karolewski, P. (1987): Odporność wybranych proweniencji sosny zwyczajnej na porażenie przez *Lophodermium pinastri* a wrażliwość ich na działanie SO_2 (The resistance of the selected provenances of Scots pine to the attack of *Lophodermium pinastri* and their susceptibility to SO_2 action). In: Siwecki, R. (Ed.) *Biological reaction of trees to industrial pollution. Materials from the II National Symposium, Kórnik, 16–19 May, 1984.* Adam Mickiewicz University Press, pp. 301–306.

Smith, H. J. and Davis, D. D. (1978): Susceptibility of conifer cotyledons and primary needles to acute doses of SO_2. *Hort. Sci.* **13**:703–704.

Szmidt, A. (1978): Zmienność katalazy w populacjach sosny zwyczajnej znajdującej się pod wpływem zanieczyszczeń przemysłowych (Differences of catalase in Scots pine populations under industrial pollution). In: Siwecki, R. (Ed.) *Reakcje biologiczne drzew na emisje przemysłowe*. Summaries. Institute of Dendrology, Kórnik, Poland, pp. 13–14.

Tingey, D. T. and Olszyk, D. M. (1985): Intraspecies variability in metabolic responses to SO_2. In: Winner, W. E., Mooney, H. A. and Goldstein, R. A. (Eds) *Sulfur dioxide and vegetation. Physiology, ecology, and policy issues.* Stanford University Press, Stanford, California, pp. 178–205.

Tzschacksch, O. (1982): Untersuchungen zur Erblichkeit der SO_2-Resistenz bei Kiefer (*Pinus silvestris* L.) und Douglasie (*Pseudotsuga menziesii* (Mirb.) Franco) mit Schlußfolgerungen für die Forstwirtschaft. *Beitr. Forstwirtsch.* **16 (3)**:103–106.

Tzschacksch, O. (1987): Zur Labor- und Feldresistenz der Kiefer (*Pinus sylvestris* L.) gegenüber phytotoxischen Stoffen und Schlußfolgerungen für die Anbauwürdigkeit von Kiefernarten in den Immissionsschadgebieten des oberen Erzgebirges. *Beitr. Forstwirtsch.* **21 (3)**:97–102.

Van Buijtenen, J. P. and Giertych, M. (1987): The genetics of stress resistance in forest trees. *Ann. Forest.* **13 (1–2)**:51–68.

VanHaverbeke, D. F. (1979): Short note: winter injury in a Scotch pine (*Pinus sylvestris* L.) clonal seed orchard. *Silvae Genet.* **28 (4)**:167–169.

Vogl, M. (1969): Orientierende Untersuchungen zur Nettoassimilation und Rauchärte zweier Provenienzen von *Pinus silvestris* L. in Abhängigkeit von der Jahreszeit und anderen Faktoren. *Arch. Forstwes.* **18 (12)**:1317–1323.

Vogl, M. (1970): Untersuchungen über Unterschiede in der relativen SO_2-Resistenz bei Kiefern-Nachkommenschaften. *Arch. Forstwes.* **19 (1)**:3–12.

Vogl, M. and Börtitz, S. (1965): Physiologische und biochemische Beiträge zur Rauchschadenforschung. 4. Mitteilung. Zur Frage der physiologisch und physikalisch bedingten SO_2-Resistenz von Koniferen. *Flora* **155 (3)**:347–352.

Weidensaul, T. C. (1979): Effects of O_3 and SO_2 on bud formation and development and needle injury on 15 Scotch pine provenances (Abstract). *Phytopathology* **69 (9)**:1040.

Weidensaul, T. C. and Darling, S. L. (1979): Effects of ozone and sulfur dioxide on the host–pathogen relationship of Scotch pine and *Scirrhia acicola. Phytopathology* **69 (9)**:939–941.

CHAPTER 18

Inheritance of wood properties

E. G. Ståhl and B. Ericson

Introduction

The economic importance of wood properties changes from time to time. For instance, naval stores were the main export products from Sweden and Norway until the 19th century, whereas nowadays they provide only a minor source of income. The wood properties of main economic interest are those that influence the strength and durability of sawn products as well as those directly correlated to the amount and properties of pulp produced. Although other properties may be just as important when considering short-term profits, changes in industrial structure or trends in consumer preferences enhance the difficulties in formulating the goals for a long-term breeding programme. In forest tree breeding programmes only wood properties with economic importance, large total variation and a high heritability should be included. The wood properties treated in this review are outlined in Table 18.1. (Not mentioned here are properties such as stem volume as well as the tapering and straightness of the stem. These properties are discussed in other sections of this book.)

In general, wood quality is the result of the effects of a number of wood properties which are correlated, positively or negatively with each other. In this chapter we have treated yield variation between sites (geographic areas) and provenance variation separately. Although between-site variation relates to environmental variation, it also includes provenance variation. The term "provenance variation" refers to the variation observed between different provenances grown together on one or more sites.

The comparisons are based mainly on field trials with planted trees. These values can differ substantially from those obtained in naturally regenerated Scots pine, as, in general, wood quality is not improved in planted stands (Uusvaara, 1974; Vuokila, 1982).

The main focus of this chapter is the inheritance of various wood properties. It should, however, be kept in mind that although the seed source is an important determinant of wood properties, conditions throughout the rotation have a major influence on the wood quality at harvest.

Table 18.1. Wood properties

Stem properties	Width of annual ring Proportions of juvenile and mature wood Proportions of early- and latewood Amount of reaction wood Amount of heart wood Amount of spiral grain
Tracheid properties	Cell wall thickness Tracheid diameter Diameter of lumen Tracheid length
Weight-related properties	Basic density Dry matter content
Branch-related properties	Knot content Spike knot frequency
Chemical properties	Content of extractives Content of lignin Content of cellulose Content of hemicellulose

Properties related to variation in stem characteristics

Variation in wood properties could be related to the following characteristics: width of the annual ring, proportions of juvenile and mature wood, proportions of earlywood and latewood within each annual ring, reaction wood content, proportions of sapwood and heartwood and spiral grain.

The width of the annual ring simply reflects the diameter growth per annum. Thus, its only direct influence on wood quality is limited to cosmetic effects on sawn timber and veneer. Nevertheless, the width of the annual ring is included in most calculations of assortments used today, since this variable is, to varying degrees, correlated with other properties directly affecting wood quality.

For *Pinus sylvestris,* positive correlation has been found between the width of the annual ring and branch diameter (Persson, 1976), knot content, lignin content and pentosane content (Ericson et al., 1973). Negative correlations have been reported between annual ring width and basic density (Ericson, 1960; Persson, 1976), percentage heartwood, percentage latewood, extractives content, pulp yield (kg/m^3) and the bursting and tearing strength of the pulp (Ericson et al., 1973). Although the correlation between wood density and annual ring width is generally negative, for annual ring widths less than one millimetre the starvation wood effect can occur (Ericson, 1966).

In general, trees with rapid radial growth tend to have a low basic density, thick branches (thus a high knot content), and low percentages of latewood and heartwood, the net result being poor quality sawn timber and a low pulp yield (measured in kg/m^3). Although it is tempting to conclude that tree improvement programmes aimed at increasing production would lead to a similar reduction in quality, the relationships are not strong. In fact, through a careful combination of selection criteria, high growth rate and good quality can be combined (Zobel and Talbert, 1984). However, some properties, e.g. branch diameter and basic density, are difficult to control when the goal is increased diameter growth. The evaluation of wood properties in breeding trials demands that adjustments are made that take into account differences in ring width or diameter between the samples tested (for example see Persson, 1972).

The change from juvenile wood to mature wood in Scots pine involves substantial alterations in wood properties and is strongly coupled to age differences between the apical meristem and the cambium (Rendle, 1960). Juvenile wood is found in the upper part of the tree and in the lower part, close to the pith.

In softwoods, according to Haygreen and Bowyer (1982), basic density, tracheid length, strength, cell wall thickness, percentage latewood and transverse shrinkage increase whereas microfibril angle, longitudinal shrinkage and moisture content decrease in mature wood. Low strength, prevalence of spiral grain (Harris, 1969) and longitudinal shrinkage render juvenile wood undesirable for use in most solid wood products (Haygreen and Bowyer, 1982). The heterogeneity in properties of sawn goods containing both juvenile and mature wood is also a problem. In pulpwood the use of juvenile wood is associated with increased costs and decreased productivity.

The number of annual rings with juvenile wood is more or less constant within a species. Thus, trees with rapid radial growth have an increased amount of juvenile wood (Zobel, 1981). The increased growth rate induced in breeding programmes will lead to a corresponding increase in juvenile wood content.

Proportions of early- and latewood within each annual ring. The annual shift in production from early- to latewood occurs around the middle of the growing season. According to Antonova et al. (1983), there are two phases of cambial activity, corresponding to the formation of early- and latewood. The two phases proceed independently and under different environmental conditions. In springtime, high auxin levels induce the production of large tracheids while available photosynthate is used for needle and shoot development. Thus, there is little photosyntate surplus for use in thickening of tracheid walls. In contrast, as the summer progresses auxin production decreases, leading to decreased tracheid dimensions while the amount of photosynthates available for cell wall formation increases (Zimmerman and Brown, 1971; Menyailo, 1985). On the basis of this relationship it can be predicted that the amount of latewood should increase with tree age since the period of shoot and needle growth becomes shorter as the tree gets older. The relationship between growth period length and earlywood/latewood formation also explains why latewood content decreases with increasing latitude and altitude (Ericson et al., 1973). Large differences in latewood content occur between and within provenances of Scots pine. In Sweden, northern provenances, with a relatively short period of shoot elongation, have less earlywood than southern provenances grown at the same site.

Earlywood is characterized by thick tracheids with thin cell walls and a proportionally large lumen. Lignin content is high and cellulose content low since the middle lamella is relatively large. Latewood is characterized by tracheids with thick cell walls and a small lumen. Basic density is a function of the amount of cell wall material and is therefore higher in latewood than in earlywood. There are indications from other pine species that the heritability of basic density could be different in earlywood than in latewood (Goggans, 1962).

Compression wood in Scots pine occurs on the lower side of a leaning stem and in the branches. In compression wood the tracheids are short, while lignin content and fibril angle are high and the cellulose content low. As sawn goods are dried, longitudinal shrinkage occurs in compression wood. The content of compression wood in Scots pine is dependent on the straightness of the tree, the inheritance of which is discussed elsewhere in this volume.

Proportions of sapwood and heartwood. The formation of heartwood starts once the active processes in the wood have stopped. The heartwood, which constitutes the inner part of the tree, provides both stability and durability. During heartwood formation the chemical content as well as other wood properties are altered. The amount of heartwood formed at a given age is dependent on the growth rate of the tree. As a consequence, forest tree improvement programmes aimed at increasing growth rate and decreasing rotation age often lead to a reduction in heartwood content.

Spiral grain. The term "spiral grain" refers to the alignment of tracheids at an angle to the axis of the tree stem. Both left and right spirals can be formed (Harris, 1981). Mild forms of spiral grain are fairly common in Scots pine while serious spiral grain is unusual. Spiral grain is apparently caused by anticlinal divisions in which new cambial cells are formed in one direction only (Haygreen and Bowyer, 1982). Spiral grain causes lumber to have diagonal grain, reduced strength and stiffness and a tendency to twist when dried. Consequently planning is difficult.

Spiral grain seems to be moderately inherited. In other pine species heritabilities around 0.5 have been found. There are also indications from a number of small trials and vegetatively propagated materials that grain abnormalities, i.e. curly grain, interlocked grain, wavy grain, are genetically determined (Harris, 1981), whether or not there is a strong nonadditive component in the genetic variation of spiral grain to be determined. In *P. radiata* Pederick (1971) found the nonadditive component to be small while Zobel et al. (1968) found that in *P. taeda* the genetic component was mainly nonadditive.

Tracheid properties

Stem tissue consists primarily of tracheids. The size and properties of tracheids are therefore major determinants of wood quality. Tracheid properties influence the quality of sawn goods as well as pulp and paper. Complicating the situation is the fact that tracheid properties vary greatly within each tree as well as within each annual ring.

Each tracheid consists of a lumen, i.e. the empty space within each cell, and a cell wall. The space between tracheids is occupied by an intercellular substance called the compound middle lamella. The cell wall comprises an outer primary wall and a secondary wall consisting of three S-layers. Development of the primary wall is initiated before the cell begins enlarging. Formation of the secondary wall begins slightly before or once final size is achieved. The main part of a cell wall consists of the S2 layer which has a high content of cellulose and hemicellulose.

In general, tracheid size increases from the pith outwards. There is also a length gradient, with the longest tracheids being found in the lowest two thirds of the stem (Ericson et al., 1973). Within each annual ring the tracheids formed in earlywood are generally thicker and have a thinner S2 layer than the tracheids in latewood. Tracheids are generally large in the root system and small in branches.

Within stands of Scots pine the taller trees tend to have longer tracheids. Ståhl (1988) found that tracheid length was positively correlated with total height, diameter and age but negatively correlated with annual increment. Basic density and tracheid length were not correlated. Between-site variation was examined in the same study. The longest tracheids were found at intermediate altitudes in southern and central Sweden while shorter tracheids were found at coastal sites, high altitudes and northern latitudes. Similar results have been reported from Sweden by Ericson et al. (1973) and in Finland by Nepveu and Velling (1986). A provenance transfer southwards and to higher altitudes increased tracheid length (Ståhl, 1988). Studies on the fibre yield and fibre mass of various provenances were made in Finland by Velling and Nepveu (1986). No significant variation in fibre yield (per cent) was found although both dry mass and fibre mass varied significantly.

Tracheid properties in other pine species show rather high heritability. Broad-sense heritability estimates for tracheid length were 0.6 in *P. elliottii* (Zobel et al., 1962) and 0.8 in *P. radiata* (Dadswell and Wardrop, 1961). Narrow-sense heritability estimates for tracheid length in *P. taeda* were 0.8 in latewood and 0.6 in earlywood while other tracheid properties ranged from 0.9 to 0.3 in latewood and from 0.8 to 0.1 in earlywood (Goggans, 1962).

Cell wall thickness should be directly correlated to basic density, as the main part of the wood substance in a tree is located in the cell wall. Nevertheless, data are scarce regarding the relationship between cell wall thickness and basic density, probably because a correct estimate of cell wall thickness is difficult to achieve and because the total amount of cell wall mass is largely determined by tracheid dimensions.

Weight-related properties

Basic density, or specific gravity, is probably the single most important property-determining quality. It affects the yield and quality of pulp and paper and is a major determinant of the durability and strength of sawn products. Density is also, as previously mentioned, correlated with a number of other properties. For this reason basic density receives great emphasis in most breeding programmes. The aim of the programme may not always be to increase basic density. For example, for certain end-products, such as tissue and high-quality printing paper, and for pine species with a short rotation the aim may be to decrease density.

Although basic density is a complex property, according to Zobel and Talbert (1984) it is primarily determined by three wood characteristics: latewood content, cell size and cell wall thickness. In spite of this, basic density is treated as a single property in most breeding programmes.

The variation in basic density within and between trees can be described as follows: Within each annual ring basic density is higher and its variation greater in the latewood. The juvenile wood formed at an early age and that formed in the crown has a lower density than mature wood. Therefore, basic density is highest in the lower part of the stem and increases gradually from pith to bark. In general, a wide annual ring or a large diameter is associated with low density. In Sweden the basic density of Scots pine decreases at high elevations and northern latitudes close to the timberline (Ericson, 1966; Ståhl, 1988). Similar altitudinal and latitudinal trends are commonly found in other pine species as well (Zobel and Talbert, 1984). In Sweden, Ståhl (1988) found the highest basic density in the central part of the country, around 60° latitude. Provenance variation in basic density seems to be fairly large, and provenance ranking does not vary much between different environments. Although the effect of transfer on basic density varies from region to region, in Sweden a moderate southward and upward transfer maintains or increases basic density levels (Ståhl, 1988).

Estimates of the genetic variation and heritability of basic density in *P. sylvestris* are high ($h^2>0.5$), often being higher than heritabilities for volume-related or phenological properties (Persson, 1972; Halupáné-Grosz and Mátyás, 1975). Unfortunately direct measurements of basic density require that increment cores or discs are removed from the tree, thereby damaging or destroying it. Indirect methods such as the use of a pilodyn instrument, have been tried (Velling and Nepveu, 1986), but they are sensitive to variation in environmental conditions and sample quality.

Branch-related properties

The properties of stem wood are to a large degree affected by the number, size, status (living or dead) and angle of branches. Quality classifications for sawn goods are often based on branch number and size—and with good reason—since the strength and beauty of the final product is generally related to knot content. In the pulp industry a reason is that the knots have a different tracheid length and density than the surrounding wood and therefore are more difficult to process, but advanced chemical processes have, in part, overcome this problem. In modern forestry the whole tree tends to be used, instead of just the stem. Branches, and in some cases the stump, are harvested and used for pulp or fuel. In these cases the properties of the non-stem wood become more important.

Quality classifications of sawn goods are generally based on assessments of the thickest branch because its size is often of crucial importance for strength. The occurrence of spike knots, i.e. large branches, emerging from the trunk at an extremely sharp angle, often caused by loss of the leader or a leading bud, is extremely detrimental to strength.

The knot content of stem wood in old virgin stands is usually less than 1% (Ericson et al., 1973) but values from artificially regenerated stands are higher. Branch diameter is closely correlated with tree diameter but not with tree height. Knot content is positively related to spacing and site quality, even for trees with the same DBH (Ericson et al., 1973; Persson, 1976). Branch diameter is related to the juvenile diameter increment of the stem (Persson, 1977).

Although the occurrence of spike knots is normally associated with injury they can also be found on trees without any external signs of damage. In Sweden, moose browsing is often the main source of injury. Spike knot frequency is determined to some degree by genetics. In trials in northern Sweden the northern, more hardy provenances were found to have a lower frequency of spike knots than southern, less hardy ones (Ståhl et al., 1986). The low degree of frost damage to buds and shoots of hardy provenances probably explains why they had few spike knots.

Genetic variation in branch diameter and knot content seems to be substantial. Still, the genetic component is difficult to distinguish, owing to the close interaction with both diameter and spacing (Remröd, 1976; Persson, 1977). Provenance variation in branch properties has been reported by Nilsson (1968), Remröd (1976) and Eriksson et al. (1987). In Sweden, northern continental provenances seem to have better branch properties than southern or coastal provenances. The same trend has been observed in Finland (Velling, 1988).

The heritability of branch-related properties is generally intermediate to high. Narrow-sense heritabilities from 0.2 to 0.6 for branch characters were reported by Nilsson (1968), Eriksson et al. (1987), Pöykkö (1982), Velling and Tigerstedt (1984) while Andersson and Hattemer (1978) reported broad-sense heritabilities from 0.8 to 0.9 for the same properties. Significant between-family variation was found by Velling (1982), Werner and Ericsson (1980), Andersson (1986), Gullberg and Vegerfors (1987) and Rusanen and Velling (1988).

Chemical properties

The chemical components of wood can be divided into four types: cellulose, hemicellulose, lignin and extractives. Cellulose is the main component of interest for the pulp and paper industry while the other components are used in the chemical industry and for energy production.

Between-site variation in the chemical properties of wood can be considerable. Ericson et al. (1973) found that the content of extractives in northern Sweden was higher than that in southern Sweden whereas the opposite was true of lignin content.

Provenance variation in extractive content is great. Bridgen and Hanover (1982) found genetic variation in chemical, physiological and anatomical traits associated with oleoresin production in Scots pine. Savchenko and Vasilevskaya (1977) found that oleoresin content was heritable, and that the oleoresin content of parent trees was positively correlated with growth rate of their offspring. Bridgen (1979) observed a large degree of interaction between seed source and environmental conditions, with 56% of the variation in total resin acids being due to this interaction. Velling and Nepveu (1986) noticed significant provenance variation as well as significant trial–provenance interaction, which was attributed to the poor adaptation of southern and coastal provenances to northern and inland conditions.

Halupané-Grosz and Mátyás (1975) found high broad-sense heritability for extractive content ($H^2=0.51$) while lignin content heritability was lower ($H^2=0.22$). For these properties the between-provenance variance component was lower than the within-provenance one, indicating that the adaptive value of these traits is only of minor importance, i.e. the selection pressure within populations was low.

In *P. taeda* the inheritance pattern for cellulose was reported to be mainly non-additive by Jett et al. (1977). The dominance component accounted for 15% of the phenotypic variance while the additive component was essentially zero.

Breeding for wood quality

The heritabilities of wood properties are generally as high or higher than heritabilities of volume-related properties. Nevertheless most breeding programmes are aimed at increasing volume production, while improvement of wood properties is given secondary priority. The reason for this lower priority is probably that the formulation of goals regarding wood properties in a long-term breeding programme is difficult. Furthermore, selection for wood properties is a complicated and expensive process, often requiring destructive sampling. The benefits to be gained by breeding for wood properties are also difficult to evaluate. An increase in volume production by 10% is a straightforward value easily used in gain evaluation. By contrast, it is often more difficult to evaluate an increase of the same magnitude in mean knot content or fibre length, especially since wood properties are often correlated with each other.

The forest managers' view is often that the primary goal of the forest tree breeder should be to increase volume production while wood properties should be controlled to the extent that they are kept at reasonable levels. Silvicultural activities can then be used to increase quality in specified stands.

In a practical breeding programme aimed at producing a specified product only some of the many wood properties can be included. For the sawmill owners, the branch properties, stem straightness and wood density are of major importance. In contrast, pulp mill owners place more value on tracheid size, basic density and chemical properties. Breeding programmes aimed at increasing wood quality in general should probably include basic density and branch properties in addition to volume production and stem straightness. To evaluate quality, plus trees should be selected in young stands. In old stands when natural pruning has reached above 4–5 m on the bottom log, evaluation of this most valuable part of the tree becomes difficult. Furthermore, the majority of stem injuries have been healed over by the end of a full rotation. In the Swedish breeding programme for Scots pine, tree selections are being made in planted stands at an age of 20 to 50 years (Werner et al., 1981). At present, the only wood property actively selected for, is basic density, with one third of the evaluated trees being rejected based on low values. For branch characters and straightness a value at or above the stand average is accepted.

A breeding programme aimed at increasing survival and wood production may also lead to improvements in wood quality (Ståhl et al., 1986). Since height growth is positively correlated with the frequency of straight stems (Prescher and Ståhl, 1986) and with fibre length (Ståhl, 1988) selecting for tall trees should yield a number of side-benefits. The negative consequences associated with rapid radial growth can be minimized if increased volume production is not used as a motivation for reducing stocking levels in silvicultural programmes (compare Dittmar, 1977).

Finally, it is important to emphasize again that a breeding programme can only produce good planting material. For a high quality seedling to manifest its full potential requires that it receives good silvicultural treatment. By combining effective forest-tree improvement programmes with sound silvicultural management future generations can be assured of having stands of Scots pine with good wood properties. A failure to combine forest tree breeding and silviculture would drastically reduce this possibility.

Literature

Andersson, B. (1986): Möjligheter till urval med avseende på virkeskvalitet i unga avkommeförsök av tall (Possibilities with selection for quality in young progeny trials of Scots pine (*Pinus sylvestris* L.). Föreningen Skogsträdsförädling, Institutet för skogsförbättring. Årsbok 1985, pp. 58–80.

Andersson, E. and Hattemer, H. H. (1978): Variation among clones and ortet-ramet relationship in grafted Scots pine (*Pinus sylvestris* L.). *Stud. For. Suec.* **148**.

Antonova, G. F., Shebeko, V. V. and Malyutina, E. S. (1983): Sezonnaja dinamika kambialnoj aktivnosti i differentsiatsii trakheid v stvole sosny obyknovennoj (Seasonal dynamics of cambial activity and tracheid differentiation in the stem of Scots pine). *Khimiya Drevesiny.* No. 1, 16–22, 120. Sukachev Inst. For. and Timber, Krasnoyarsk, USSR.

Bridgen, M. R. (1979): Genetic variation of Scotch pine oleoresin physiology. *Agricult. Forest. Wildlife*

Bridgen, M. R. and Hanover, J. W. (1982): Genetic variation in oleoresin physiology of Scotch pine. *Forest Sci.,* **28**:582–589.

Dadswell, H. E. and Wardrop, A. (1961): Some aspects of wood anatomy in relation to pulping quality and to tree breeding. *J. Austr. Pulp and Paper Ind.* **13**:161–173.

Dittmar, O. (1977): Die Ergebnisse der internationalen Kiefern-Provenienzversuche Chorin 85 und Finowtal 145 nach 66- bzw. 36jähriger Beobachtungszeit. *Beitr. Forstwirtsch.* **3**:116–119.

Ericson, B. (1960): Studies of the genetical wood density variation in Scots pine and Norway spruce. Forest Research Institute of Sweden. Department of forest yield research, Reports No. 4.

Ericson, B. (1966): Effect of thinning on the basic density and content of latewood and heartwood in Scots pine and Norway spruce. Royal College of Forestry, Department of forest yield research, Research notes No. 10.

Ericson, B., Johnson, T. and Persson, A. (1973): Wood and sulphate pulp of Scots pine from virgin stands. Royal College of Forestry, Department of forest yield research, Research notes No. 25.

Eriksson, G., Ilstedt, B., Nilsson, C. and Ryttman, H. (1987): Within- and between-population variation of growth and stem quality in a 30-year-old *Pinus sylvestris* trial. *Scand. J. For. Res.* **2**:301–314.

Goggans, J. F. (1962): The correlation, variation and inheritance of wood properties in loblolly pine (*Pinus taeda* L.). N. C. State Coll. Sch. For. Tech. Rep. 14.

Gullberg, U. and Vegerfors, B. (1987): Genotype-environment interaction in Swedish material of *Pinus sylvestris. Scand. J. For. Res.* **2**:417–432.

Halupané-Grosz, Z. and Mátyás, C. (1975): Major observations in examination of characteristics of wood of graft clones of Scotch pine. *Erdészeti Kutatások* **71**:149–161.

Harris, J. (1969): On the causes of spiral grain in corewood of radiata pine. *New Zealand J. Botany* **7**:189–213.

Harris, J. M. (1981): Spiral grain formation. In: Barnett, J. R. (Ed.): *Xylem cell development.* Caste House Publ. Ltd, Kent, pp. 256–274.

Haygreen, J. G. and Bowyer, J. L. (1982): *Forest products and wood science—an introduction.* The Iowa State University Press, Ames, Iowa.

Jett, J. B., Weir, R. J. and Barker, J. A. (1977): *The inheritance of cellulose in loblolly pine.* TAPPI Conference Papers, Forest biology wood chemistry conference, pp. 159–162.

Menyailo, L. N. (1985): Role of the needles in hormonal regulation of xylogenesis in pine. *Soviet plant physiology* **32**:95–100.

Nepveu, G. and Velling, P. (1986): Effet du milieu et de la provenance sur le rendement en fibres chez le pin sylvestre (*Pinus sylvestris* L.) en Finlande. *Ann. Sci. For.* **43 (2)**:227–238.

Nilsson, B. (1968): Studies of the genetical variation of some quality characters in Scots pine (*Pinus silvestris* L.). Royal College of Forestry, Department of forest genetics, Research Notes No. 3.

Pederick, L. A. (1971): Inheritance of spiral grain in young radiata pine. Presented to XVth Congress IUFRO, Gainesville, Florida.

Persson, A. (1972): Studies on the basic density on mother trees and progenies of pine. *Studia For. Suec.* **96.**

Persson, A. (1976): The influence of spacing on the quality of sawn timber from Scots pine. Royal College of Forestry, Department of forest yield research, Research Notes No. 42.

Persson, A. (1977): Kvalitetsutveckling inom yngre förbandsförsök med tall (Quality development in young spacing trials with Scots pine). Skogshögskolan, Inst. för skogsproduktion, Rapporter och uppsatser nr 45.

Prescher, F. and Ståhl, E. G. (1986): The effect of provenance and spacing on stem straightness and number of spike knots of Scots pine in South and Central Sweden. *Studia For. Suec.* **172.**

Pöykkö, T. (1982): Genetic variation in quality characters of Scots pine. An evaluation by means of the heritability concept. *Silva Fenn.* **16**:135–140.

Remröd, J. (1976): Choosing Scots pine (*Pinus sylvestris* L.) provenances in Northern Sweden—analysis of survival, growth and quality in provenance experiments planted 1951. Royal College of Forestry, Department of forest genetics, Research Notes No. 19.

Rendle, B. J. (1960): Juvenile and adult wood. *J. Inst. Wood Sci.* **5**:58–61.

Rusanen, M. and Velling, P. (1988): Harvest index in young Scots pine progeny tests, variation and correlation with growth and quality traits. *Fol. Forest.* **708.**

Savchenko, A. I. and Vasilevskaya, L. S. (1977): Sozdanie zhiznestojkikh sosnovykh nasazhdenij (Creation of healthy resistant pine stands), *Lesnoe Khozyaistvo* **5**:28–31.

Ståhl, E. G. (1988): Transfer effects and variations in basic density and tracheid length of *Pinus sylvestris* L. populations. *Studia For. Suec.* **180.**

Ståhl, E. G., Persson, A. and Persson, B. (1986): Timber quality in northern latitudes—implications for forest tree improvement. In: *Provenances and forest tree breeding for high latitudes. Proceedings of the Frans Kempe symposium in Umeå, June 10–11 1986.* Swedish University of Agricultural Sciences, Department of Forest genetics and plant physiology, pp. 149–158.

Uusvaara, O. (1974): Wood quality in plantation grown Scots pine. *Comm. Inst. Forest. Fenn.* **80**:2.

Velling, P. (1982): Genetic variation in quality characteristics of Scots pine. *Silva Fenn.,* **16 (2)**:129–134.

Velling, P. (1988): The relationships between yield components in the breeding of Scots pine. Thesis. University of Helsinki, Department of Plant Breeding.

Velling, P. and Nepveu, G. (1986): Männyn puuaineen laadun ja tuotoksen vaihtelu suomalaisessa provenienssikoesarjassa (Variation of wood quality and yield in a Finnish series of provenance trials on Scots pine). *Silva Fenn.* **20 (3)**:211–231.

Velling, P. and Tigerstedt, P. M. A. (1984): Harvest index in a progeny test of Scots pine with reference to the model of selection. *Silva Fenn.* **18 (1)**: 21–32.

Vuokila, Y. (1982): Metsien teknisen ladun kehittäminen (The improvement of technical quality of forests). Institutum Forestale Fenniae, Helsinki. *Fol. Forest.* **523.**

Werner, M. and Ericsson, T. (1980): Virkeskvalitetsstudier i avkommor från en tallfröplantage (Wood quality studies in progenies from a Scots pine seed orchard). Föreningen skogsträdsförädling, Institutet för skogsförbättring, Årsbok 1979, pp. 40–69.

Werner, M., Hadders, G. and Rosvall, O. (1981): Förädlingsprogram för tall. gran och contortatall (Breeding programme for Scots pine, Norway spruce and lodgepole pine). Föreningen Skogsträdsförädling, Institutet för skogsförbättring, Årsbok 1980, pp. 59–111.

Zimmermann, M. H. and Brown, C. L. (1971): *Trees: structure and function.* Springer, New York.

Zobel, B. J. (1981): Wood quality from fast-grown plantations. *Tappi* **64 (1)**:71–74.

Zobel, B. J., Cole, D. and Stonecypher, R. (1962): Mimeograph of paper read to tree improvement meeting of the Society of American Foresters in Macon, Ga., October 25.

Zobel, B. J., Stonecypher, R. W. and Browne, C. (1968): Inheritance of spiral grain in young loblolly pine. *For. Sci.* **14**:376–379.

Zobel, B. J. and Talbert, J. (1984): *Applied forest tree improvement.* John Wiley, Chichester.

CHAPTER 19

Inheritance of tree form

M. Giertych

Traditional opinions

Ever since the earliest days of tree breeding it was believed that selections of Scots pine should focus on trees displaying a good stem form. Based on intuition rather than on anything else tree breeders have recommended that in breeding programmes only such trees be considered which have straight, clean stems with fine branches projecting from the stem as far as possible at right angles. Often, particularily in Scandinavian countries it was recommended that selected trees demonstrate narrow crowns. This latter requirement proved impossible on the continent of Europe because the best trees were not narrow crowned and the Scandinavian ideal was not to be found.

It was, of course, recognized from the very start that this form depends very much on the environment, on the tending practices, stand density and the like. Instead of numerical values, trees should have more desirable characteristics in relation to others growing in the same environment. For example, Lindquist (1948) recommended that in windy areas where many trees are deformed the selections should be restricted to trees which are "dominating, have absolutely straight stems and slender branching".

From experience with seed transfers it was recognized that there is a considerable racial variation in tree form. For example, Rohmeder and Schönbach (1959) admit that Scots pine races from southwestern Germany have a large percentage of crooked trees, both locally and when planted elsewhere. According to them the good form of trees from northern and northeastern Europe as well as from the mountains of Central Europe is maintained only in an oceanic climate. This is best observable in comparative plantations.

Trees have a natural tendency to grow straight. The opposite case is a consequence of some stress they experienced during their life. Thus, when selecting straight trees, in fact, we are selecting for resistance against the factor that locally is most likely to deform trees (snow, wind, insects, browsing). In a different environment such selections may prove useless. The characteristic of coarse branches which is typical of the races from southwestern Germany, appears less dependent on the environment.

Clonal tests

Initially the breeding programmes consisted primarily of selecting plus trees, propagating them by grafting. It was considered useful to evaluate the grafts as a means of testing the genetic value of the plus trees. It was observed, for example, (Arnborg and Hadders, 1957) that ramets of some clones consistently have short branches and slender crowns while others have broad crowns and thick stems. The clones were found to differ in average branch angle and in the tendency of branches to grow straight or curve upwards. In a test on different sites (Johnsson, 1965), on clay and on a poor sandy soil at 3 localities in South Sweden, the clones were ranked in the same order for branch characteristics. Ramets of ortets with narrow crowns were also narrow crowned. The response of some traits to different environments was in the same direction for all genotypes. The clonal averages for branch angle were unaffected by differences in the soil and climate. In a trial including different spacings and fertilizer treatments the branch angle of 5-year-old grafts remained unaffected (Ehrenberg, 1970). Lindgren (1985) found high correlations for various quality characteristics of clones between 3 field tests in south, central and northern Sweden. All of this indicates that the genotype x environment interaction effect is small at the clonal level for crown characteristics.

In general, while clonal differences do exist, the correlations between ortets and ramets, except for branch angle, are not very high (Andersson and Hattemer, 1978). It is said of clonal comparisons, or tree shows, as they used to be called, that much may depend on the rootstock, on the quality of the graft union and on topophytic effects. This is true of growth rate which is the main objective of tree breeding. Thus, clonal comparisons based on grafts are not considered very reliable and nowadays are seldom used, though for form traits they could still be useful. The utility of clonal tests would markedly improve if reliable methods of rooting were found or otherwise propagating Scots pine for clonal forestry.

Progeny comparisons

Parent–progeny relations

The crucial point is whether the traits of interest are passed onto the generative progeny. It was claimed by Lindquist (1948) that in progenies of Scots pine plus trees at age 2 "in many cases a tendency to slender branch formation has already been revealed". Rohmeder and Schönbach (1959) report that when seeds from trees with poor stem form are collected there will be a fair proportion of trees with crooked stems in the progeny.

In special studies devoted to the inheritance of tree form (Ehrenberg and Gustafsson, 1957), where controlled crosses were made between plus and minus Scots pine trees and the young seedling progenies were compared with free pollinations, the relation between parents and offspring was demonstrated at an early age. The open pollinated progeny of one minus tree showed a high percentage of forking, abnormal bud formation and proleptic shoots. Progenies of both of the minus trees studied had long, thick, acute-angled branches. The progenies of the plus trees were more vigorous as well as of better form than those of the minus trees. In crosses of plus x minus trees the progeny displayed more forkings and prolepses than in crosses between the plus trees, suggesting that these characters are dominant or semi-dominant. The negative effect appears, regardless of whether the minus tree is used as the male or the female parent.

Later on in the same experiment it was reported that at age 5 plus tree progenies tended to have more slender crowns and fewer branches per whorl, but the relationship between parents and offspring in branch traits was less well defined (Ehrenberg, 1966). Still later it was reported that progenies of parent trees with forkings and fasciations of annual shoots also had these traits (Ehrenberg, 1970). While at age 10 some minus tree progenies were among the best in height growth, they had much poorer stem form (Ehrenberg, 1974). In this particular study the minus trees had exceptionally bad form. Thus, the fact that these negative traits were passed onto the progeny does not necessarily imply that a normal selection for good form would improve the progeny much over the stand average.

There are numerous freak trees with several odd characteristics that seem to be genetically controlled and linked to tree form. For example, trees are known with stunted or wolf growth, stem fasciations and abundant cone production (cone disease). These features are passed onto the progeny (Schröck, 1957; Simak, 1971).

There are also exceptional plus trees as for example, tree no. E 1101, the so-called "Kanerva pine" found in Finland which is very narrow-crowned and thin-branched and this quality appears to be controlled by a single dominant gene since it is passed onto 50% of its progeny in the form of a reduced number of branches per unit height and lower branch basal area per stem basal area. Unfortunately this E 1101 factor appears to be associated with stem crookedness (Velling, 1982; Kärki, 1984). It is also linked to some needle characteristics such as greater frequency of 3-needled fascicles and the tendency for needles to be at a wide angle to the shoot axis (Mikola, 1985).

The phenotype of progenies of these exceptional trees with single dominant gene effects does not necessarily offer any evidence for successful selections of traits under polygenic control.

On the other hand, a general comparison of plus tree progenies with standard material indicates that plus tree selection is effective in improving stem form traits (Velling, 1982; Andersson, 1986). Also a comparison of branch and stem characteristics in progeny tests with those of parent clones in a clonal archive were rather good (Andersson, 1986).

Comparisons between progenies

Selection based on comparisons of progenies is usually more effective than phenotypic selection of parent trees. It was shown that selfing of Scots pine trees resulted in progenies with poor stem form while outbreeding (interprovenance crosses) demonstrated that the crooked stems of southwest-German races is a dominant character while the poor stems of the French provenances is not (Scamoni, 1950). In a 20-year-old test plantation of Scots pine half-sib progenies it was shown that significant differences occurred in stem taper (Georgi and Schröck, 1967).

Open-pollinated 6–15-year-old progenies of plus trees of Scots pine from various seed stands indicated that there were no significant differences between the progenies but considerable ones between the stands. Branch number per whorl depended on soil and climate but less on stand or mother tree. Frequency of large forks and of large and small bayonette type malformations did not significantly differentiate the progenies nor the stands. On the other hand, branch angle and frequency of small forks were different from stand to stand. Fortunately small forks are less frequent on fast-growing trees which makes joint selection for the two traits easier (Nanson, 1969).

In a study of 12-year-old half-sib progenies from 36 plus trees it was found that they differed in the number of branches in the last whorl and in slenderness (the ratio of tree height to branch length) (Palmberg, 1970).

In another study, plus tree progenies differed in straightness and in the frequency of spike knots (Werner and Ericsson, 1980). The latter trait represented a genetic component in a 10-year study for both male and female parents while in a 23-year-old study there was no significant progeny variance component (Gullberg and Vegerfors, 1987).

In a 19-year-old trial in Finland most of the branch characteristics observed differentiated the provenances significantly on a uniform site, but less so on a more variable one. A younger, 9-year-old trial also revealed large progeny differences in branch characteristics (Velling, 1982). A 13-year study in Sweden on two locations gave similar results for both branch characteristics and stem form, including crookedness and spike knots (Andersson, 1986). In both trials, however, the correlations of desirable form characteristics with growth traits were not very strong.

Thus, regardless of whether the progenies are from parents selected for form traits or not there exist genetic differences that can readily be used for tree form improvement.

Heritabilities

In Table 19.1 heritabilities of various form traits for Scots pine are compiled. Since the values were obtained in various conditions the environmental variance had to be different and therefore discrepancies should be expected quite apart from the

Table 19.1. Heritabilities of form traits (taken from Faulkner 1969, expanded)

Trait	Age of material	Heritability b. s.	Heritability n. s.	References
No. of branches per stem length	19	–	0.08	Nilsson, 1968
	4	–	0.03–0.23	Schrum and Gerhold, 1970
	5	–	0.06–0.15	Schrum and Gerhold, 1970
	6	–	0.11–0.36	Schrum and Gerhold, 1970
No. of branches per whorl	19	–	0.60&0.28–0.56	Nilsson, 1968
top whorl	10	–	0.18&0.34	Ehrenberg, 1963
2nd whorl	10	–	0.09&0.33	Ehrenberg, 1963
3rd whorl	10	–	0.09&0.41	Ehrenberg, 1963
4th whorl	10	–	0.43&0.14	Ehrenberg, 1963
Branch angle	4	–	0.23–0.37	Schrum and Gerhold, 1970
	5	–	0.47–0.50	Schrum and Gerhold, 1970
	6	–	0.24–0.55	Schrum and Gerhold, 1970
	12	–	very low	Palmberg, 1970
top 3 whorls	6–9	0.71	–	Arnborg and Hadders, 1957
2nd whorl	10	–	0.39&0.36	Ehrenberg, 1963
3rd whorl	10	–	0.26&0.96	Ehrenberg, 1963
4th whorl	10	–	0.14&0.81	Ehrenberg, 1963
4th whorl	19	–	0.08&0.16	Nilsson, 1968
clonal		0.88–0.85	–	Andersson and Hattemer, 1978
		0.69–0.80	–	Andersson and Hattemer, 1978
Branch length	4	–	0.21–0.50	Schrum and Gerhold, 1970
	5	–	0.37–0.41	Schrum and Gerhold, 1970
	6	–	0.35–0.42	Schrum and Gerhold, 1970
Crown length	19	–	0.58&0.34–0.68	Nilsson, 1968
Crown taper	4	–	0.13–0.30	Schrum and Gerhold, 1970
	5	–	0.14–0.54	Schrum and Gerhold, 1970
	6	–	0.20–0.46	Schrum and Gerhold, 1970
Ht./branch length				
top whorl	10	–	0.14&0.87	Ehrenberg, 1963
2nd whorl	10	–	0.25&0.30	Ehrenberg, 1963
3rd whorl	10	–	0.74&0.68	Ehrenberg, 1963
4th whorl	10	–	0.83&0.89	Ehrenberg, 1963
Stem straigthness clonal		0.87&0.72	–	Andersson and Hattemer, 1978
Deviation from vertical	1	–	0.19	Bolland, 1967
	2	–	0.05	Bolland, 1967
	3	–	0.02	Bolland, 1967
	4	–	0.22	Bolland, 1967
	5	–	0.56	Bolland, 1967

high

fact that there may be genetic differences between the material investigated by various authors. It is therefore rewarding to see that there is so much agreement between the heritability estimates of various investigations.

Branch length appears to have the highest heritability as well as the traits dependent on branch length such as the height/branch length ratio. The same is true of crown length. Branch angle is of medium heritabilitiy except for the very low value reported by Palmberg (1970). Crown taper being dependent on branch angle has a similar heritability. The number of branches per whorl is more heritable than the number of branches per unit height. Finally the study on the degree of leaning of pine seedling stems appears to give very variable heritability values depending on the year of study. For stem straightness only a clonal estimate of broad sense heritability is available, indicating a very high heritability.

In general, it can be said that compared with growth parameters form traits are more clearly heritable. Andersson and Hattemer (1978) claim that taller trees are rating lower for straightness. The correlation between branch thickness and stem straightness can be either positive or negative depending on site.

Genotype x environment interactions

It is usually reported that for form traits the site is not likely to affect rankings of clones or progenies (Johnsson, 1965; Ehrenberg, 1970; Fries, 1984; Lindgren 1985; Andersson, 1986; Gullberg and Vegerfors, 1987). In the last two studies quoted some genotype x locality interaction was reported for spike knot frequency. However, except for this trait the genotype x environment interaction can be considered as practically non-existent. This is also true of a man-made environmental variable such as stand density (spacing), which though in itself affecting form characters, does not induce any genotype x spacing interactions (Ehrenberg, 1970; Fries, 1984; Gullberg and Vegerfors, 1987).

In contrast to the predominating evidence referred to above Nanson (1969) found that for branch angle the ranking of populations can be significantly different on different sites.

Juvenile–mature correlations

Bolland (1967) claims that there are good correlations between young one-year-old seedlings and the same trees five years later in stem and branching characteristics. I have not found any more specific data on these correlations, however, most authors seem to assume that since heritabilities are high and parent–progeny as well as occasionally ramet-ortet agreements are good there is sufficient justification to evaluate form traits early. Obviously, however, this cannot be done before the factors causing stem deformation have acted. Thus, if deformations are caused by snow damage, the evaluations cannot be done until some years after the trees grew above the regular snow cover level so as to select trees resistant to snow pressure.

Provenance variation

Perhaps the best evidence for genetic control of stem quality comes from provenance experiments. Let me mention two examples.

In a study conducted in Michigan (USA) a number of Scots pine provenances have been tested at five different locations. An evaluation of the percentage of crooked stems at 12 years of age has been published (Wright et al., 1976) pooled for "varieties" as defined by Wright (1976). This presentation of data for groups of provenances is unfortunate because there exists the possibility of various populations from one region differing in tree form. However, the data are interesting and are presented in Table 19.2 after normalisation (conversion to units of standard deviation from location means). The "varieties" are arranged in sequence

Table 19.2. Differences in frequency of crooked trees in 19 varieties of Scots pine as defined by Wright 1976 grown for 11–12 years at 5 locations in Michigan, USA

Variety	Country of origin from Wright, 1976	Locations					Mean
		Kellogg	Allegan	Rose L.	Dunbar	Linden L.	
Lapponica	FIN, SWE, SIB	1.16	1.49	0.67	0.09	1.22	0.93
Scotica	SCO	0.36	–	1.44	0.51	1.22	0.88
Septentrionalis	FIN, NOR, SWE	0.89	1.26	0.90	0.30	0.77	0.82
'Krasnoyarsk'	SIB	1.55	0.56	–	0.93	0.09	0.80
Rhodopaea	GRE, BUL	0.76	0.79	1.21	0.09	0.54	0.68
Mongolica	SIB	1.16	0.56	0.67	0.58	−0.03	0.59
Rigensis	LAT, SWE	0.36	0.79	0.67	−0.05	0.65	0.48
Uralensis	URA	0.36	−0.15	0.90	0.16	0.99	0.45
Illyrica	YUG	−1.29	0.32	0.52	1.28	0.43	0.25
Armena	GEO, TUR	−0.30	0.09	−0.25	0.37	−0.37	−0.09
Polonica	POL	−0.56	0.09	−1.41	0.72	0.43	−0.18
'E. Anglia'	ENG	−0.82	–	–	0.44	–	−0.19
Carpatica	E CZE	−1.09	−0.38	−0.25	−0.05	0.31	−0.29
Hercynica	FGR, GDR, CZE	−0.83	0.09	−0.56	−0.47	−0.82	−0.52
Subillyrica	ITA	−0.56	–	−1.64	0.09	−0.59	−0.67
Aquitana	S FRA	−0.59	−1.55	−0.72	−0.26	–	−0.72
Pannonica	HUN	−0.82	−2.25	−0.72	−0.47	−1.28	−0.89
Iberica	SPA	−0.03	−0.61	0.29	−3.60	−1.05	−1.00
Haguenensis	BEL, FRA, FRG	−2.28	−1.08	−1.72	−0.67	−2.52	−1.65

The data are taken from Wright et al., 1976, but are converted into units of standard deviation from location means, i.e. normalized (negative values indicate crooked trees).

Table 19.3. Proportion of crooked trees in various provenances of Scots pine from the IUFRO 1938 trial at various locations

IUFRO no.	Provenance	IUFRO location no. and country								
		8 FRA	9 FRA	10 IT	12 NOR	20 USA	21 POL	22 HUN	23 CZE	Mean
30	Modum, Norway					2.01			1.76	1.88
16	Kandersteg, Switzerland		1.60							1.60
29	Bromarv, Finland					1.45				1.45
48	Vitsand, Sweden				0.67	1.46		2.11	1.34	1.39
34	Tinoava, Romania					2.46		0.56	1.07	1.36
6	Asnes, Norway			2.45	0.72		0.38	0.95	0.59	1.02
4	Tonset, Norway				0.99		0.56			0.77
32	Griva, Lat. SSR					0.71		0.79		0.75
49	Axamo, Sweden				0.45	0.64			0.96	0.68
12	Susk, Ukr. SSR			0.89					0.38	0.64
3	Sääminki, Finland			0.87	0.64	1.05	−0.14	0.36		0.55
36	Langenbrand, FRG					0.51				0.51
22	Ruciane, Poland	0.58		0.46		−0.03	1.60	0.29	−0.10	0.46
45	Vys. Tatry, Czechoslovakia					0.80			0.06	0.43
38	Supraśl B, Poland					0.40				0.40
11	Vecmokas, Lat. SSR			0.76	−0.47	0.63	1.42	−0.53	0.38	0.36
41	Susk, Poland	0.80				−0.18				0.31
37	Supraśl A, Poland					0.23				0.23
39	Bolewice A, Poland					−0.19			0.59	0.20
8	Voxna, Sweden			−0.20	0.51					0.15
55	Luboml, Ukr. SSR		0.15	−0.18		−0.38	−0.14	0.66	0.22	0.06
31	Val di Fiemme, Italy		−0.93	0.03				1.05		0.05
21	Göddenstedt, FRG			0.01		−0.55	1.42	−0.20	−0.52	0.03

53	Mustejki, Lit. SSR			0.58		−0.50	1.25	−1.55	0.32	0.02
2	Rovaniemi, Finland				0.45		−0.48			−0.01
54	Rychtal, Poland			0.27		−0.70	−0.14			−0.19
9	Tönnersjöheden, Sweden				−0.30	−0.09				−0.19
44	Trebon, Czechoslovakia					−0.23				−0.23
7	Svanoy, Norway	0.78	0.34		−1.97		−0.66		0.16	−0.27
20	Brody, Poland	−0.86	0.52	0.08		−0.94	−0.14	−0.17	−0.58	−0.30
28	Les Angles, France	−1.30		−0.39					0.75	−0.31
42	Kurivody, Czechoslovakia								−0.32	−0.32
14	Talmacel, Romania		−0.21	0.48		0.59		−2.18		−0.33
43	Tisnov, Czechoslovakia								−0.37	−0.37
18	Herselt, Belgium			−0.15			−1.00	0.06	−0.47	−0.39
40	Bolewice B, Poland					−0.47				−0.47
1	Inari, Finland						−0.48			−0.48
19	Diever, The Netherlands			−0.06		−1.37	0.04	−0.56		−0.49
25	Lenti, Hungary			−0.61		−0.62		−0.53		−0.59
17	Glen Gary, Scotland				−1.68		0.04		−0.47	−0.70
33	Breda, The Netherlands					−0.51			−0.90	−0.71
13	Presov, Czechoslovakia		−1.47	−0.09						−0.78
26	Ambert, France			−1.42						−1.42
35	Langensteinbach, FRG					−1.49				−1.49
23	Elmstein, FRG			−1.12		−1.38	−1.87	−0.89	−2.45	−1.54
24	Zellhausen, FRG			−2.67		−1.37	−1.69	−0.20	−2.23	−1.63
Literature:		1	1	2	3	4	5	6	7	

The data are normalized (negative values indicate crooked trees).
Literature: 1. Lacaze, 1964; 2. Tocci, 1976; 3. Dietrichson, 1964; 4. Schreiner et al., 1962; 5. Giertych, 1986; 6. Magyar, 1964; 7. Vincent, 1965.

on the basis of average deviation from the mean. It can be seen that the straight stemmed trees are most common among populations from northern and generally extreme parts of the species range (Scandinavia, Scotland, Siberia, the Balkans). On the other hand, most crooked trees are to be found in the northwestern part of continental Europe.

The average performance appears to have good confirmation on almost all sites. There appears to be very little "variety" x environment interaction. Only a few "varieties" with a generally medium stem crookedness have widely different scorings at different locations, e.g. *illyrica, polonica.*

Wright's data come from not too distant locations, all in Michigan, and are generally managed similarly. In Table 19.3 I have made a similar compilation of results for a more heterogeneous set of locations of the IUFRO 1938 Scots pine provenance experiment. The table is based on data from various authors, gathered at different ages of the trees and following different management practices. Again the data were normalized. For locations 20, 22 and 23, where the original data were in the form of percentage of trees or percentage volume in different tree trunk classes, the percentages were multiplied by the classes and summed over all classes to obtain an estimate of provenance crookedness (or straightness depending on definition of classes). The provenances are ranked in descending order of average straightness. This sequence has to be treated with caution with respect to underrepresented provenances. The overall picture is, however, very similar to that discussed for the Michigan series. The straight-stemmed provenances are primarily from Scandinavia, the most crooked ones from FRG, Holland and Czechoslovakia. Polish and western USSR provenances are generally of medium crookedness and include some truly interactive provenances, e.g. Mustejki, Ruciane. Also among the medium provenances is the very interactive Italian provenance, Val di Fiemme, which gave crooked trees in France and straight ones in Hungary.

Comparing the IUFRO 1938 data with the results obtained in the Michigan series the only obvious discrepancy concerns the outlier Scottish population, straight in Michigan and crooked in Norway and Czechoslovakia. Obviously different populations were tested and it is hardly surprising that among the Scottish outlier populations there is much genetic variability. Another example of large genetic variability among relatively adjacent populations, confirmed on several sites, is the straight-stemmed population Tinoava from Romania and the crooked populations Talmacel (Romania) and Lenti (Hungary).

For the major regions of Scots pine occurrence, however, the agreement between the two series is very good. Also in a provenance series studied in Poland which includes additionally some Swedish sources the variation in crookedness showed that the latter, though slow growing, have straight stems. Of the Polish provenances the straightest stems are in provenances from the Masurian Lake District in northern Poland while the populations from central Poland have the most crooked stems (Giertych, 1989).

The generally good agreement between the estimates of tree form at provenance level indicates that selection for this trait can be very effective genetically.

Literature

Andersson B. (1986): Möjligheter till urval med avseende pa virkeskvalitet i unga avkommeför sök av tall (Possibilities with selection for quality in young progeny trials of Scots pine). Föreningen skogsträdföradling, Inst. f. Skogsförbättring. Årsbok 1985; 58–80.

Andersson E. and Hattemer H. (1978): Variation among clones and ortet-ramet relationship in grafted Scots pine (*Pinus sylvestris* L.). *Studia Forest. Suec.* **148**:1–31.

Arnborg T. and Hadders G. (1957): Studies of some forestry qualities of clones of *Pinus sylvestris. Acta Horti Gotoburgensis, Medd. Göteborgs Botan Trädgärd* **21** **(3)**:125–157.

Bolland G. (1967): Ein Frühtest zur Verbesserung der Schaftform der Kiefer. *Arch. Forstwes.* **16** **(6/9)**:659–662.

Dietrichson J. (1964): Proveniensproblemet belyst ved studier av vekstrytme og klima (The provenance problem illustrated by studies of growth-rhythm and climate). *Medd. Det. norske Skogforsoksvesen* **19** **(71)**:499–656.

Ehrenberg C. E. (1963): Genetic variation in progeny tests of Scots pine (*Pinus silvestris* L.). *Studia Forest. Suec.* **10**:1–135.

Ehrenberg C. E. (1966): Parent-progeny relationship in Scots pine (*Pinus silvestris* L.): results from three progeny tests with plus and minus tree progenies in southern Sweden. *Studia Forest. Suec.* **40**:1–54.

Ehrenberg C. E. (1970): Breeding for stem quality. *Unasylva* **24** **(2–3)**: 23–31.

Ehrenberg C. E. (1974): Progeny tests of Scots pine (*Pinus sylvestris* L.), IUFRO excursion 1974. 09. 03 to Röskär, Point 7. (mimeogr.).

Ehrenberg C. and Gustafsson A. (1957): Plus-och minusträd: urval och avkommeprövning (Plus and minus trees: selection and progeny trials). *Skogen* **44** **(22)**:722, 724, 727–728.

Faulkner R. (1969): Some characters of secondary importance to stem straightness in the breeding of conifers. 2nd World Consultation on Forest Tree Breeding, FAO-FO-FTB-69–3/2: 1–12.

Fries A. (1984): Spacing interaction with genotype and with genetic variation for production and quality traits in a trial of seedlings and grafted clones of Scots pine (*Pinus sylvestris* L.). *Silvae Genet.* **33** **(4–5)**:145–152.

Georgi E. and Schröck O. (1967): Ergebnisse des Kiefernherkunftsversuches Lieberose. Beilage zu *Forstwirtsch.* **17** **(11)**:1–8.

Giertych M. (1986): Provenance variation of Scots pine (*Pinus sylvestris* L.) on a 46-year-old international experiment in Poland. *Arboretum Kórnickie* **31**:183–193.

Giertych M. (1989): Interakcja genotypu ze środowiskiem oraz z wiekiem polskich proweniencji sosny zwyczajnej (*Pinus sylvestris* L.) (Genotype interaction with environment and age for Polish provenances of Scots pine). *Arboretum Kórnickie* **33** (in print)

Gullberg U. and Vegerfors B. (1987): Genotype–environment interaction in Swedish material of *Pinus sylvertris.* In: Gullberg, V. (Ed.) Studies on the utilization of adapted forms in the breeding of *Pinus sylvestris.* SLU Inst. för Skogsgenetik, Research Notes 38, 1–21.

Johnsson H. (1965): Miljöns och genotypens inflytande pa tallens växtform i experimentel belysning (Environment and genotype influences on Scots pine stem form in experimental plots). Fören. Skogsträdsföradling, Årsbok 1964, 115–125.

Kärki L. (1985): Genetically narrow-crowned types combine high timber quality and high stem wood production at low cost. In: Tigerstedt, P. M. A., Puttonen, P. and Koski, V. (Eds) *Crop physiology of forest trees.* University of Helsinki, Department of Plant Breeding, pp. 245–256.

Lacaze J. F. (1964): Comparison de diverses provenances de pin sylvestre (*Pinus sylvestris* L.) representees dans les arboretums forestiers. *Rev. For. Fr.* **8/9**:658–677.

Lindgren D. (1985): Undersökningar med tallympar (Investigations by aid of Scots pine grafts). In: Nilsson, J-E. (Ed.) Skogsodlinsmaterial för Harda Norliga Lägen. SLU Inst. skogling genetik och växtfysiologi, Rapport 4, 181–187.

Lindquist B. (1948): Genetics in Swedish Forestry Practice. Svenska Skogsvardsföreningens Förlag, Stockholm, p. 173.

Magyar P. (1964): Erdeifenyő-származási kísérletek Bugacon (The Bugac Scots pine provenance experiment). *Erdészeti Kutatások* **60 (1/3)**:5–31.

Mikola J. (1985): Relationships between height growth differences of Scots pine full-sib families and variation in seed size, annual growth rhythm and some foliage characteristics. In: Tigerstedt, P. M. A., Puttonen, P., and Koski, V. (Eds) *Crop physiology of forest trees.* University of Helsinki, Department of Plant Breeding, pp. 233–243.

Nanson A. (1969): Tests de descendances de pin sylvestre. Stat. Rech. Eaux et Forets, Groenendaal-Hoeilaart, Belgium, Trav. Ser. E No. 3: 1–51.

Nilsson B. (1968): Studier av nagra kvalitetsegenskapers genetiska variation hos tall (*Pinus silvestris* L.) (Studies of the genetical variation of some quality characters in Scots pine). Inst. för Skogsgenetik, Skogshögskolan, *Res. Note.* **3**:1–117.

Palmberg C. (1970): Estimating heritability in progenies of open-pollinated plus trees of *Pinus sylvestris. Silv. Fenn.* **4 (3)**:202–218. In: *For. Abs.* 1971 32 No. 2067.

Prescher F. and Ståhl E. C. (1986): The effect of provenance and spacing on stem straightness and number of spike knots of Scots pine in south and central Sweden. *Studia For. Suec.* **172**:1–12.

Rohmeder E. and Schönbach H. (1959): Genetik und Züchtung der Waldbäume. Paul Parey, Hamburg and Berlin, p. 338.

Scamoni A. (1950): Über die weitere Entwicklung künstlicher Kieferkreuzungen in Eberswalde. *Züchter* **20 (1/2)**:39–42.

Schreiner E. J., Littlefield E. W. and Eliason E. J. (1962): Results of 1938 IUFRO Scots pine provenance test in New York. Station Paper NE Forest Exp. Sta. No. 166: 1–23.

Schröck O. (1957): Beobachtungen an der Nachkommenschaft einer Zapfensuchtkiefer. *Silvae Genet.* **6**:169–178.

Schrum, G. M. and Gerhold, H. D. (1970): Heritability estimates for Scotch pine. In: Namkoong, G. and Stern, K. (Eds) *Proceedings of the 2nd meeting of working group on quantitative genetics.* Sec. 22 IUFRO, Raleigh N. C., USA, USDA For. Ser., New Orleans, pp. 87–89.

Simak M. (1971): Kottesjuka hos tall (Cone clusters on Scots pine). *Sveriges Skogsvardsförbunds Tidskrift* **6 (71)**:609–619.

Tocci A. V. (1976): Esperienze sulle provenienze del pino silvestre (*Pinus sylvestris* L.). *Ann. Inst. Sperimentale per la Selvicoltura* **7**:203–232.

Velling P. (1982): Genetic variation in quality characteristics of Scots pine. *Silva Fenn.***16**:129–134.

Vincent G. (1965): Die Ergebnisse der Kiefernprovenienzversuche in der CSSR. Tagungsberichte Nr. 75 d. Dt. Akad. d. Landwirtsch. wiss. zu Berlin: 125–135.

Werner M. and Ericsson T. (1980): Virkeskva litetsstudier i avkommer fran en tallfroplantage (Studies on wood quality in progeny of a Scots pine seed orchard). Foreningen Skogstradsföradling Institut för skogsförbrättring, Àrsbok 1979, Uppsala: 40–69.

Wright J. (1976): *Introduction to forest genetics.* Academic Press, New York, p. 463.

Wrigth J. W., Lemmien W. A., Bright J. N., Day M. W. and Sajdak R. L. (1976): Scotch pine varieties for Christmastree and forest planting in Michigan. Res. Pap. 293, Natural Resources, Mich. State Univ. Agric. Exp. Sta., East Lansing: 1–15.

CHAPTER 20

The value of early testing

J.-E. Nilsson

Introduction

Because the value of early testing lies in the ability to forecast the future development of trees, reliable early predictions, i.e. high juvenile–mature correlations are needed to shorten generation times in breeding programmes, while poor predictions might delay or even prevent breeding progress. Traits regulated by single genes and less influenced by the environment can easily be assessed very early. However, such traits are rare in forest trees. Instead nearly all economically important traits are regulated by large gene complexes, interacting with tree age and with the variable environmental conditions during rotation time.

Genetic gain from early testing

The relative gain at mature age obtained from juvenile (J) selection as compared with mature (M) selection is by definition (Falconer, 1960):

$\mathrm{Gain}_{M/J}/\mathrm{Gain}_M = i_J h_J r_G / i_M h_M$, where i_J, i_M = selection intensities at juvenile and mature age; h_J, h_M = square roots of heritabilities at juvenile and mature age; r_G = juvenile–mature genetic correlation, i. e. correlation between breeding values at juvenile and mature age.

In practical work the same selection intensity can usually be used both in juvenile and mature selection. The heritability often varies considerably from juvenile to mature age indicating large changes in genetic versus environmental effects during stand development (Namkoong and Conkle, 1976; Franklin, 1979). As short-term experiments can usually be designed to reduce the experimental errors compared with long-term experiments, juvenile heritability can usually be at least as high as the mature heritability. This means that the expected gain from early selection highly depends on the magnitude of the juvenile–mature correlation (Franklin, 1977).

The gain or loss at mature age obtained from early compared with mature selection must be related to the gain in time which can be substantial in long-lived species and override possible non-optimal decisions based on early testing. Lambeth (1980) concluded from studies of various pine species that early selection for height is more efficient than mature selection in terms of gain per year in breeding. However, too early selection is not recommended and the most effective selection time increases with the rotation age.

Experimental considerations in early testing

Design of experiments

Short-term field experiments for early testing can be designed more cost-efficiently and experimentally more attractively than long-term experiments by the use of a narrow spacing. In a narrow spaced experiment the competition among trees relatively soon resembles that of a normally spaced stand or field experiment and hopefully increases the juvenile–mature correlation and precision of early selection.

Greenhouses and climate chambers are useful locations for short-term experiments where the advantages of both close spacing and accelerated growth can be obtained. As the environment in these experiments is mostly specialized and different from outdoor conditions, it is necessary to follow up the results of these experiments with ordinary field experiments under natural conditions.

Non-genetic influences on early testing

Non-genetic factors like physiological seed effects (e.g. seed weight) and environmental conditions influence the early performance of plants. Although seed weight is partly genetically determined (by the genetic composition of the mother), it is also strongly influenced by the physiological environment of the cone during seed ripening. Low temperatures during seed development negatively affect seed size and consequently the growth of young pine seedlings. However, there is a large confusion concerning the duration of seed effects on the development of Scots pine from one year (Mikola, 1980) to more than 16 years (Giertych, 1974).

Reduction of seed effects must be a main objective in early testing. The earlier an experiment is to be evaluated the heavier are the demands on physiological homogeneity of the seed material. It is recommended that seedlings for early testing should be raised from seeds matured in the same year and, if possible, on the same locality.

Indirect selection

Traits are often more or less correlated, either causally or through the dependence on other traits, and the correlation patterns can vary at different ages or in different environments (c.f. climate chamber and field experiment), emphasizing the need of verifying that early selection does not result in unacceptable consequences on correlated traits at later ages.

Traits for observation

Growth

Stem volume is not easily measured on young trees. However, height at young ages is positively correlated with stem volume of older trees. Prescher (1985) found that the correlation between height after 12 years and stem volume at age 30 was mostly $0.7 < r < 0.8$ for various Swedish Scots pine provenances indicating that tree height at young ages can be advantageously used to predict later stem volume. For older trees, when height observations are laborious, stem diameter is an alternative predictor of stem volume (Eriksson and Ryttman, 1986).

From studies on various pine species Lambeth (1980) suggested that juvenile–mature correlation for height growth in the field could be estimated from the age/age ratio ($r = 0.30$ after 1/10 and $r = 0.50$ after 1/5 of rotation) and concluded that early selection for height in field experiments is more efficient than mature selection in a breeding programme in terms of gain per year.

The heritabilities for height observed in field experiments usually increase with time (Eriksson and Ryttman, 1986), but field experiments of Scots pine where the development of trees has been observed from young to mature age are rare, and our knowledge of juvenile–mature height correlations is, therefore, limited. The phenotypic juvenile-mature correlations vary considerably among some of the oldest provenance experiments of Scots pine:

Giertych, 1974	−0.43	6 vs 59 y (52° lat N)
—	−0.83	16 vs 59 y (52° lat N)
Lundmark and Persson, 1985	0.75	16 vs 38 y (within families)
Nanson, 1968	−0.19	40 vs 59 y (45–57° lat N)
—	0.83	5 vs 58 y (45–57° lat N)
Petrini, 1959	0.35	5 vs 54 y (57–63° lat N)
—	0.81	10 vs 54 y (57–63° lat N)
Ståhl et al., 1981	0.64	21 vs 39 y (56–69° lat N)
Wiedemann, 1930 vs Patlaj, 1973	0.75	8 vs 58 y (50–61° lat N)

Studies on height correlations relying on ages of 15–25 years and less are more common:

Giertych, 1974	0.77	6 vs 16 y (52° lat N)
Hadders, 1976	0.39	3 vs 16 y (plus tree fam., 62–63° lat N)
Johnsson, 1955	0.94	8 vs 17 y (57–65° lat N)
Krusche et al., 1980	0.40	3 vs 11 y (plus tree fam.)
Petrini, 1959	0.35	5 vs 22 y (57–63° lat N)
Remröd, 1976	0.75	6 vs 27 y (59–68° lat N)
Saatcioglu, 1967	0.93	5 vs 25 y (39–69° lat N)
Stefansson and Sinko, 1967	0.56	6 vs 20 y (59–68° lat N)
Wright and Baldwin, 1957	0.86	4 vs 17 y (45–69° lat N)

The magnitude of the age–age height correlation is strongly influenced by the genetic diversity of the tested entries. For this reason the correlations calculated for provenance tests tend to be higher than suggested by Lambeth (1980) indicating ten years as a sufficient age for provenance height selection.

For small-range materials (e.g. plus tree selection for a breeding zone) the situation is more uncertain due to the lack of experiments of relevant ages. Usually positive juvenile–mature correlations are observed (Hadders, 1976; Krusche et al., 1980; Lundmark and Persson, 1985) but in exceptional cases also negative correlations (Giertych, 1974). For long-term breeding of Scots pine from a small geographical area it seems advisable not to make height selection in progeny test before 15–20 years or 3 m height (Lindgren, 1984).

Climate chamber experiments. For experimental reasons the performance of plants during the first few years is sometimes better studied under more controlled artificial conditions, like climate chambers or greenhouses than in ordinary field experiments. However, the different growth pattern of young Scots pine seedlings (free growth) compared with older trees (predetermined growth with all growth primordia present in the last year's bud) complicates height selection on juvenile plants. Only the observation of predetermined growth data can supply a reliable basis to make a meaningful selection for mature growth possible. Because at least two growth seasons per year can be achieved in a climate chamber, the juvenile period of the plants is quickly passed and the predetermined growth of non-juvenile plants takes over.

Good

Unfortunately, little is known about the genotype x environment interaction of growth in artificial compared with natural environments. Therefore, the reliability of the climate chamber results must be thoroughly verified by comparison with older field growth before used in practical early testing. A breeding programme based on selection for early height growth in artificial conditions cannot be recommended as yet. However, should future experimentation prove the possibility of successful height selection in artificial environments, it may become a valuable method of intensifying breeding programmes.

Survival

Natural mortality in Scots pine is caused by a large number of stress factors, e.g. low temperature, snow, ice, wind, drought, insects, pests, animals, air pollution or their interactions. Good adaptation to climate and weather stresses largely prevents damage from secondary calamities. Therefore, most genetic research and breeding efforts on the survival of Scots pine have been directed towards climate and weather stresses. When planning a field experiment, the unpredictability of weather stresses causes a substantial uncertainty in deciding at what age a reliable survival ranking can be made or which stress factors will cause the future damage. On locations with high snow cover many winter stresses are avoided before the trees are exposed to them above the snow cover. Therefore, 10 years is often considered as a minimum age for survival ranking in field trials. Ståhl and Andersson (1985) found that mortality due to weather stresses occurs at least till 30 years of age, however, with no change in ranking among provenances compared to 10 years earlier.

Field experiments with low occurrence of damage or mortality are not ideal for ranking or selection of the genetic entries. This is avoided by replicating field experiments on locations with varied exposure to weather and climate stresses. If intensive survival selection is the main goal, field experiments with high natural mortality rates can be of great value. However, if severe damage occurs at a very early age, certain stresses may not have occurred resulting in non-optimal selection.

High mortality is mostly a problem on northern and high altitude locations with short summers and cold winters. Trees not adapted to initiate cold hardening during the short night conditions in late summer are severely stressed during the following cold winter period and thereafter susceptible to secondary calamities. On northern sites with a severe climate, volume production is more affected by survival than by tree height (Remröd, 1976; Eriksson et al., 1980) and a good ranking of Scots pine provenances for mature volume production is obtained from the survival × height product. Marklund (1981) found a correlation $r=0.75$ for volume production between ages 13 and 70, Ståhl et al. (1981) $r=0.64$ between ages 21 and 39 and Persson (1975) $r=0.42$ and $r=0.91$ between ages 13 and 59. In all cases survival was by far the more significant factor in volume production.

Early testing of adaptation to specific weather or climate stresses (e.g. low temperature, temperature fluctuations, long summer nights, drought) is more easily and accurately performed in controlled laboratory or climate chamber conditions than in field experiments. Artificial freezing tests of young Scots pine seedlings in late summer show positive correlations between freezing damage at juvenile age and field mortality after 10–20 years for provenances (Nilsson and Eriksson, 1986), seed orchards (Andersson, 1986) and families (Nilsson and Andersson, 1987). Andersson (1986) also found that the correlations were higher for sites in severe climate than in more southern locations where other factors, e.g. insects and voles, contribute relatively more to the field mortality.

Resistance

Resistance to pathogens largely depends on the physiological status of the host at the time of attack. When selecting for disease resistance, it is, therefore, important that the plants or trees studied are grown under as uniform conditions as possible and have equal physiological status when tested. These preconditions are mostly better fulfilled in controlled environments, like greenhouses, than in the field. On the other hand, it is difficult to obtain an artificial climate comparable to that in nature and to grow plants in greenhouse until they reach the size when attacks usually occur in nature. Martinsson (1987) found slightly positive correlations between damages from pine twist rust on 2-year-old plus tree families in the greenhouse and on 8-year-old families in field trials. However, he noted also maternal effects in the greenhouse experiment, indicating that studies on artificial inoculation are preferably performed on young trees in natural forest habitats (c.f.

Bedker and Blanchette, 1988) rather than on seedlings in artificial short-term experiments.

Resistance to pollutants can be tested on young seedlings in short-term experiments, in the field or under laboratory conditions with controlled exposure to high concentrations of pollutants. Oleksyn and Białobok (1986) found a positive correlation (r=0.44) between needle injury of young seedlings of Scots pine provenances exposed to SO_2 *in vitro* and in the field. Garsed and Rutter (1982) found poor correspondence between Scots pine resistance to acute and chronic air pollution exposures. As with the problem of pathogen resistance there is an uncertainty about the relevance of evaluating young forest tree seedlings after exposure to often high concentrations of chemicals for short periods as short-cuts for long-term field studies.

Quality

Many quality traits are strongly affected by the environment. A wide spacing usually negatively affects quality traits (Persson, 1976), e.g. wood density, branch size, spike knots and natural branch pruning. Quality selection on sites with irregular spacing and competition, e.g. in severe climates, can therefore be hazardous. Quality traits are at present less suitable for very early testing in controlled artificial environments than hardiness and growth-related traits.

Heritabilities estimated from clone, provenance or progeny experiments of various ages show that similar genetic selection gains can be obtained for quality traits as for growth traits (Andersson, 1985*a*). However, the heritability estimates vary considerably between studies and between quality traits.

Field studies indicate that the effect of genotype × location interaction on quality is smaller than the individual genetic and location effects (e.g. Johnsson, 1964). Selection for quality traits can consequently be made earlier in environments that favour the development of early competition, e.g. on sites with good growth conditions and dense spacing (Lindgren, 1985).

Other applications

Early testing can be used in applications other than provenance or family selection. By comparing the early performance of seedlots of uncertain origin, or uncertain genetic composition, with genetically well-known reference materials the utility area of the unknown material can be determined (Fig. 20.1). One application is the determination of the potential planting zone of early-age seed orchard crops by comparing their rate of cold acclimation in artificial freezing tests with that of plants of known origin and hardiness (Andersson, 1985*b*). Similarly, the effect of supplementary mass pollination may be studied. Because these conclusions are put into practice directly, seed effects and other non-genetic effects must be thoroughly eliminated from the results. Otherwise there is a substantial risk that non-genetic factors will influence results of early tests to an unacceptable degree. If the comparisons are made on juvenile plants the minimum requirement should be that the seeds of reference plants are matured in the same environment as the tested entries.

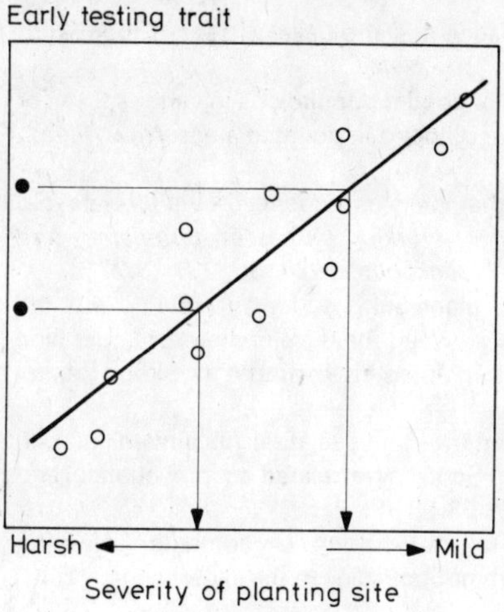

Fig. 20.1. Principle for early determination of the planting area of a population with uncertain properties and/or genetic constitution (filled circle) by early testing of young seedlings. After fitting the regression line to the entries with known planting areas (open circles) the potential sites of the unknown entries can be estimated by plotting their values on the early test trait (vertical axis) and projecting them on the horizontal axis

Early testing techniques also have direct economic applications in commercial plant production. As an example the determination of plant hardiness for cold storage in late summer (Rosvall-Åhnebrink, 1985) can be mentioned. If fast and reliable early tests can be developed, elimination of plants in poor physiological condition or of poor genetic constitution before planting in the field might reduce the future mortality on harsh sites.

Literature

Andersson, B. (1985*a*): Möjligheter till urval med avseende på virkeskvalitet i unga avkommeförsök av tall (Possibilities of selection with respect to timber quality in young progeny trials of Scots pine). Föreningen Skogsträdsförädling, Institutet för Skogsförbättring, Årsbok 1985:58–80.

Andersson, B. (1985*b*): Frystest av norrländska plantageskördar (Freezing tests of northern Scots pine seedlings from seed orchards). In: Nilsson, J. E. (Ed.) *Reforestation material for harsh northern sites. Proceedings of a seminar at Umeå, February 19–20, 1985.* Swed. Univ. Agr. Sci., Dept. For. Genetics and Plant Physiol., Report 4:42–56.

Andersson, B. (1986): Freezing tests of Scots pine (*Pinus sylvestris* L.) seed orchard crops. In: *Provenances and forest tree breeding for high latitudes. Proceedings of the Franz Kempe symposium in Umeå, June 10–11, 1986.* Swed. Univ. Agr. Sci., Dept. For. Genetics and Plant Physiol., Report 6:99–111.

Bedker, P. J. and Blanchette R. A. (1988): Mortality in Scots pine following inoculation with the pinewood nematode. *Bursaphelenchus xylophilus. Can. J. For. Res.* **18**:574–580.

Eriksson, G., Andersson, S., Eiche, V., Ifver, J. and Persson, A. (1980): Severity index and transfer effects on survival and volume production of *Pinus sylvestris* in northern Sweden. *Studia For. Suec.* **156**:1–32.

Eriksson, G. and Ryttman, H, (1986): Kvalitet och volym i ett 30-årigt proveniensförsök med tall i södra Lappland (Quality and volume in a 30-year-old provenance trial of Scots pine in southern Lapland). Swed. Univ. Agr. Sci., Skogsfakta, Biologi och Skogsskötsel 33:1–4.

Falconer, D. S. (1960): *Introduction to quantitative genetics.* Ronald Press, New York, p. 356.

Franklin, E. C. (1977): Juvenile-mature correlations. In: *Proccedings of the joint IUFRO Workshop,* S2.02–08 and S2.03–01. *Brisbane, 1977,* pp. 205–212.

Franklin, E. C. (1979): Models relating levels of genetic variance to stand development of four North American conifers. *Silvae Genet.* **28**:207–212.

Garsed, S. G. and Rutter, A. J. (1982): Relative performance of conifer populations in various tests for sensitivity to SO_2 and the implications for selecting trees for planting in polluted areas. *New Phytol.* **92**:349–367.

Giertych, M. (1974): Inadequacy of early tests for growth characters as evidenced by a 59-year-old experiment. In: *Proceedings of the IUFRO joint meeting of working parties on population and ecological genetics, breeding theory and progeny testing. Stockholm, 1974,* pp. 237–242.

Hadders, G. (1976): En undersökning av samband mellan plant-och ungdomsutveckling hos tall (*Pinus sylvestris* L.) (An investigation on the relationship between the development of Scots pine at the seedling stage and youth stage). Föreningen Skogsträdsförädling Institutet för Skogsförbättring 1976 Årsbok, pp. 170–188.

Johnsson, H. (1955): Utvecklingen i 15-åriga försöksodlingar av tall i relation till proveniens och odlingsort (Development in 15-year-old experiments of Scots pine related to provenance and location of use). *Svenska Skogsvårdsföreningens Tidskrift* **53**:58–88.

Johnsson, H. (1964): Miljöns och genotypens inflytande på tallens växtform i experimentell belysning (The influence of environment and genotype on the growth of Scots pine in the experimental light). Fören. Skogsträdsförädling. Årsbok 1964: 115–125.

Krusche, D., Das, B. L. and Stephan, B. R. (1980): Results of a progeny test with *Pinus sylvestris* and estimation of genetic gains from different selection methods. *Silvae Genet.* **29**:122–129.

Lambeth, C. L. (1980): Juvenile-mature correlations in *Pinaceae* and implications for early selection. *For. Sci.* **26(4)**:571–580.

Lindgren, D. (1984): Prediction and optimization of genetic gain with regard to genotype × environment interactions. *Studia For. Suec.* **166**:15–24.

Lindgren, D. (1985): Undersökningar med tallympar (Investigations by aid of Scots pine grafts). In: Nilsson, J. E. (Ed.) *Reforestation material for harsh northern sites, Proceedings of a seminar at Umeå, February 19–20, 1985.* Swed. Univ. Agr. Sci., Dept. For. Genetics and Plant Physiol. Report 4:181–187.

Lundmark, L. and Persson, T. (1985): Höjden i ung ålder som prognosinstrument (Tree height at an early age as a predictor). Swed. Univ. Agr. Sci. (Unpublished thesis).

Marklund, E. (1981): Äldre tallproveniensförsök ger underlag för produktionsprognoser (A production prognosis based on older provenance trials of Scots pine). *Svenska Skogsv. förb. Tidskr.* **5**:9–14.

Martinsson, O. (1987): Scots pine resistance to pine twist rust — conformity between the resistance found in artificial environment and field trials. *Silvae Genet.* **36 (1)**:15–21.

Mikola, J. (1980): The effect of seed size and duration of growth on the height of Scots pine (*Pinus sylvestris* L.) and Norway spruce (*Picea abies* (L.) Karst) provenances and progenies at the nursery stage. *Silva Fenn.* **14**:84–94.

Namkoong, G. and Conkle, M. T. (1976): Time trends in genetic control of height growth in ponderosa pine. *For. Sci.* **22**:2–12.

Nanson, A. (1968): La valeur des tests prècoces dans la selection des arbres forestiers en particulier au point de vue de la croissance. Stn. de Recherches des Eaux et Forêts, Groenendaal-Hoeilaart, Dissert. Facult. de Sci. Agron., Gembloux.

Nilsson, J.-E. (1985): Härdiga arvsanlag längst i norr (Hardy genes at the timber line in the far north?). In: Nilsson, J.- E. (Ed.) *Reforestation material for harsh northern sites. Proceedings of a seminar at Umeå, February 19–20, 1985.* Swed. Univ. Agr. Sci., Dept. For. Genetics and Plant Physiol. Report 4:84–92.

Nilsson, J.–E. and Andersson, B. (1987): Performance in freezing tests and field experiments of full-sib families of *Pinus sylvestris* L. *Can. J. For. Res.* **17**:1340–1347.

Nilsson, J.-E. and Eriksson, G. (1966): Freeze testing and field mortality of *Pinus sylvestris* L. in northern Sweden. *Scand. J. For. Res.* **1 (2)**: 205–218.

Oleksyn, J. and Białobok, S. (1986): Net photosynthesis, dark respiration and susceptibility to air pollution of 20 European provenances of Scots pine, *Pinus sylvestris* L. *Environ. Pollut. Series A* **40**:287–302.

Patlaj, I. N. (1973): Geograficheskie kultury sosny obyknovennoj v Ukrainskoj SSR (Scots pine provenance studies in the UkrSSR). *Proceedings of the IUFRO symposium on genetics of Scots pine.* Warsaw-Kórnik, pp. 1–26.

Persson, A. (1975): Äldre tallproveniensförsök (Older provenance trials of Scots pine). In: *Transfer of Scots pine* (*Pinus sylvestris* L.) *seed.* Swed. Univ. Agr. Sci., Dept. For. Genetics. Research Notes **17**:81–90.

Persson, A. (1976): Förbandets inverkan på tallens sågtimmerkvalitet (The influence of spacing on the quality of sawn timber from Scots pine). Royal Coll. Forestry, Dept. Forest Yield. Res., Research Notes 42. p. 122.

Petrini, S. (1959): De två äldsta svenska tallproveniensförsöken (The two oldest Swedish Scots pine provenance trials). Meddelande från statens skogsforsknings institut **48 (11)**:1–49.

Prescher, F. (1985): Framtida förädlingsstrategi för tall (Future breeding strategy for Scots pine). Slutredogörelse till Skogs-och Jordbrukets Forskningsråd. Swed. Univ. Agr. Sci., Dept. For. Yield. Res., Department Note.

Remröd. J. (1976): Val av tallprovenienser i norra Sverige-analys av överlevnad, tillväxt och kvalitet i 1951 års tallproveniensförsök. (Choosing Scots pine (*Pinus sylvestris* L.) provenances in northern Sweden-analysis of survival, growth and quality in provenance experiments planted in 1951). Royal Coll. For., Dept. For. Gen., Research Notes 19.

Rosvall-Åhnebrink, G. (1985): Invintring av plantor för höstplantering eller vinterlagring. (Hardening of plants for autumn planting or winter storage). In: Skogsskötsel i södra Sverige. Swed. Univ. Agr. Sci., Skogsfakta Konferens 7:33–37.

Saatcioglu, F. (1967): Results of the 25 years provenance experiment established by using 16 Scotch pine of European and 1 of native provenances in Turkey. *Silvae Genet.* **16 (5–6)**:172–177.

Stefansson, E. and Sinko, M. (1967): Försök med tallprovenienser med särslkild hänsyn till norrländska höjdlägen (Experiments with provenances of Scots pine with special regard to high-lying forests in northern Sweden). *Studia For. Suec.* **47**:1–108.

Ståhl, E. and Andersson, S.(1985): Avgång och skadefrekvens i 30-åriga tallproveniensförsök i hårda klimatlägen (Mortality and frequency of damage in 30 year old Scots pine trials in severe climatic conditions). In: Nilsson, J. E. (Ed.) *Reforestation material for harsh northern sites. Proceedings of a seminar at Umeå, February 19–20, 1985.* Swed. Univ. Agr. Sci., Dept. For. Genetics and Plant Physiol. Report **4**:66–74.

Ståhl, E. G., Andersson, S. and Öh, K. A. (1981): Ett proveniensförsöks utveckling under perioden 20 till 40 år (The development of a Scots pine provenance trial during the period 20–40 years). Skogsgenetisk forskning 1980. Sw. Univ. Agr. Sci., Dept. For. Gen. Rapporter och Uppsatser 31:15–22.

Wiedemann, G. (1930): Die Versuche über Einfluss der Herkunft des Kiefernsamens. *Zeitschr. Forst. Jagdwes.* **62**:809–836.

Wright, J. W. and Baldwin, H. I. (1957): The 1938 IUFRO Scotch pine test in New Hampshire. *Silvae Genet.* **6 (1–2)**:2–14.

CHAPTER 21

Utilization of improved material: a survey

J. Mikola

Scandinavia

Finland

Scots pine is by far the most important species in forest regeneration in Finland. During the 1980s it has been planted or direct-seeded on 100 000–120 000 ha annually (70% planting, 30% seeding). Its proportion in the total area of artificial regeneration has been around 80%. The area of Scots pine-dominated forests is 12.4 million ha. Natural regeneration has been practiced on an area about equal to direct seeding (Anon., 1986*a*, 1987).

The annual utilization of Scots pine seed has been around 3 t in forest nurseries and 12 t in direct seeding. Planting is fairly evenly distributed over the whole country, with about 160 million plants in total per year, but direct seeding predominates in the north.

Within the framework of a national programme, a total of 3 022 ha of first-generation clonal seed orchards of Scots pine were set up in Finland between the years 1952 and 1976. The programme aimed at the orchard production of all seed needed in artificial reforestation, the bulk of which should be used in direct seeding. The climate in North Finland is very adverse to flowering and seed ripening. Therefore, Scots pine seed orchards for the northern half of the country (1 800 ha in total) were established in southern and central Finland. In the southern half of the country orchard-produced seed is now available to satisfy all nursery needs. Direct seeding, with its great expenditure of seed (0.3 to 1.0 kg per ha), still relies mainly on unselected material harvested from identified natural stands or from specified regions. During the last few years orchard seed has been used for direct seeding as well, up to 1 000 kg annually (Valli and Venäläinen, both pers. comm.).

The production of improved material for the northern half of the country still faces severe problems. The northern orchards produce considerable amounts of seed, but due to pollen contamination from the surrounding southern forests, the material harvested so far has proved to be unadapted to the harsh climatic conditions of the native habitats of the mother trees. Fortunately, the northern × southern hybrid material arising in these orchards at the young age has been shown to be quite applicable for reforestation in intermediate areas halfway between the origins of mother trees and the locations of seed orchards (Kylmänen, 1980; Nikkanen, 1982).

In connection with plus tree selection, more than 600 stands with a total area of 4 200 ha have been selected and registered for seed production purposes, but their utilization is, however, negligible.

In a recent survey of all Scots pine progeny tests in southern Finland which had reached an age of 10 years or more (12 experiments, 240 families in total), the progenies of seed orchard clones showed an average superiority of 5% in height growth as compared to unselected control material from natural stands (Mikola, unpublished data). Also quality characters like branch thickness and branch angle have been observed to be slightly improved in the orchard-produced material (Velling, pers. comm.).

Sweden

In Swedish forestry Scots pine is also the most important tree species, although not as predominantly as in Finland. Its annual regeneration area is about 130 000 ha, i.e. 50% to 60% of the total reforestation and it divides roughly into 68% planting, 2% direct seeding and 30% natural regeneration.

The establishment of clonal seed orchards of Scots pine was started in the 1940s. In 1984, their total area was 477 ha, and 94 ha were still being built up. Between 1974 and 1983, 20 413 kg of seed in total was harvested from these orchards, i.e. about 2 t per year on the average. The total need of Scots pine seed has been estimated to be between 3.3 and 3.8 t annually, assuming that all regeneration in the future will be carried out by planting. The average production of the orchards is expected to be about 7 kg of seed per year per hectare (Werner et al., 1981). Forty percent of the demand is mainly fulfilled with seeds collected from felled trees in selected stands. At present, however, there is a serious shortage of any kind of Scots pine seed for the northernmost areas (Nilsson, pers. comm.).

Progeny testing has been initiated for almost all of the existing orchards (Hadders and Samuelson, 1985). The results obtained to date indicate that orchard progenies surpass unselected material of respective provenances by 5% to 10% in height growth and by 10% to 15% in diameter growth. This suggests that the genetic gain in volume production would be around 15% to 20%. In many timber quality traits, however, the relative superiority of first-generation seed orchard material has proved to be low or negligible (e.g. Werner and Ericsson, 1979; Andersson, 1985). For the orchards still being established, plus tree selection has been shifted from natural stands to young plantations, in which the equal age and spacing of the trees should provide better conditions for an efficient phenotypic selection (Werner et al., 1981).

Norway

Scots pine is the second most important tree species in Norway, with an annual planting of about 8.5 million seedlings, constituting 10% to 20% of the total reforestation. The need for seed orchards has been calculated to be

60 ha but only 14 ha have been established and no expansions are being planned (Veie, 1987).

The existing clonal seed orchards have not proved as useful as expected. All orchards have been transferred into a more favourable climate at lower elevations. This has created problems with background pollination. The material produced at low elevations has not been hardy at the higher elevations of the mother tree habitats (Jonsen, pers. comm.).

Western Europe

United Kingdom

In the British Isles, Scots pine is the only native conifer of commercial importance. In the early years of British tree breeding, started in the late 1940s, most of the breeding effort was devoted to this species. Since then, ideas about the potentials of different tree species in British forestry have changed rapidly. During the last ten years Scots pine has received about 15% of the resources devoted to tree breeding. Sitka spruce, lodgepole pine, Douglas-fir and larch surpass it in priority (Faulkner, pers. comm.).

The establishment of clonal seed orchards of Scots pine began in the early 1950s. Up to 1987, 71 clonal orchards with a total area of 52 ha had been established but many of these have been withdrawn. Cone collections are still made in 22 orchards, totalling 23 ha. Only two orchards (8.8 ha) have been established since 1970 and these are composed of clones tested in half-sib progeny tests. All existing orchards are of the clonal type but the present breeding strategy of the species is based on both clonal and seedling seed orchards (Samuel, pers. comm.).

The Netherlands

Scots pine is the most important forest tree species in the Netherlands, covering about 45% of the forested land area. Almost all of the present stands, totalling about 100 000 ha, are descendants of foreign populations (Squillace et al., 1975). The recent planting area of the species has been about 300 ha annually, making up 20% to 30% of total reforestation (de Vries, pers. comm.). The area of Scots pine is expected to decrease in the future by replacement with Douglas-fir (Kriek, 1980). An extensive selection and breeding programme was initiated in the early 1960s. Two clonal seed orchards based on the best parents from half-sib progeny tests have been established. In the early 1980s these orchards, totalling 13 ha in area, produced 50 to 60 kg seed annually, covering two thirds of the needs of Dutch forestry (Kriek, 1983).

France

There are about 1.1 million ha of Scots pine forests but recently the annual planting area of the species has been only a few hundred hectares (Roman-Amat, 1987). Separate breeding programmes have been set up for two indigenous provenances and a single foreign (Polish) one. Three new seed orchards, with clones selected on the basis of progeny performance, totalling 30 ha, have been established between 1985 and 1990. About 12 000 ha of stands, from 19 different provenance regions, have also been selected and registered for seed collection purposes (Chollet, 1987; Roman-Amat, 1987).

Federal Republic of Germany

Scots pine dominates an area of about 1.2 million ha in the FRG, constituting 15% to 20% of the total forest area. A great part of the present forests originate from cultivation with non-autochthonous provenances.

At present the reproductive material of Scots pine is mainly harvested from selected stands. More than 1 700 stands with a total area of 12 600 ha have been approved for this purpose in the whole FRG (Hoffmann, pers. comm.). The situation of Scots pine breeding and seed procurement in different lands will be briefly surveyed in the following.

In Lower Saxony Scots pine is the main tree species, dominating 50% of the total forest area of 928 200 ha. Most reforestation needs of the species are fulfilled with selected material from seed collection stands. Five seed orchards with a total area of 19 ha have been established.

In Bavaria Scots pine covers 19% of the area. The annual regeneration area amounts to 560 ha. Seed orchard area of the species is 11 ha. Since 1980 orchard seed has fulfilled 60% of seed needs in the state forests, and 40% have been collected from registered seed stands (Köhler-Maier, pers. comm.).

In Hesse the area of Scots pine forests is 155 000 ha. The area of seed orchards is 12.5 ha. In the 1970s seed orchards produced 28% of the total needs of Scots pine seed (Weisgerber, 1980) but in the last few years the amount of seed collected from seed orchards has been very low (Rau, pers. comm.). Registered seed stands form the main commercial seed supply of Scots pine.

In Baden–Württenberg the area of Scots pine in state and corporate forests is 85 800 ha (11%). Some 120 ha are annually planted with the species, and natural regeneration is practiced on about 40 ha. Two seed orchards totalling 13 ha cover 95% of the seed needs for artificial reforestation in the state forests (Franke, pers. comm.).

Some 15 000 ha are planted annually with Scots pine in the (former) GDR requiring 25 million plants and 6 000 kg of seed. Some 13 000 ha of Scots pine stands have been selected and registered for seed collection purposes. The area of first-generation clonal seed orchards is 127 ha. Moreover, 7 clonal orchards with an area of 55 ha have been established with clones selected on the basis of progeny test results.

Untested seed orchards of Scots pine have recently produced 900 to 1 400 kg of seed annually, satisfying 15% to 23% of the needs in the country. Tested orchards presently produce less than 5% but by the year 2000 they are expected to cover at least 30% of the total seed utilization. Nowadays still up to 80% of reforestation seed is harvested from registered seed stands (Bolland, 1981; Schütze, pers. comm.).

Other European countries

Poland

Scots pine is the most important tree species in Poland, occupying 72% of the state's forest area of 8.5 million ha.

Current demand of Scots pine seed in Poland is almost completely met by stand collections. There are two categories of seed collection stands, i.e. selected seed stands and temporary seed stands.

The total areas of temporary seed stands and selected stands are 180 000 and 3 880 ha, respectively. All new plantations established with material from selected stands are registered as particular progeny plantations. These plantations are specially developed to provide seed for further reforestation. In 1980 the total area of registered progeny plantations was about 4 500 ha (Kocięcki, 1980).

Both clonal and seedling seed orchards have been established since 1970. At present their total areas are 247 and 91 ha, respectively (Patalas, pers. comm.). Up to 1990 the area of clonal orchards will be enlarged to 500 ha. Seedling seed orchards from generative progeny of the plus trees will be established at a faster rate, with the goal of an area of 700 ha in 1990 (Kocięcki, 1980). Scots pine is planted annually on an area of 40 000 to 45 000 ha.

Economic seed stands form the main seed supply with an annual harvest of about 15 000 kg. About 1 000 kg per year are collected from selected seed stands. Because of the young age of the seed orchards, very little seed has been available from them. So far cones have not been collected in seedling seed orchards but clonal orchards provided 190 kg of seed in 1986 (Patalas, pers. comm.).

Hungary

The species covers about 160 000 ha in Hungary, corresponding to about 9% of the total forested area. The present seed orchard area of the species is 140 ha; 54 ha of tested orchards are expected to produce certified strains which promise a yield gain of 8 to 10% in cultivation. Seed production in the Hungarian Scots pine seed orchards has been remarkably high, up to 10 kg per hectare and year already at the age of 10 to 15 years. Roughly half of the present demand is covered by material from untested seed orchards, and one fourth by material from tested orchards (Mátyás, pers. comm.).

Czechoslovakia

In Bohemia Scots pine occupies 465 000 ha and in Slovakia 150 000 ha, i.e. 18% and 8%, respectively, of the total forest area. As a result of artificial regeneration, the species now covers large areas outside its original range. In recent years the annual regeneration area of Scots pine has been around 8 000 ha in the whole country. More than 95% of the regeneration is by planting. The present areas of registered seed collection stands and seed orchards of Scots pine are about 22 000 ha and 170 ha, respectively. The area of seed stands is planned to be kept around 20 000 ha and that of seed orchards to be increased to 210 ha. At present about 90% of the demand of Scots pine seed is met by seed collection stands and the rest by seed orchards. The coverage of orchard-produced seed is to be increased to about 30% in the near future (Ŝindelár, pers. comm.).

Bulgaria

In Bulgaria Scots pine forests cover 570 000 ha, which is 17% of the total forest area and 52% of the coniferous forests. About 1 500 kg of Scots pine seed is used for reforestation annually. At present seed is mainly harvested from selected stands. The number of selected permanent seed stands of Scots pine is 361, and their total area is 9 640 ha. Also 46 ha of clonal seed orchards and 140 ha of seedling seed orchards of the species have been established (Alexandrov, pers. comm.).

The natural range of Scots pine covers almost all countries in Europe but in countries that have not been discussed in the foregoing it is a species of minor importance (e.g. for Ireland see Mooney, 1986; for Belgium see Nanson, 1974). Obviously very little breeding work of Scots pine has been done in many countries of central, south-eastern and southern Europe, where the species occurs only at high altitudes. However, selection of seed stands has been reported from Yugoslavia (Jovanović and Tucović, 1974) and from Italy (Magini, 1974). Also in Greece the species is included in the breeding programme (Panetsos, 1974).

Soviet Union

Scots pine is the most important tree species of the country, and it has a very central position in seed production programmes. The annual requirement for Scots pine seed in reforestation and afforestation is several hundreds of tons (Chebotarev, 1972).

In 1971, work begun in state forest companies to establish a permanent seed supply basis for all important tree species for the whole country. Up to 1978, a forest inventory of more than 13 million ha had been carried out for breeding purposes. The inventory resulted in the formation of more than 140 000 ha of permanent seed collection stands, in the selection of 7 000 ha of plus stands and of about 15 000 plus trees and in the establishment of 9 500 ha of clonal seed orchards. For Scots pine, the inventorized area was 4.8 million ha, the area of permanent seed stands was 76 000 ha, the area of plus stands was 3 600 ha and that of clonal seed orchards 5 200 ha (Novoseltseva, 1980).

The long-range plan for the establishment of a permanent supply of genetically improved seed was revised in 1981. It was considered necessary to increase the total area of clonal seed orchards of the main tree species by 67 500 ha, of which 18 500 ha should be allocated to Scots pine (Nikolayuk and Yarkin, 1984). Up to 1983, more than 19 million ha of forests had been surveyed by the selection inventory, and the total area of seed orchards and permanent seed stands of all species has increased to 11 600 and 160 000 ha, respectively (Chebotarev, 1984; Petrov, 1987).

Established permanent seed stands and seed orchards have been estimated to supply about 1% of the seed needs of the whole USSR in 1985. In 1990 and in 2000 their contribution is expected to be 11% and 25%, respectively. The situation is different, however, in various regions. In the Baltic states and in White Russia the improved seed sources already produce more than is needed (Nikolayuk and Yarkin, 1984). In the Latvian SSR seed orchards alone already fully meet the demand for seed. In the Lithuanian SSR the same situation will be reached in 1990 (Chebotarev, 1984).

Other countries

Turkey

In Turkey Scots pine forests cover about 740 000 ha which is nearly 4% of the total forest area. The planting stock produced for 1987 was 37 million. Planting material currently originates from selected seed stands. Their total area is 4 320 ha and the annual seed harvest has been 2 to 3 t in recent years. Also 12 clonal seed orchards with a total area of 64 ha have been established but their seed production is still very low (Boydak, pers. comm.).

People's Republic of China

Scots pine (*Pinus sylvestris* var. *mongolica*) is one of the main tree species in the north-eastern provinces of the People's Republic of China, although its coverage of the total forest area of the country is only about 0.004%. There are 386 200 ha of Scots pine forests in China, of which about 30 000 ha are man-made. The annual regeneration area of the species is nearly 100 000 ha and it consists of 70% planting and 30% direct seeding.

In the 1970s, about 477 ha of Scots pine seed orchards were established in the provinces of Heilongjiang, Jilin, and Liaoning. Their seed production is still very low, some 50 kg per year in total. At present almost all seed needed in reforestation is harvested from forest stands. Some 2 000 kg of Scots pine seed are annually collected from selected stands, the total area of which is 1 090 ha. There are also about 6 700 ha of special seed collection areas of the species and some 30 000 kg of seed can be collected from them annually (Hong, pers. comm.).

United States of America

Scots pine was one of the first exotic tree species introduced to North America. It has been cultivated as an ornamental for over 300 years. In the states bordering the Great Lakes it has been planted for timber production but because of poor stem form in many older plantations it has not gained much favour as a timber tree (Wright, 1980). Some breeding efforts from the forestry standpoint have been undertaken, however, including selection of seed collection areas and establishment of clonal seed orchards (Schreiner, 1974).

Since 1950 Scots pine has been planted extensively for the production of Christmas trees. This is the largest current use of the species. In the 1970s the annual harvest of plantation-grown Christmas trees in the United States was about 30 million trees, and two thirds of this amount was Scots pine. In the state of Michigan, for instance, about 3 million trees were harvested annually as Christmas trees, and more than 6 million were planted for the same purpose (Wright, 1980).

Most of the seed are collected from plantations growing in the United States. Several domestic plantation sources are used as named varieties. Many plantations have been specially developed for seed production (Gerhold, pers. comm.). The most advanced project is a seed orchard developed from a full-sib progeny test in Pennsylvania. Its seed was released in 1984 for commercial plant production as a genetically improved variety. The new "Pennspanish" variety has clearly better crown form and colour than other commonly cultivated varieties and it grows faster than the Spanish material from which it was derived (Gerhold, 1985). Another successful seed orchard project of much the same kind has been carried out in eastern Nebraska (van Haverbeke, 1981*a, b*).

Canada

In the early 1900s Scots pine was widely planted for timber production in the eastern provinces of Canada. Several thousand hectares of production plantations exist in Quebec and Ontario but because of poor stem quality in these older plantations and the generally better productivity of native pines, Scots pine is not planted as a timber tree at present. However, it is extensively used for Christmas tree plantations as in the neighbouring areas in the United States.

In Ontario, between 1.5 and 2 million Scots pine seedlings a year are grown for Christmas tree production. In Quebec the nurseries of the Ministry of Energy and Resources produced some 340 000 Scots pine seedlings for Christmas tree growers in 1985; ten years earlier their annual production of Scots pine plants was 2.7 million. At that time 70% of the seed utilized were collected from the best production plantations of the province and the rest was imported from France and Turkey. In 1989 the Ministry established 10 ha of clonal seed orchards with material selected from old local plantations (Beaudoin, pers. comm.).

In the prairie provinces a breeding programme based on sources from the Soviet Union has been started and a seed orchard for the production of an improved strain of Scots pine for shelterbelt use is being established (Anon., 1986*b*).

Conclusions

The results of tree breeding are already widely utilized in the artificial regeneration of Scots pine all over the world. In many countries the sale of unselected material is virtually forbidden and only material from officially approved sources, mainly from selected seed collection stands and from untested first-generation seed orchards, is used in reforestation practice. In some Central-European countries improved material from tested seed stands and seed orchards is already available.

The total area of selected seed collection stands or areas of Scots pine in the world, excluding the 180 000 ha of temporary seed stands of Poland, is nearly 160 000 ha. The permanent seed stands of the Soviet Union make up nearly one half of the total area.

First-generation clonal or seedling seed orchards of Scots pine have been established on an area of about 10 000 ha in the whole world. (Summarized data on the area of seed orchards by countries are presented in Chapter 10.) Again, the Soviet Union alone contributes about one half to the total figure and its seed orchard programme with Scots pine is still expanding rapidly. The next four big Scots pine countries in the world, with regard to seed orchard area of the species, are Finland, Sweden, the People's Republic of China and Poland. Sweden and Finland are the leading countries in the utilization of orchard-produced Scots pine seed. In both countries it presently covers more than half of the annual needs for artificial reforestation of the species.

The establishment of improved first-generation seed orchards, with progeny-tested clones (i.e. so-called 1.5-generation orchards), and second-generation orchards will become topical in many countries before the end of this century. Other breeding methods, aimed at the production and vegetative propagation of superior individuals, are not likely to displace the seed orchard approach in the near future. However, in the longer run new findings in biotechnology may completely change the situation.

Literature

Anderson, B. (1985): Möjligheter till urval med avseende på virkeskvalitet i unga avkommeförsök av tall (Possibilities with selection for quality in young progeny trials of Scots pine). Föreningen Skogsträdsförädling, Institutet för skogsförbättring, Årsbok 1985, 58–80.

Anonymous (1986*a*): Yearbook of forest statistics 1985. (Official statistics of Finland XII A:17.) *Fol. Forest* **660.**

Anonymous (1986*b*): Report of the PFRA Tree Nursery, Indian Head, Saskatchewan, Agriculture Canada, Prairie Farm Rehabilitation Administration.

Anonymous (1987): Yearbook of forest statistics 1986. (Official statistics of Finland XVII A:18.) *Fol. Forest* **690.**

Bolland, G. (1981): Entwicklungsstand und Perspektiven der Samenplantagen in der Deutschen Demokratischen Republik. *Soz. Forstwirtsch.* **10/1981**:306–308.

Chebotarev, I. N. (1972): Organization of the forest seed production in the USSR. In: *Proceedings of the International symposium on forest tree breeding, genetics and seed production of coniferous tree species, Novosibirsk, June 19–25, 1972,* Part I, pp. 7–18.

Chebotarev, I. N. (1984): O dal'neishem razvitii rabot po sozdaniyu postoyannoj bazy na selektsionnoj osnove (Future development work in the establishment of a permanent seed supply of improved material). *Lesnoe Khozyaistvo* **2**:33–36.

Chollet, F. (1987): Les vergers à graines forestiers. Amélioration génétique des arbres forestiers. *Rev. Forest. Franc.* Numéro spécial 1986:74–80.

Gerhold, H. D. (1985): 'Pennspanish' Scotch pine released for production. *Am. Christ. Tree J.* **29 (2)**:30–33.

Gerhold, H. D. (1987): (Personal communication.)

Hadders, G. and Samuelson, K.-R. (1985): Forest tree seed orchards in Sweden, 1985. Föreningen Skogsträdsförädling, Institutet för Skogsförbättring, Årsbok 1984, 65–86.

Jovanović, M. J. and Tocović, A. (1974): Forest tree breeding in Yugoslavia. In: Toda, R. (Ed.) *Forest tree breeding in the world,* Meguro, Tokyo, pp. 102–108.

Kocięcki, S. (1980): Directions and performance of Scots pine selection in Poland. In: *Proceedings of the IUFRO working party symposium on Scots pine forestry of the future, September 29–October 4, 1980, Kórnik, Poland,* pp. 173–184.

Kriek, W. (1980): Superior Scots pine for the Netherlands. *Proceedings of the IUFRO working party symposium on Scots pine forestry of the future, September 29–October 4, 1980, Kórnik, Poland,* pp. 158–172.

Kriek, W. (1983): Naar een betere genetische kwaliteit van het Nederlandse grovedennenbos (Towards a better genetic quality of the Dutch Scots pine forests). *Nederlands Bosbouwtijdschrift* **55 (7/8)**:314–329.

Kylmänen, P. (1980): Preliminary results concerning the usability of North Finland x South Finland hybrid seed born in young Scots pine seed orchards. *Fol. Forest.* **423**:1–16.

Magini, E. (1974): Breeding forest trees in Italy. In: Toda, R. (Ed.) *Forest tree breeding in the world.* Meguro, Tokyo, pp. 91–101.

Mooney, O. V. (1986): Scots pine—the forgotten species of Irish forestry? *Irish Forestry* **43 (1)**:7–13.

Nanson, A. (1974): Outlines of forest tree breeding in Belgium. In: Toda, R. (Ed.) *Forest tree breeding in the world.* Meguro, Tokyo, pp. 40–45.

Nikkanen, T. (1982): Survival and height growth of North Finland x South Finland hybrid progenies of Scots pine in intermediate areas. *Fol. Forest.* **527**:1–31.

Nikolayuk, V. A. and Yarkin, N. P. (1984): Puti razvitiya lesnogo semenovodstva v SSSR (The ways of the forest seed development in the USSR). *Lesnoj Zh.* **1**:5–10.

Novoseltseva, A. I. (1980): On the application of forest genetics and forest tree breeding research to practical forest seed production. *Silva Fenn.* **14 (1)**:28–31.

Panetsos, K. P. (1974): Forest tree breeding in Greece. In: Toda, R. (Ed.) *Forest tree breeding in the world.* Meguro, Tokyo, pp. 109–115.

Petrov, S. A. (1987): The matters of gene resources revealing and preservation of main forest species. In: *Proceedings of the 18th IUFRO world congress, 7–21 September, 1986, Ljubljana,* Division 2, Vol. II, pp. 614–621.

Roman-Amat, B. (1987): Pin sylvestre. Amélioration génétique des arbres forestiers. *Rev. Forest. Franc.* Numéro spécial 1986:132–134.

Schreiner, E. J. (1974): Tree breeding in United States forestry practice. In: Toda, R. (Ed.) *Forest tree breeding in the world.* Meguro, Tokyo, pp. 190–201.

Squillace, A. E., la Bastide, I. G. A. and van Vredenburch, C. L. H. (1975): Genetic variation and breeding of Scots pine in the Netherlands. *For. Sci.* **21 (4)**:341–352.

van Haverbeke, D. F. (1981*a*): Seed orchards—a way to better Christmas trees. Part I. Establishment. *Am. Christ. Tree J.* **25 (2)**:16–19.

Veie, B. (1987): Fro fra skogfroplantasjer: Situasjonsbeskrivelse—Norge (Seed from forest tree seed orchards—Report of the situation in Norway). Dokumentation från Nordiska skogbrukets frö- och planträds seminarium 28–29 april 1987, p. 2.

Weisgerber, H. (1980): Saatguterträge aus Samenplantagen. *Allgem. Forst- Jagdz.* **35 (26)**:673–675.

Werner, M. and Ericsson, T. (1979): Virkeskvalitetsstudie i avkommor från en tallfröplantage (Wood quality studies in progenies from a Scots pine seed orchard). Föreningen Skogsträdsförädling, Institutet för Skogsförbättring, Årsbok 1979, 40–69.

Werner, M., Hadders, G. and Rosvall, O. (1981): Förädlingsprogram för tall, gran och contortatall (Breeding programme for Scots pine, Norway spruce and Lodgepole pine). Föreningen Skogsträdsförädling, Institutet för Skogsförbättring, Årsbok 1980, 59–111.

Wright, J. W. (1980): Choosing Scotch pine for planting in the Northeastern United States. In: *Proceedings of the IUFRO working party symposium on Scots pine forestry of the future, September 29–October 4, 1980, Kórnik, Poland,* pp. 128–140.

Index